Production Networks in Southeast Asia

This book answers the recently topical questions of how China's processed trade affects the trade of Southeast Asia. What is Southeast Asia's role in Factory Asia, the region's complex of cross-border supply chains? What is Southeast Asia's involvement in building or joining production networks in the region? And, most important, how can Southeast Asia increase the value added of its products and improve its competitiveness?

This book provides rigorous analysis of how trade policy affects value added, highly disaggregated at the firm and product level, of the six Southeast Asian countries – Indonesia, Malaysia, the Philippines, Singapore and Viet Nam – and combines this with thorough examinations of their trade, industrial and labour policies.

Lili Yan Ing is a Senior Economist at Economic Research Institute for ASEAN and East Asia (ERIA).

Fukunari Kimura is Chief Economist at Economic Research Institute for ASEAN and East Asia (ERIA).

Routledge–ERIA Studies in Development Economics

1 Economic Consequences of Globalization: Evidence from East Asia
Edited by Shujiro Urata, Chin Hee Hahn and Dionisius Narjoko

2 Real and Financial Integration in Asia
Edited by Shandre Mugan Thangavelu and Aekapol Chongvilaivan

3 Energy Market Integration in East Asia: Deepening Understanding and Moving Forward
Edited by Yanrui Wu, Fukunari Kimura and Xunpeng Shi

4 Rebalancing Economies in Financially Integrating East Asia
Edited by Jenny Corbett and Ying Xu

5 Strengthening Social Protection in East Asia
Edited by Mukul G. Asher and Fukunari Kimura

6 Economic Integration in East Asia: Production Networks and Small and Medium Enterprises
Charles Harvie, Dionisius Narjoko and Sothea Oum

7 ASEAN and Regional Free Trade Agreements
Edited by Christopher Findlay

8 Age Related Pension Expenditure and Fiscal Space: Modelling Techniques and Case Studies from East Asia
Edited by Mukul G. Asher and Fauziah Zen

9 Production Networks in Southeast Asia
Edited by Lili Yan Ing and Fukunari Kimura

Production Networks in Southeast Asia

Edited by Lili Yan Ing
and Fukunari Kimura

Routledge
Taylor & Francis Group
LONDON AND NEW YORK

First published 2017 by Routledge

2 Park Square, Milton Park, Abingdon, Oxfordshire OX14 4RN
52 Vanderbilt Avenue, New York, NY 10017

Routledge is an imprint of the Taylor & Francis Group, an informa business

First issued in paperback 2019

British Library Cataloguing-in-Publication Data
A catalogue record for this book is available from the British Library

Library of Congress Cataloging-in-Publication Data
Names: Ing, Lili Yan, editor. | Kimura, Fukunari, editor.
Title: Production networks in Southeast Asia / edited by Lili Yan Ing and Fukunari Kimura.
Description: Abingdon, Oxon ; New York, NY : Routledge, 2017. | Series: Routledge-ERIA studies in development economics ; 9 | Includes bibliographical references and index.
Identifiers: LCCN 2016040050 | ISBN 9781138222830 (hardback) | ISBN 9781315406787 (ebook)
Subjects: LCSH: Southeast Asia—Commerce. | Southeast Asia—Foreign economic relations. | Manufacturing industries—Southeast Asia. | Economic development—Southeast Asia.
Classification: LCC HF3790.8.Z5 P76 2017 | DDC 338.8/7—dc23
LC record available at https://lccn.loc.gov/2016040050

ISBN: 978-1-138-22283-0 (hbk)
ISBN: 978-0-367-35043-7 (pbk)

Typeset in Galliard
by Apex CoVantage, LLC

Contents

Figures

Tables

Contributors

Rafaelita Aldaba is a Senior Research Fellow and the Acting Vice President at the Philippine Department of Trade and Industry (PIDS).

Trinh Bui is a Research Consultant at the General Statistics Office (GSO), Viet Nam.

Jose Caraballo is an Assistant Professor of Economics at the University of Puerto Rico, Cayey.

Xiaomin Cui is a Research Assistant at the China Center for Economic Research (CCER) at Peking University.

Hal Hill is a Professor at Australian National University (ANU).

Lili Yan Ing is a Senior Economist at the Economic Research Institute for ASEAN and East Asia (ERIA).

Patarapong Intarakumnerd is a Professor at the National Graduate Institute for Policy Studies (GRIPS).

Xiao Jiang is an Assistant Professor of Economics at Denison University, Granville.

Andrew Jia Yi Kam is a Research Fellow at the Institute of Malaysian and International Studies (IKMAS) at Universiti Kebangsaan Malaysia.

Fukunari Kimura is the Chief Economist at the Economic Research Institute for ASEAN and East Asia (ERIA).

Archanun Kohpaiboon is an Assistant Professor at Thammasat University.

Przemyslaw Kowalski is a Senior Trade Policy Analyst at the Organisation for Economic Co-operation and Development (OECD).

Javier López-González is a Trade Policy Analyst at the Organisation for Economic Co-operation and Development (OECD).

Siwage Dharma Negara is a Researcher at the Economic Research Center of the Indonesian Institute of Sciences (P2E–LIPI) and visiting fellow at the Institute of Southeast Asian Studies (ISEAS).

Anh Duong Nguyen is a Deputy Director at the Central Institute for Economic Management (CIEM), Viet Nam.

Chandra Putra is a PhD student at Australian National University.

Ben Shepherd is a Principal at Developing Trade Consultants Ltd.

Siew Yean Tham is Adjunct Professor at Institute of Malaysian and International Studies, Universiti Kebangsaan Malaysia and Senior Fellow at ISEAS-Yusuf Ishak Institute, Singapore.

Mun-Heng Toh is an Associate Professor at National University of Singapore (NUS).

Tri Thanh Vo is the Vice President at the Central Institute for Economic Management (CIEM), Viet Nam.

Miaojie Yu is a Professor at the China Center for Economic Research (CCER) at Peking University.

Preface

This book was initiated by growing discussions in Southeast Asia on how a developing country can increase its value added. We invited distinguished scholars to write chapters on Malaysia, Indonesia, the Philippines, Singapore and Viet Nam. We also complement the country analyses by adding analyses on Factory Asia, how China affects Southeast Asian trade, and the lessons learnt – the good and the bad – from regional production network in other regions such as Germany-Eastern Europe, and Latin America.

The key message of this book is that trade is growing, production is sliced, and tasks are fragmented: more openness in trade and investment may tend to pull down the ratio of domestic value added but level up the total amount of domestic value added.

Acknowledgements

We thank all of our students who keep plying us with challenging questions. To you, please keep asking questions, and for you, this book is presented.

This book has benefited from constructive discussions and comments from distinguished scholars, particularly Martin Richardson, Olivier Cadot, Mari Pangestu, Lee Sung Joon, Shujiro Urata, Hank Lim, Simon Tay, Peter Drysdale, and Inkyo Cheong. This book is also enriched by discussions with our scholar and government official counterparts.

We thank Fadriani Trianingsih, Rizqy Anadhika, and Abigail Ho for their excellent assistance. We also thank our editorial team: Maria Priscila del Rosario and Stefan Weisak, Samantha Phua and Yong Ling Lam. Elisa Ayu provided generous administrative support. We are grateful to the Economic Research Institute for ASEAN and East (ERIA) for making this project possible.

The last but most important, thank you goes to my husband, Lee Jae Hun, who always provides kind support for every single of my piece, and to my two angels, Michelle Lee and Han Na Lee, who keep reenergizing me with their non-stop questions and laughs. I hope this book will still be valid in their generation.

1 Introduction

Lili Yan Ing and Fukunari Kimura

1. Introduction

East Asia Pacific (EAP) economic growth has outperformed global growth for the past two decades, except from 1997 to 1999, due to the Asian Financial Crisis. Compared with other regions, EAP had the highest growth in 2014, and growth in developing EAP remained strong in 2014, although slowing somewhat following the global financial and economic crisis. Most East Asian economies are well positioned to weather renewed volatility.

The drivers of global economic growth are shifting toward East Asia, and Emerging East Asian economies (EEA) are well situated to benefit from the growing power of Asian economies. Developing East Asia grew by 6.2 percent in 2014, with slower expansion in China pulling down much of the regional aggregate (World Bank 2015). The emerging Southeast Asian Economies, Indonesia, Malaysia, the Philippines, Thailand and Viet Nam experienced strong growth in the past two decades – among the highest in the world. The average growth of these economies was 5.2 percent from 2005 to 2015. Despite the recent global crisis in 2008, the five ASEAN countries grew at an average of 5 percent in 2015 (EIU 2013). The growth of Emerging East Asia is estimated to be on a par with China.

East Asia has achieved sustained economic growth by applying development strategies that have aggressively exploited the mechanics of global value chains, and thus it is meaningful to analyze the industrialization process with trade in value added, input–output and firm-level data merged with product-level data.

Section 2 explains trade in a new paradigm. Section 3 draws the new things from this book and outlines short explanations of Chapters 2 to 13.

2. Trade in a new paradigm

2.1 The mechanics of production networks and trade in value added

Trade-in-value-added (TiVA) data provide novel and insightful information on how a country's economy is connected with the rest of the world. We notice that in the globalization era, tight links with the world economy are essential to economic prosperity, and government policies are the key for it. However, we still do not know exactly what sort of connectivity is effective for economic development and what kind of policies are needed to achieve the goal. TiVA data provide us a novel angle of looking at the nature and characteristics of global value chains (GVCs).

Many scholars and practitioners have tried to develop various methods for exploiting TiVA data, and the development process is still ongoing. Because various indicators, figures and tables generated by TiVA data are so stimulating, there is obviously a great temptation to jump into easy though sometimes misleading conclusions. We would like to list three caveats that we should address, based on the development of analytical tools for TiVA data up to this moment.

First, TiVA data present international industrial linkages, but do not directly tell us the nature of such linkages. Since the mid-1980s, we have observed the emergence of international production/distribution networks (Jones and Kierzkowski 1990; Ando and Kimura 2005) or the second unbundling (Baldwin 2011). These are the international division of labor in terms of production processes or tasks in contrast with the traditional industry-by-industry international division of labor or the first unbundling. With such a new type of international division of labor, international trade in intermediate products and parts and components has explosively increased in contrast to traditional international trade in raw materials and final products.

ASEAN and East Asia are the regions in which international production networks have most advanced in the world, particularly in machinery industries, including general machinery, electric machinery, transport equipment and precision machinery. Of course, production networks are observed in industries such as textiles and garment, cut flowers, software industry and others. However, in these industries, most of the activities still belong to slow value chains or the first unbundling. The difference between slow value chains and production networks or between the first unbundling and the second unbundling ultimately resides in the way of organizing production/distribution chains, rather than being based on industry-by-industry differences.

Second, the decomposition of value added in produced/exported goods into domestic and foreign value added is useful, but we have to be careful in drawing policy implications from it. In the context of a country's economic growth, the value added of the country is certainly a prime target, together with domestic employment. The higher the value added, the higher the national welfare as far as economic efficiency is concerned. We have to note, however, that what is important is the amount of value added rather than the ratio of domestic value added. The trick is that the international division of labor allows countries to import what each country is relatively poor at producing and export what it is relatively good at producing. With enhancing international trade, the world total welfare goes up, and at the same time, in most of the cases, each country's welfare also enhances. Trade and investment liberalization tend to pull down the ratio of domestic value added but push up the amount of domestic value added. In general, the ratio of domestic value added depends on many factors such as (1) size of the domestic economy, (2) the existence of exporting upstream industries such as agriculture and mining, (3) general openness to trade of the economy concerned, (4) the degree of participation in international production networks or the second unbundling, and (5) the extent of industrial agglomeration. (1) and (2) are supposed to enhance the ratio of domestic value added, which must be adjusted for international comparison. (3) and (4) would lower the ratio of domestic value added, which should be interpreted as rather a good phenomenon. (5) may increase the ratio of domestic value added. The issue in (5) depends on the form of industrial agglomeration. In the old days, many developing countries tried to build up industrial agglomeration by conducting the so-called import substitution strategy with trade barriers and mostly ended up with big failures. We should not follow such steps. In the current ASEAN and East Asian developing economies, industrial agglomeration is formed together with tight and thick connectivity with international production networks. In such a case, the core of industrial agglomeration is the inter-firm (arm's length) division of labor supported by intra-firm supplies from international production networks (Kimura and Ando 2005). Industrial agglomeration provides local firms to participate in production networks, enjoy technology transfer and spillover, and eventually upgrade their capability from process to product innovation. Policies to support the formation of such industrial agglomeration, which is quite different from the import substitution strategy in the past, may result in a higher ratio of domestic value added.

Third, TiVA data are constructed based on international input–output (IO) tables, and thus their reliability depends on the quality of the original statistical information. Each country's IO tables are so-called secondary statistics estimated from "primary" statistics such as production statistics of each industrial sector and other fragmented information. Missing is comprehensive information on the production structure of services subsectors. Although analyses on services ingredients in production are now one of the focal points in the TiVA literature, we must be somewhat conservative in interpreting results of data analyses, particularly for disaggregated services subsectors. In addition, international IO tables, and thus TiVA data, are in nominal prices. We thus have to take into consideration over-time price changes, particularly for fluctuations of energy prices, when making time-series comparisons. Or, even in one-shot analysis, we should notice that price differences across countries are not taken care of. These three caveats are already well recognized among researchers and would eventually be overcome, at least partially. However, we should bear them in our minds for the moment.

3. The new information from this book

This book provides insights on global value chains of Southeast Asia's products by painting an overall picture of it using the available OECD TiVA data, which are complemented by country analyses using input–output and merged firm-level and product-level data. This is also enriched by lessons learned of improving value added from central Eastern Europe and Latin American countries. This book is concluded by examining thorough industrial policies.

In Chapter 2, Javier López-González and Przemyslaw Kowalski explain the position of Southeast Asia in Asia's factory. The authors demonstrate that the Association of Southeast Asian Nations (ASEAN) has embraced the new opportunities global value chains (GVCs) offer. While Brunei's domestic value added in exports originates overwhelmingly from the primary sector of oil and gas, Singapore is highly specialized in services. The Philippines, on the other hand, has significantly increased its domestic value added embodied in exports of electrical equipment, which is accompanied by significant use of foreign inputs. This complementarity between the use of foreign inputs and domestic specialization is key to understanding global value chain (GVC) participation. In the case of ASEAN, it is shown that there is an element of synchronization in terms of the growing importance of domestic and foreign value added in ASEAN where for all countries, except for Cambodia, a positive correlation emerges between changes in the overall importance of domestic and foreign value added across sectors. This suggests that the sectors, which have a growing share of domestic value added to total exports, tend to have a growing share of foreign value added in inputs. In other words, there is an indication that imported inputs may help developing domestic capacity.

The chapter also challenges what is referred to in the existing literature as the retention of domestic value added and upgrading in GVCs: this should no longer be framed in the context of increasing the *share of the pie* that is occupied by domestic value added in the production of an industry's exports, but rather in the context of how further engagement offers important opportunities for growing the total *value of the pie*. The main drivers of participation in GVC, while mainly structural, such as the size of the economy and the distance to manufacturing hubs, trade and investment openness, as well as logistics performance, hard and soft infrastructure and good governance, also play significant roles. Much progress has been made in the process of completing the ASEAN Economic Community (AEC), but with competitive pressures rising, as other countries increasingly look to join GVCs, there is a strengthened case for continuing the process of reform through further trade and investment openness and domestic regulatory reform.

In Chapter 3, Miaojie Yu and Xiaomin Cui analyze the impacts of the processed trade of China on Southeast Asia's trade. This chapter employs the trade data at the Standard International Trade Classification (SITC) one-digit level from the UN Comtrade database to study the impact of China's trade on ASEAN's trade, and understand how ASEAN countries can improve the value added of their exports and their trade competitiveness. First, they find that on average ASEAN had the strongest revealed comparative advantage (RCA, hereafter) in animal and vegetable oils, fats and waxes, where the RCA index was about 6.2 during the period 2000–2013. ASEAN had an increasing comparative advantage in miscellaneous manufactured articles in the period 2000–2013 and also suffered a gradual loss in comparative advantage in manufacturing machinery and transport equipment from 2000 to 2011, even though the RCA index recovered slightly in the period 2012–2013. Conversely, from 2000 to 2013, China did not have comparative advantages in agriculture products and crude materials on average. This reflects the structural complementarity of exports between China and ASEAN. However, China remained competitive in manufacturing, with an RCA index for manufacturing machinery and transport equipment increasing steadily from 0.8 to 1.45 from 2000 to 2011. Until 2013, ASEAN also had weak RCA in manufacturing, where China and ASEAN may compete severely.

Second, when it comes to the question of what the impacts are of China's trade on ASEAN's trade, the empirical results show that both complementarity and substitution/competition effects exist. With the gradual deepening and broadening of regional cooperation between China and Southeast Asia, an increase in exports from China to ASEAN will tend to promote ASEAN's exports. However, China's increasing exports to the rest of the world will tend to crowd out ASEAN's own exports. When it comes to the effect of China's exports on ASEAN's value added of exports, the results work in the opposite direction. If ASEAN imports more goods from China, its value added of exports will tend to be lower. Conversely, ASEAN's value added of exports will tend to be higher due to the substitution/competition effect if China's exports to the rest of the world increase. In addition, ASEAN countries can improve their value added of exports and trade competitiveness through trade liberalization and by increasing investment in R&D and training programs for skilled labor.

Chapters 4 to 9 provide insights on the value added of the six Southeast Asian countries: Indonesia, Malaysia, the Philippines, Singapore, Thailand and Viet Nam.

In Chapter 4, Lili Yan Ing and Chandra Putra present that imported inputs increase Indonesia's firm productivity via better Indonesian product quality and variety. They argue that trade evolves and production is sliced. Much of production is based in production networks. Imports are largely used as inputs for exports. Many countries are engaged directly and indirectly in producing final products. The development of global production chains, with an increased use of imported inputs, caused a reduction of the domestic value-added content for each unit of manufacturing production and exports. They provide a comprehensive analysis of how imported imports affect Indonesia's product varieties and quality. Their analysis argues that reductions in input tariffs will increase value added via product variety and quality. It shows that a reduction in input tariffs will increase value added, not only via its interaction with importing firms, but also with exporting firms that use imported products as their inputs.

Using the very disaggregated merged Indonesian firm- and product-level data from 2000 to 2010, the findings show that a reduction of 1 percent in input tariffs will increase value added by 0.2 percent, not only via its interaction with importing firms, but also with exporting firms that use imported products as their inputs. A 1 percent reduction in input tariffs will increase product variety and quality by 3.5 percent and 1.5 percent, respectively. Exporting firms tend to have a higher value added than the average of all firms, and they also tend to increase variety and quality of products. Foreign firms also tend to have a relatively higher value added than the general average, but they do not necessarily have increased product variety and higher quality.

In Chapter 5, Siew Yean Tham and Jia Yi Kam examine the imported contents of Malaysia's manufacturing sector by using Hummel's vertical specialization method employing the input–output tables of Malaysia from 2000 to 2010. The results are then compared with the OECD trade in value-added data for Malaysia for the years 1995, 2000, 2005, 2008 and 2009. The study finds that a decrease in direct and indirect imported inputs based on Hummel's vertical specialization index is interpreted as an increase in the local content of Malaysia's exports. Analyses based on both data sources, input–output and the OECD TiVA data by sectors, show that the highest foreign inputs are found in non-resource manufacturing sectors such as basic metals, machinery and equipment and electrical and optical equipment as compared to resource-based sectors such as food products and wood and paper products. The compressed input–output table showed a decrease in the information, communication and technology (ICT) sector's use of imported inputs. Using a more detailed classification of the electronics sector, two main shifts can be observed in electronics exports from 2000–2013. First, there was a shift toward finished goods in terms of the number of products, but not in terms of export values. Second, there was a shift from electronics manufacturing service (EMS) to semiconductor manufacturing service (SMS) activities. The contraction in the ICT and electronics sector has led to a greater focus on semiconductor manufacturing. The chapter also presents a case study on the ICT sector, and finds, by using revealed comparative advantage (RCA) analysis, that the product groups with the highest RCA are auto electronics and finished goods, followed by electronic data-processing (EDP) and components/devices semiconductor parts.

In Chapter 6, Rafaelita Aldaba examines the extent and depth of participation of the Philippines in the electronics global GVC using trade-in-value-added (TiVA) and extensive margin indicators. The Philippines remains strong in semiconductors, but is lagging behind other ASEAN countries. According to the TiVA database, the level of participation of the Philippines in the electronics GVC increased substantially between 1995 and 2009. The extensive margins show that the Philippines has been regaining its position in regional production networks as indicated by the rising number of exported products to the region. Based on the RCA analysis, it is indicated that the Philippines has remained strong in semiconductors but has been lagging behind Singapore, Malaysia and Thailand. The extensive margin analysis also shows that the Philippines has been regaining its position in the regional production networks as indicated by the rising number of exported products to the region. The foreign inputs in Philippine electronics exports have increased from 51 percent in 1995 to 61 percent in 2000, but this declined to 53 percent in 2008 and 50 percent in 2009. The Philippine electronics industry has been largely dominated by the semiconductor sector. Its participation in global/regional production networks has relied mainly on mature and legacy products and processes focusing on semiconductor assembly and test services (SATS), which is the back-end part of the semiconductor manufacturing services (SMS). The gradually declining trend in the number of foreign inputs indicates the need to diversify and upgrade the industry's GVC participation through market upgrading characterized by moving from semiconductors to electronic manufacturing services (EMS), particularly in areas with high growth potential such as auto electronics, power electronics, electronic data processing and consumer electronics. The upgrading process will require human resources development, establishing an innovation ecosystem, efficient logistics and infrastructure and developing a parts, supplies and materials sector to support the industry.

In Chapter 7, Mun-Heng Toh reviews the cluster-based development strategy adopted in Singapore. Singapore's development strategy is predominantly based on trade openness and liberal inflows of foreign direct investments. That strategy has enabled Singapore to be plugged into GVAs to benefit from inflows of foreign investment and participation in international trade. The author makes use of information and indicators of GVC participation from the TiVA database with specific reference to Singapore. The results affirm that Singapore has provided an example that success in pursuing a strategy of trade-led economic growth is translated into successful trade in integrated global markets and into upgrades within the GVCs. Participation in the GVCs facilitated by a relatively large network of regional and bilateral FTAs has helped to bring gain to Singapore in terms of improved competitiveness, and better access to global markets and expansion of production and jobs. It has also helped Singapore to increase productivity and avoid the middle-income trap.

The TiVA analysis indicates that services play an important role in global trade, contributing, on average, more than 30 percent of the total value added in manufactured goods. Liberalization of the services trade would allow for more efficient and higher-quality services to enhance the competitiveness of manufacturing activities. In this regard, Singapore's concerted efforts in promoting trade liberalization in services relating to transport, logistics, finance and communication, via the regional and bilateral FTAs, are steps in the right direction. New concepts and the implications of trade in value added are appraised to provide new perspectives of and prospects for continued sustainable growth of Singapore's economy.

In Chapter 8, Patarapong Intarakumnerd illustrates the development of domestic value added in Thailand's automotive industry. Thailand's automotive sector has become a part of the regional production networks of many car manufacturers. Completely built-up (CBU) vehicles and completely knocked-down (CKD) kits are built by local producers, and have been a major export product since 2000. Lately, it has become more 'product-specific'. Rather than producing and innovating varieties of vehicles as in the past, product champions like one-ton pick-up trucks and, subsequently, eco cars have emerged in the past 15 years.

By using the Grubel-Lloyd index calculated based on UN Comtrade data in 1992, 1996, 2002 and 2011, the chapter examines inter-industry trade (one-way trade) and intra-industry trade. The findings show that the degree of intra-industry trade between Thailand and its main auto and auto parts trading partners

(Australia, China, India, Indonesia, Japan, Korea, Malaysia, the Philippines and Viet Nam) has been increasing. The intra-industry trade pattern was mainly one-way trade prior to 1997, and the degree of intra-industry trade increased significantly in 1999. In general, Thailand tends to have higher degrees of IIT in automobile products with Japan and Indonesia. This is because both Thailand (for sedans) and Indonesia (for eight-seat cars) became important bases of Japanese car makers in the region.

Intarakumnerd also explains the different types of government policies in the auto industry, which can be divided into three phases: phase I, 1960–1970, the import substitution regime, which was mainly an import substitution policy and revised investment promotion law to encourage automotive assembly in Thailand; phase II, 1970–1996, which consisted of significant FDI and export promotion; phase III, 1997–2015, which was predominantly further liberalization for FDI and technological upgrading.

In Chapter 9, Tri Thanh Vo, Anh Duong Nguyen and Trinh Bui seek answers to two questions – (1) whether Viet Nam is depending more on imported products to meet its export growth; and (2) whether export growth brings sufficient benefits for the domestic economy. Apart from reviewing trade data and existing literature, the authors calculate domestic value added in Viet Nam's exports by sector, using input–output tables for 2007 and 2011. They found that the inducement impact of exports to the economy and income have decreased in 2007–2011. The domestic value-added content of exports exhibited a complicated pattern, though overall exports increased. The gross benefits are even higher due to the drastic increase in exports.

Instead, export growth stimulated more rapid increase of imports. The domestic value-added content of exports fell almost continuously from 78.9 percent in 1995 to 63.7 percent in 2011, except for a minor, one-off increase in 2009. Still, such a fall was overwhelmed by more drastic increases in gross exports averaged at 21.5 percent per annum during 2002–2006 and 18.0 percent per annum during 2007–2014. This relieves the concern about decreasing benefits from trade. Viet Nam should thus focus more on ensuring better diffusion of exports to domestic aggregate economic activity and return to various production factors. Moreover, Viet Nam should deepen linkages between multinational enterprises and local firms and adopt a more targeted industrial policy.

In Chapter 10, Hal Hill and Archanun Kohpaiboon survey Southeast Asia's experience with industry policy. 'Industry policy', defined as non-neutral inter-industry (and sometimes inter-firm) incentives, remains a contested field. There is much general agreement in the development economics and political economy literatures about the factors that underpin rapid economic development. Key policies include macroeconomic stability; openness to trade, investment and technology; a stable and business-friendly commercial environment; mechanisms that ensure broad-based, 'inclusive' development; and investment in supply-side capabilities, ranging from infrastructure to human capital. The contestation focuses on whether these general, economy-wide approaches are sufficient, or whether in addition there is a role for sector-specific interventions. While the earlier literature focused on the merits of import-substitution versus export orientation, recognizing the inevitability of global economic integration, the contemporary literature has focused on whether there is a case for 'smart industry policy' in broadly open economies.

Against this backdrop, Hill and Kohpaiboon survey and analyze Southeast Asia's experience with industry policy. They suggest designing policies for industrial progress, rather than 'industry policy'. They adopt a case study approach, investigating cases of sector-specific successes in several countries, and the factors underpinning the success. Examples include the Thai automotive industry, Malaysian higher education, the Philippine BPOs and the Cambodian garments industry. The special case of Indonesia is also examined, a country that has achieved episodes of rapid industrialization, but in which the general policy environment rather than sector-specific factors has been the major driver of growth.

In spite of the significant country and sector differences, several key points emerge from these case studies. The first is openness. All four cases involved export-oriented goods or services, in which foreign investors also played a significant role in transmitting technology and international market information. Second, government policies contributed significantly to the successes in one way or another. Often this simply took the form of a major deregulation that enabled countries to exploit their latent comparative advantages. Third, there were specific country features in all cases that, combined with open policies, underpinned success. For Cambodian tourism, Angkor Wat was obviously the key, combined with the restoration of peace and open policies toward FDI, labour and civil aviation. In the Philippines, it was English language proficiency, aided by an outward-looking and cosmopolitan culture. Malaysia's cultural and linguistic openness to Islam, to Chinese, Indian and Malay communities, and the widespread use of English all contributed to its educational success. Fourth, a measure of good luck (and timing) was present in most of the cases, and it enabled governments to build on reforms that were already under way: like with BPOs in the Philippines, where the government had recently liberalized a chronically repressed telecoms industry just in time to link in to the wave of technologically driven international outsourcing of services. Thailand's liberalization occurred when its three potential Southeast Asian auto rivals had policy regimes that effectively precluded their international participation in the industry. Malaysia opened up a hitherto state-dominated tertiary education system just as the internationalization of higher education was gathering pace.

In Chapter 11, Xiao Jiang and Jose Caraballo conduct a multiregional input–output analysis to estimate the employment outcomes of global value chain participation in the form of trading intermediates inputs for the six Asian economies, namely China, Indonesia, India, Japan, the Republic of Korea and Taiwan, included in the World Input–Output Database (the selection of these countries was mainly due to data availability). They claim that during 1995 to 2008, many Asian countries experienced what they termed 'value-added erosion' to describe the phenomenon of the decline in the sectoral shares of domestic value added in a country's exports as the country becomes more integrated into the global value chains (GVCs).

Following related theoretical models, they argue that the decline of domestic value-added share in a country's exports is likely to be caused by the expansion of high value-adding activities performed by foreign lead firms in the upper stream of the GVCs. By applying econometric methods to the multiregional global input–output model, they found that there is a relatively robust relationship between the surge in foreign high-skilled labor embodied in exports and value-added erosion, as validated with two cross-sectional dimensions (across industries within many countries and across countries within many manufacturing industries). The decline of domestic value-added share is, to a lesser extent, also related to other variables such as capital intensiveness in industry and labor productivity.

These Asian countries face the challenge of shifting toward production of higher value-added goods and services, bringing them into more direct competition with advanced economies, while at the same time their competitiveness in lower value-added goods and services is being eroded by the increasing presence of lower-labor-cost countries in global markets. In addition, all of these countries, despite their developmental heterogeneity, failed to increase their share of value added in the period 1995–2009.

In Chapter 12, Siwage Dharma Negara explains the endogenous growth theory that innovation is one of the key drivers of technological progress and productivity growth of a country. Technological improvements stemming from firms' innovative activities can contribute to a country's overall productivity and export competitiveness. For innovation to flourish, an environment conducive to firms conducting risky innovative activities is needed. Studies show that public policies, including labor market policies, can influence the operating conditions and institutional structures of firms to foster innovation that leads to productivity gains. However, empirical evidence is mixed on the impact of labor market policies on firms' incentives to innovate.

Negara constructs a set of panel data of 32 countries, including seven ASEAN countries, namely, Cambodia, Indonesia, Malaysia, the Philippines, Singapore, Thailand and Viet Nam (note: data for Brunei Darussalam, Lao PDR and Myanmar are not available), from 2009 to 2013. This study uses scores data for innovation competitiveness published by the World Economic Forum (WEF) since 2005 as a proxy for innovation capacity. The indicators are derived using a standardized survey targeted to more than 14,000 business executives in 144 countries. In addition to innovation scores data, the chapter uses alternative proxies for innovation, including the number of patents residents filed at the national office, the number of venture capital deals and the number of people employed in knowledge-intensive services from the Global Innovation Index published by Cornell University, INSEAD and the World Intellectual Property Organization (WIPO). It uses scores data for labor market efficiency published by the GCI and cost of redundancy dismissal from the World Bank's Doing Business, as a proxy for labor market policies. Using this balanced panel data, Negara offers simple empirical models to measure the relationship between labor market policies and innovation capacity; and between innovation capacity and trade competitiveness. The main findings are that countries with more flexible labor market policies have higher levels of innovation competitiveness and that there is a positive correlation between innovation competitiveness and trade competitiveness. The quality of higher education is positively associated with a country's innovation competitiveness. On trade aspects, this chapter finds preliminary evidence that past innovation is positively associated with trade competitiveness. This is in line with previous studies that find a positive link between innovation and exporting.

The final chapter, by Ben Shepherd, illustrates lessons learned from regional production networks in other regions such as central Eastern Europe and Latin America. The author demonstrates that there is clear evidence of increased internationalization of ASEAN value chains, as well as industrial growth. Regional production networks in Latin America have been largely driven by NAFTA, and its trade between Latin America and the United States, and changes have typically been more rapid in Central and Eastern Europe than in ASEAN, probably due to economic transition and joining the European Union. In terms of the cross-regional comparison that is the analytical focus of this chapter, one finding stands out in sharp relief: value chain development has been very intense in CEE, but relatively limited in LAC. This result needs to be interpreted with caution, because the TiVA data are much more complete in their coverage of CEE than of LAC. Based on the available data, however, it appears that regional production networks in LAC remain considerably less than in ASEAN or CEE, and the dynamic of change is much slower.

The story of regional production networks in CEE provides an interesting insight for Southeast Asia. In CEE production networks, Germany plays an important role as an anchor economy – a source of final demand and of technology-rich investment – and ASEAN will need to continue looking to regional economies such as Japan and the Republic of Korea in this light. The emphasis in most ASEAN countries will now need to shift to 'moving up' to higher value-added activities, such as research and development, which have positive spillovers for the rest of the economy.

References

Ando, M and Kimura, F 2005, 'The Formation of International Production and Distribution Networks in East Asia', in *International Trade in East Asia (NBER–East Asia Seminar on Economics, Volume 14)*, eds. T Ito and AK Rose, University of Chicago Press, Chicago, pp. 177–213. The former version was distributed as NBER Working Paper No. 10167, National Bureau of Economic Research, Inc., December 2003.

Baldwin, R 2011, '21st century regionalism: filling the gap between 21st century trade and 20th century trade rules', *Centre for Economic Policy Research (CEPR) Policy Insight* No. 56 (May), Centre for Economic Policy Research.

Economist Intelligent Unit data http://country.eiu.com/AllCountries.aspx;www.eiu.com/home.aspx.

Jones, RW and Kierzkowski, H 1990, 'The Role of Services in Production and International Trade: A Theoretical Framework', in *The Political Economy of International Trade: Essays in Honor of Robert E. Baldwin*, eds. RW Jones and AO Krueger, Basil Blackwell, Oxford, pp. 31–48.

Kimura, F and Ando, M 2005, 'Two-dimensional fragmentation in East Asia: conceptual framework and empirics', *International Review of Economics and Finance* (special issue on 'Outsourcing and Fragmentation: Blessing or Threat', ed. Henryk Kierzkowski), vol. 14, no. 3, pp. 317–348.

World Bank World Development Indicators http://data.worldbank.org/data-catalog/world-development-indicators

2 Global value chain participation in Southeast Asia

Trade and related policy implications[1]

Javier López-González[2] and Przemyslaw Kowalski

1. Introduction

Global value chains (GVCs) are a defining feature of twenty-first-century trade with three key production hubs emerging in Europe, North America, and Asia. It is the latter that is the concern of this chapter, which delves more deeply into the internal and external linkages of the region in an effort not just to characterize the evolution of production sharing between regional and non-regional partners, but also to identify the key sectors in which specialization has taken place and the policies that can best help Association of Southeast Asian Nations (ASEAN) countries benefit from participation in GVCs.

To this end, the discussion is framed along three dimensions. The first is an assessment of the evolution of ASEAN participation in GVCs. The second is a discussion seeking to explain why participation might be important and how the 'upgrading' debate should be framed. The last relates to the role that policy can play in shaping participation.

Throughout, traditional metrics for value chain participation, which are derived from Hummels et al. (2001) indicator of 'vertical specialization' and its refinement by Koopman et al. (2011), are used. Participation is defined in terms of the value added embodied in exports both looking backward and forward from a reference country: backward when it comes to foreign value added embodied in exports, and forward when it refers to domestic value added that is used as an input to produce the exports of the destination country.

While both of these measures are expressed as shares of the reference country's exports, in fact they measure very different forms of engagement. For example, a country that is predominantly assembling products into final goods and subsequently exporting these will have a strong backward participation index but a very weak forward participation. Conversely, a country that predominantly supplies intermediates to an assembler will have a highly developed forward participation indicator but a small backward participation measure. These participation measures, therefore, give a metric of engagement in the form of buying from (backward) and selling to (forward) GVCs or the demand and supply sides of value chain activity.

2. Mapping participation in Asia

The term *value chain* refers to the different processes, activities, firms, or workers engaged in the production of goods and services from conception to end-use (Gereffi and Fernandez-Stark 2011). Although they are not new, they have recently taken on a more global dimension through their increased expansion towards emerging economies. Firms in developed countries can now more readily combine their high-tech know-how with lower wage labour in developing countries to produce at lower costs, and this has led to a growing redistribution of economic activity towards the developing world, and towards Asia in particular. This chapter looks at Asia's position in the context of the global evolution of value chain activity, as well as at the intra- and extra-regional dynamics that underpin Asia's evolving participation.

2.1 The global context

There is a strong regional dimension to value chain activity (Figure 2.1); the use of value added to produce exports is intensive with respect to neighbouring countries. For example, 13 per cent of the value of Chinese exports comes from neighbouring Asian countries, Mexico relies on 13 per cent value added from the United States (US), and Germany sources 14 per cent of its value added in exports from neighbouring European countries.

However, the global element is also important. Germany, for example, is a strong supplier of value added to many countries outside of the European Union (EU) (such as Turkey, South Africa, China, or Korea) as is the United States, which supplies a significant (more than 1.5 per cent) share of the value added to the exports of all countries except Russia. These countries, which can be seen to transcend regional boundaries or which coordinate regional production, can be thought of as 'headquarter' economies, whereas those that use rather than sell their value added can be likened to 'factory' economies (Baldwin and Lopez-Gonzalez 2013).

There is a significant heterogeneity in the share of domestic value added embodied in exports. For example, natural resource-rich countries such as those in South America, Russia, and Australia tend to have higher (lower) domestic (foreign) value added in their exports. But so do economies such as the United States and Japan, since they can draw on larger domestic markets for their intermediates and are more technologically advanced. In contrast, smaller countries and 'factory' economies or regions tend to have lower domestic content of exports: for example, in the Rest of EU and the Rest of Asia groupings, 69 per cent and 63 per cent, respectively, of the value added in exports is domestic.

Changes in the foreign value-added content of exports since 1995 (Annex Figure 2.1) reveal that, for most countries, there has been a growing dependence on foreign value added. For some, like China or Korea, the increases are quite significant (21 and 17 percentage points, respectively), whereas for others, such as the UK, Russia, and Canada, there have been declines – pointing to a growing domestic content share of exports. Notably, most of the bilateral declines appear

Backward participation by origin 2009	*France*	*Germany*	*UK*	*RoEU*	*Turkey*	*South Africa*	*Russia*	*India*	*China*	*Korea*	*Japan*	*RoAsia*	*Australia*	*New Zealand*	*South America*	*Mexico*	*USA*	*Canada*	*RoW*
France		2%		2%															
Germany	4%		2%	4%	2%	2%			2%	2%									3%
UK		2%		2%															
RoEU	8%	10%	5%	8%	4%	2%	2%	3%	3%	3%		3%		2%		2%			5%
Turkey																			
South Africa																			
Russia				2%	4%														
India																			
China								2%		5%	2%	4%				3%			
Korea									3%			2%				2%			
Japan									4%	5%		4%				2%			
RoAsia								2%	6%	5%	2%	6%	3%	2%		2%			
Australia										2%				3%					
New Zealand																			
South America									2%						2%				
Mexico																			
USA	3%	3%	3%	3%	2%	2%		2%	4%	5%	2%	5%	2%	2%	2%	13%		9%	2%
Canada																			
RoW	4%	4%	3%	5%	4%	5%		7%	5%	10%	3%	6%	2%	2%	2%	2%	3%	3%	2%
Domestic	75%	73%	83%	69%	78%	84%	93%	78%	67%	59%	85%	63%	87%	82%	88%	70%	89%	80%	82%
Imported	**25%**	**27%**	**17%**	**31%**	**22%**	**16%**	**7%**	**22%**	**33%**	**41%**	**15%**	**37%**	**13%**	**18%**	**12%**	**30%**	**11%**	**20%**	**18%**

Figure 2.1 Origin of value added in exports – backward participation (2009)

Note: Entries below 1.5 per cent are omitted to improve readability. Each entry identifies the share of column nation value added that the row nation uses to produce a unit of gross exports (the backward linkage). Where row nation France meets column nation Germany, the 4 per cent gives us the share of value added that France uses from Germany to produce a unit of gross exports. Countries are geographically clustered and darker colours within column nations identify larger sources of foreign value added. RoEU includes the remaining 24 EU countries in the trade-in-value-added (TiVA) database, and RoAsia includes the ASEAN-8 group as well as Taiwan and Hong Kong.

Source: Authors' calculations based on OECD TiVA database.

to have concentrated around headquarter economies such as Germany, Japan, and the United States, with important growth in China and the Rest of Asia grouping. Clearly, the geography of global production is shifting towards Asia.

The matrix of forward participation (Figure 2.2) shows how the sales of value added of each country are distributed as a share of gross exports (read across the rows).[3] Strong forward linkages are apparent not just for 'headquarter' economies such as the United States, Germany, or Japan, but also for the natural resource-rich countries, with Russia contributing 45 per cent of its gross exports to other countries' production of exports. Some nations/regions are important destinations of value added for the production of exports, that is, the Rest of EU, China, and the Rest of Asia grouping. These are the 'factory' economies, which source much of their value added from abroad to produce exports.

2.2 Looking inside Factory Asia

In Asia GVC participation is growing (Figure 2.3), with six of the eight ASEAN countries in the sample having prominent and growing backward linkages.[4] Singapore has the largest backward participation, with 50 per cent of the value added in its exports in 2009 of foreign origin. Cambodia, Malaysia, the Philippines, Thailand, and Viet Nam showed values of around 30 per cent. Brunei Darussalam and Indonesia are different – they showed a more developed forward linkage with quite limited backward interactions (only 10 per cent to 15 per cent of the value added in their exports was foreign in 2009). Nevertheless, the combined changes in these linkages confirm that participation in GVCs in the region is growing.

The sources of this growing participation are fairly concentrated (Figure 2.4).[5] ASEAN-8 reliance on the EU is high and ranges from 27 per cent of total foreign value added used to produce exports in Brunei to 12 per cent in Thailand and Viet Nam. Equally, the North American Free Trade Agreement (NAFTA) is seen as a key source of intermediate imports, mainly for Malaysia and Brunei (each 21 per cent). However, regional sourcing is also important and is growing. Brunei, Cambodia, and the Philippines, respectively, rely on intra-regional sources for 22 per cent and 16 per cent, respectively, of their total foreign value added used to produce exports. Thailand and Viet Nam have a lower, albeit non-negligible reliance on intra-regional value added (13 per cent each). To put matters into perspective, regional sourcing (intra-ASEAN) in 2009 was comparable to that from the North American Free Trade Agreement (NAFTA) region and indeed higher, in many instances, than that from China or Japan.

ASEAN-8 sales of value added into the production of other countries' exports underscore the importance of the region as a supplier of intermediate value added to other countries/regions. For example, in 2009, 10 per cent of total foreign value added used in China to produce exports came from ASEAN-8 partners, with this figure close to 12 per cent in the case of Japan. Where changes in extra-regional destinations of value added are concerned, there have been important increases in sales of value added to other Asian countries (rising by two or three percentage points in the period 1995 to 2009) and particularly to Australia–New Zealand (AUS–NZL) (rising by 13 percentage points).

	France	Germany	UK	RoEU	Turkey	South Africa	Russia	India	China	Korea	Japan	RoAsia	Australia	New Zealand	South America	Mexico	USA	Canada	RoW	Total
France		4%		9%																21%
Germany				11%															2%	23%
UK		4%		11%															3%	25%
RoEU		4%		8%																21%
Turkey		3%		5%															3%	16%
South Africa				3%					4%											17%
Russia	2%	5%		19%					4%			3%							4%	45%
India				5%					3%			4%							3%	20%
China				2%								3%								13%
Korea				3%					9%			5%								24%
Japan				4%					9%	3%		8%					2%			33%
RoAsia				3%					7%			6%								25%
Australia				3%					8%	4%	3%	6%								31%
New Zealand				2%					3%			2%	2%							16%
South America				5%					6%			3%							3%	28%
Mexico																	5%			11%
USA		2%		7%					3%			4%				2%		2%	2%	29%
Canada																	6%			15%
RoW	4%	7%	3%	21%				3%	9%	6%	3%	10%					6%		4%	32%

Figure 2.2 Destination of value added in exports – forward participation (2009)

Note: Entries below 1.5 per cent are omitted to improve readability. They show how the sales of value added of each row nation are distributed as a share of gross exports (read across the rows). The entry where column Germany meets row Rest of EU shows that 11 per cent of Germany's gross exports is composed of value added that is sold to the Rest of EU grouping for it to produce exports. Countries are geographically clustered and darker colours within column nations identify larger sources of foreign value added. RoEU includes the remaining 24 EU countries in TiVA, and RoAsia includes the ASEAN-8 group as well as Taiwan and Hong Kong.

Source: Own calculations using TiVA database.

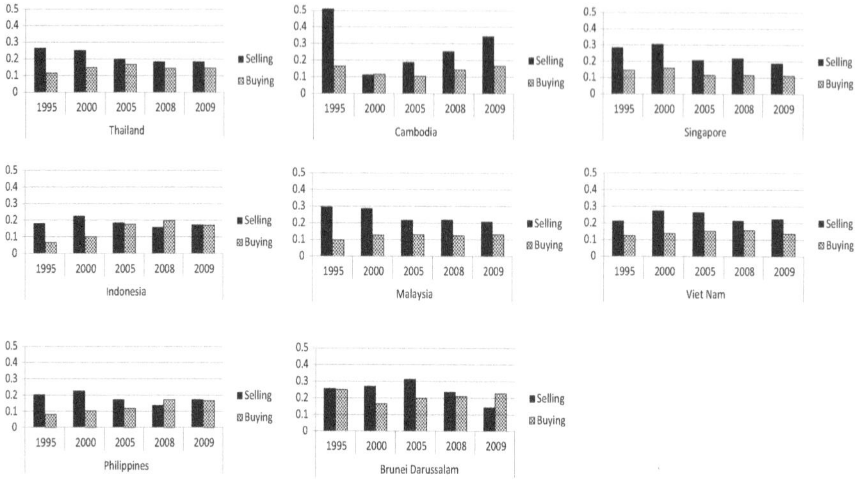

Figure 2.3 Backward and forward participation in ASEAN-8 countries, 1995–2009

Note: Black bars show the backward participation indicator (foreign value-added content of exports). Pattern bars show forward participation (the value added exported to other countries for them to produce gross exports).

Source: Authors' calculations based on OECD–WTO TiVA database.

There were significant changes in sourcing of foreign value added from 1995 to 2009 (see bottom panel of Figure 2.5). Although the EU and NAFTA remain two of the largest sources for most ASEAN-8 countries, there has been a rapid decline for all countries except Singapore. In fact, there appears to be an important switching of sources of foreign value added away from more traditional source countries or regions, such as Japan, NAFTA, and the European Union (EU), towards intra-regional and Chinese sourcing.[6] This is indicative of the important process of regionalization of economic activity occurring in Southeast Asia, with China playing an increasingly pivotal role.

Patterns of specialization in the region, seen through the sectoral decomposition of the domestic value added embodied in exports in Figure 2.5, are fairly heterogeneous. Brunei's domestic value added overwhelmingly comes from the primary sector (74 per cent of total domestic value added in 2009), with this sector also being the most important for Viet Nam (51 per cent), Cambodia (40 per cent), and Indonesia (40 per cent). In contrast, Singapore is much more service sector–oriented (61 per cent), with the Philippines (49 per cent), Malaysia (36 per cent), and Thailand (29 per cent) following closely.

Changes over time also show diverse patterns of specialization.

- The share of primary sector value added rose by 12 percentage points (from 1995 to 2009) in Brunei, with half of these points shaved off domestic service sector value added in exports.

2009	EU	India	Brunei Darussalam	Cambodia	Indonesia	Malaysia	Philippines	Singapore	Thailand	Viet Nam	China	Chinese Taipei	Hong Kong, China	Korea	Japan	AUS-NZL	NAFTA	ROW
EU	51%	21%	27%	21%	14%	19%	13%	23%	12%	12%	16%	9%	20%	13%	14%	20%	19%	47%
India	1%	0%	1%	1%	2%	2%	1%	3%	1%	2%	2%	2%	3%	1%	1%	1%	1%	1%
ASEAN-8	2%	8%	22%	16%	15%	12%	16%	10%	13%	13%	10%	11%	10%	9%	12%	21%	4%	3%
China	4%	8%	4%	17%	9%	9%	11%	6%	10%	16%	0%	13%	20%	12%	11%	8%	8%	4%
Chinese Taipei	1%	1%	1%	5%	2%	3%	7%	3%	4%	6%	7%	0%	3%	3%	3%	1%	2%	1%
Hong Kong, China	1%	1%	1%	5%	1%	1%	3%	1%	1%	1%	2%	1%	0%	1%	1%	1%	0%	1%
Korea	1%	3%	2%	3%	4%	4%	8%	4%	4%	8%	9%	6%	3%	0%	4%	3%	3%	2%
Japan	3%	4%	11%	5%	10%	11%	12%	8%	15%	11%	13%	16%	7%	13%	0%	7%	7%	3%
AUS-NZL	1%	4%	2%	1%	4%	3%	2%	2%	3%	3%	4%	4%	3%	5%	7%	0%	1%	1%
NAFTA	12%	12%	21%	16%	13%	21%	16%	18%	11%	9%	13%	12%	16%	13%	17%	16%	33%	14%
Rest of the World	24%	37%	8%	10%	24%	14%	11%	21%	24%	19%	24%	27%	15%	31%	30%	21%	22%	14%

Figure 2.4 ASEAN-8 sourcing of value added in exports (2009)

Note: Top panel: numbers show the value of column nation sales to row nation as a share of globally traded value added in final demand; values below 3 per cent are not presented. Bottom panel: changes below 1.5 percentage points are omitted.

Source: Authors' calculations based on OECD–WTO TiVA database.

	Changes 2009–1995																
	EU	*India*	*Brunei Darussalam*	*Cambodia*	*Indonesia*	*Malaysia*	*Philippines*	*Singapore*	*Thailand*	*Viet Nam*	*China*	*Chinese Taipei*	*Hong Kong, China*	*Korea*	*Japan*	*AUS-NZL*	*NAFTA*	*ROW*
EU	–9%	–7%	–5%	–24%	–8%	–2%	–4%	0%	–5%	–16%	3%	–7%	–1%	–2%	–3%	–11%	–1%	–9%
India	1%	0%	1%	1%	1%	2%	0%	3%	1%	2%	1%	1%	2%	1%	0%	1%	1%	1%
ASEAN-8	0%	2%	–3%	0%	9%	3%	8%	–4%	2%	0%	3%	3%	2%	3%	1%	13%	1%	1%
China	3%	6%	3%	15%	7%	7%	9%	3%	8%	12%	0%	10%	5%	7%	7%	6%	6%	3%
Chinese Taipei	0%	0%	0%	4%	–2%	0%	1%	0%	0%	–2%	–3%	0%	–3%	1%	0%	–1%	0%	0%
Hong Kong, China	0%	0%	0%	4%	0%	0%	–2%	0%	0%	–2%	–2%	–1%	0%	0%	0%	0%	0%	0%
Korea	0%	1%	1%	1%	–5%	1%	2%	0%	1%	–4%	–1%	2%	–2%	0%	–1%	1%	1%	1%
Japan	–2%	–4%	2%	0%	–7%	–12%	–9%	–11%	–12%	0%	–7%	–13%	–9%	–12%	0%	–4%	–6%	–1%
AUS-NZL	0%	1%	0%	1%	–1%	–1%	–1%	0%	1%	1%	1%	1%	0%	0%	1%	0%	0%	0%
NAFTA	–1%	–6%	2%	–8%	–2%	–1%	–4%	0%	–6%	–3%	–4%	–10%	1%	–8%	–12%	–12%	–11%	–2%
Rest of the World	7%	7%	–1%	7%	8%	5%	–1%	10%	11%	12%	8%	13%	5%	11%	7%	7%	9%	7%

Figure 2.4 (Continued)

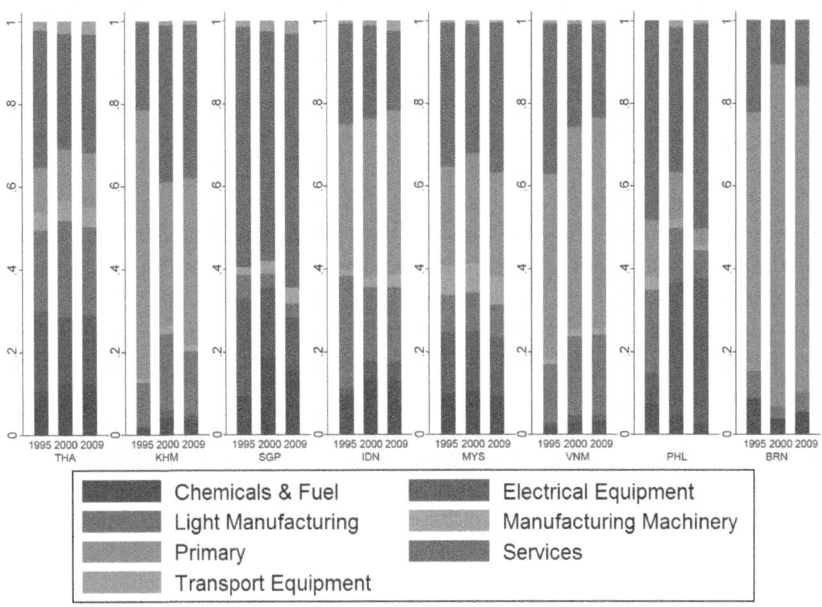

Figure 2.5 Share of domestic value added embodied in exports by source sector, 1995–2009

Note: Bars show the distribution of domestic value added across originating sector.

Source: Authors' calculations based on OECD–WTO TiVA database.

- In Viet Nam, the six- and five-percentage-point relative increases in primary and light manufacturing value added in exports, respectively, also came at the expense of an important relative decline in service sector value added.
- Indonesia's declining sectors are light manufacturing (–9pp) and services (–5pp), giving way to increases in the relative share of domestic value added in the primary sector (5pp) and, to a lesser extent, chemicals (3pp) and electrical equipment (3pp).
- Cambodia witnessed a very important shift in its domestic value added in exports with primary sectors falling by 25 percentage points and services increasing their share by 16 percentage points. Light manufacturing also increased by five percentage points.
- There were only small changes in Malaysia, with very slight increases in primary and services value added (1pp each) and corresponding falls of one percentage point in electrical equipment, light manufacturing, and chemicals.
- In contrast, the Philippines witnessed a very dramatic relative increase in its electrical equipment value added in exports, which went up by 28 percentage points and displaced light manufacturing (–13pp) and primary (–10pp) value added in exports.

- Singapore saw an 11-percentage-point relative decline in its electrical equipment value added, with increases in the chemical and fuel sector (6pp) and the service sector (3pp).
- Specialization in Thailand favoured the primary sector and light manufacturing (2pp increases in each from 1995), with relative declines in service sector value added (–4pp) and electrical equipment (–3pp).

Other sectors are notable for their absence in terms of shares in overall value added in exports, as well as changes in time such as the 'manufacturing and machinery' and the 'transport equipment' sectors. These sectors are often identified as highly 'fragmentable', and therefore it is somewhat surprising that ASEAN-8 countries have not engaged more widely in these, particularly when considering the location of large motor vehicle and machinery manufacturers in the vicinity (Japan, Korea, and China).

In terms of the composition of *foreign* value added in exports (Figure 2.6), the main import sector of value added in exports is services for all ASEAN countries, although in some instances this declined quite fast in 1995–2009 (i.e., by 19pp in Cambodia, 17pp in Viet Nam, and 13pp in Thailand).[7] In contrast, all countries

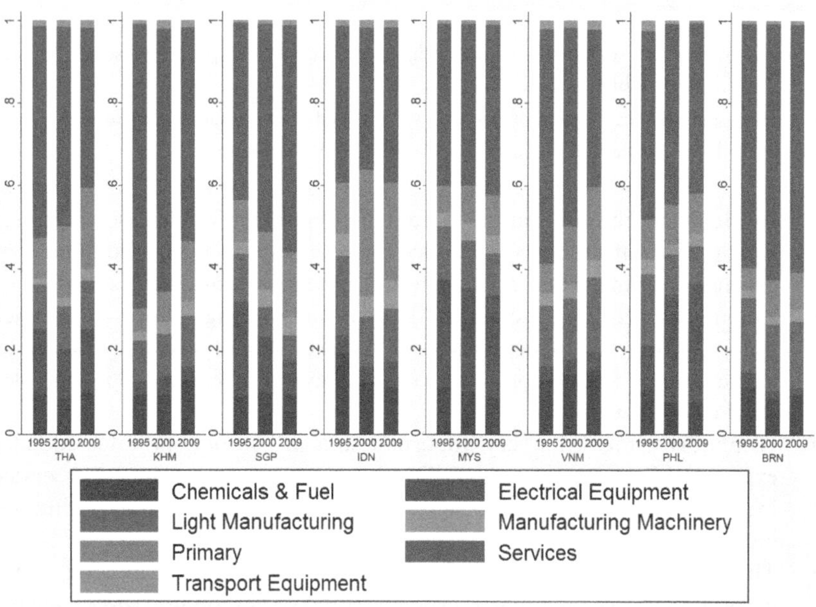

Figure 2.6 Share of foreign value added embodied in exports by source sector, 1995–2009

Note: Bars show the distribution of foreign value added across originating sector.

Source: Authors' calculations based on OECD–WTO TiVA database.

increased their use of foreign value added from the primary sector. This was most pronounced in Indonesia and Viet Nam (11pp increase), as well as in Thailand (10pp) and Cambodia (9pp). Other notable changes arise in the electrical and optical equipment sectors, where the Philippines increased its foreign sourcing by 18 percentage points, whereas Singapore reduced its foreign sourcing by 15 percentage points.

In 1995, the sources of foreign value added were concentrated (see Annex Figure 2.4). Japan was the top source in all manufacturing sectors except 'manufacturing and machinery', where the EU dominated, as it did in services, with NAFTA also playing a very strong role. Intra-regional sourcing was important in primary products and to a lesser extent chemicals and fuels. Most striking is the small role China played in 1995.

By 2009, sourcing patterns had changed dramatically. Sourcing from Japan had significantly declined with accompanying increases in Chinese and regional sourcing. ASEAN-8 partners became the top origin of 'chemicals and fuel' value added and the second source of 'electrical equipment' value added. China was the top origin for 'light manufacturing' value added; Japan lost ground to NAFTA as a source of 'transport equipment' and the EU maintained its importance as a source of services value added in exports. These patterns further corroborate the important regionalization of ASEAN economic activity and underscore the importance of neighbouring Asian partners or, in other words, the consolidation of Factory Asia as an entity.

3. How do countries benefit from participation?

Measurement and analysis of outcomes of participation in GVCs has been couched in terms of economic and social 'upgrading'. A recent review of literature on this topic by Milberg and Winkler (2010) suggested that *economic upgrading* is usually defined in terms of efficiency of the production process or characteristics of the product or activities performed, while *social upgrading* often refers to outcomes related to employment and pay, gender, and the environment. Focusing on economic upgrading, Humphrey and Schmitz (2002) distinguished between the following types of upgrading:

1 Process upgrading – where firms are seen to gain in terms of efficiency in producing a given type of output.
2 Product upgrading – where firms engage in the production of more sophisticated products.
3 Functional upgrading – acquiring new functions (such as, for example, service provision) within a given value chain.
4 Chain upgrading – where firms move into different value chains.

The *process-upgrading path* explicitly refers to efficiency, whereas *product*, *functional*, and *chain upgrading* refer to the type of activities performed in value chains without an explicit reference to value creation or productivity. Somewhat curiously, the latter three upgrading paths have been interpreted in the upgrading literature to imply that targeting specific 'sophisticated' products or production stages can be a viable strategy for upgrading in value chains. For example, the 'smiley' curve exposition, which suggests that the value added generated at the product design and marketing stages may be higher than that at the assembly or manufacturing stages, has been interpreted to imply that it may be beneficial to move away from the assembly or manufacturing parts of the chain.

Recently, and perhaps mistakenly, the concept of upgrading has therefore been seen as the need to capture a growing share of domestic value added in exports. This idea is partly spurred by the oft-cited iPad case study (Xing and Detert 2010; Kraemer et al. 2011), which underlined the low value-adding share that assembly occupies in the production process (less than 5 per cent of the sale value of the iPad remains in China), which has been used to justify policy objectives that seek to increase the *share of* domestic value added. This narrow view of upgrading may, however, miss the point that the volume of the activity may matter just as much as the domestic share of the value of the product. Although it is indeed true that assembly activities often represent a very small share of the value of the final products being assembled, it is also true that important benefits can be derived from specializing in assembly activities and performing them on a large scale.

For example, specialization in assembly of electronic devices has served several firms in Asia well. They have become assemblers par excellence and attracted clients such as Apple, Dell, Amazon, Nokia, and Samsung. From a firm perspective, adding 5 per cent of the value of these firms' leading products adds up to a relatively large sum.[8] As an alternative business development strategy, these assemblers could have instead launched a new mobile phone to rival the larger smartphone producers and to enter the higher end of the smiley curve to capture larger shares of the value of the final product, but they would at the same time have to capture a significant market share from the other electronic device producers. What is important, from the point of view of the firm, is therefore the value that is created from its economic activities and not the share that the firm occupies in the value of the final product.

Indeed, it is unlikely that profit-maximizing firms would switch to alternative products or value chain stages if this is not accompanied by higher productivity.[9] The relevant economic concept here is the principle of comparative advantage, which teaches us that some firms – or some countries – will be able to obtain higher overall gains from GVC participation by specializing in what could be considered less sophisticated products, tasks, or functions within the chain. Which segments of the value chain will be profitable will be determined jointly by the characteristics of the production process as well as relative skills and resource endowments of the firms and countries in question (i.e. the comparative advantage). What this suggests is that productivity may be a preferred unifying characteristic of upgrading in GVCs.

A good illustration of some of the pitfalls associated with defining actual upgrading is the case of China's production of electric and optical equipment. As can be seen in Figure 2.7, the domestic content of China's electrical and

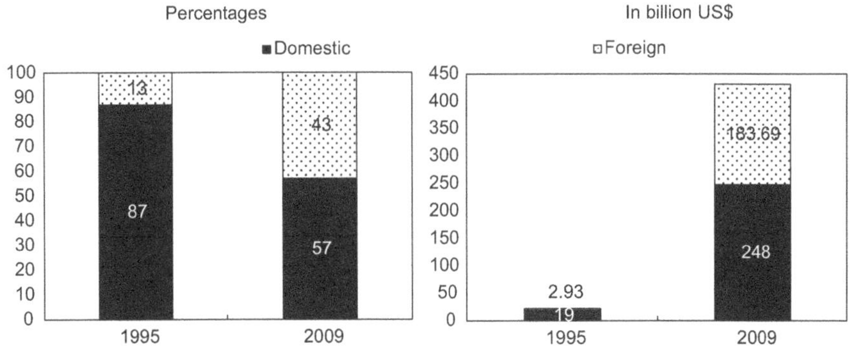

Figure 2.7 Enjoying a smaller share of a larger pie: electrical and optical equipment in China

Source: OECD (2015).

optical equipment exports fell from 87 per cent to 57 per cent between 1995 and 2009, but the volume of domestic value added embodied in exports increased more than tenfold. This also led China to grow its domestic share of global value added in exports of electrical and optical equipment from 3 per cent in 1995 to 22 per cent in 2009. While certain other developments in the Chinese and global economies not accounted for in the figure may have played a role, these figures provide prima facie evidence that firms operating in China have increased the foreign content of their products while multiplying their overall sales and profits, as well as the wage bill for the workers they employ.

This tends to support the findings from OECD research based on the TiVA database that trade is increasingly *complementary* and that access to competitive intermediate products is likely to have played an important role in helping China to gain such a large share of global value added in electrical and optical equipment. The question is whether a similar phenomenon has been at work across the ASEAN region.

There is an element of synchronicity in terms of the growing importance of domestic and foreign value added in ASEAN (see Figure 2.8), where for all countries, except for Cambodia, a positive correlation emerges between changes in the overall importance of domestic and foreign value added across sectors. This in turn suggests that the sectors which are performing best in the economy, in terms of growing domestic value added in exports, tend to be those where foreign value added has also grown the most, or, in other words, where there is some evidence that importing may help develop domestic productive capacity.

3.1 What does the empirical evidence suggest?

Recent OECD analysis has shed more rigorous empirical light on how GVC participation may influence the economic performance of countries, by assessing the

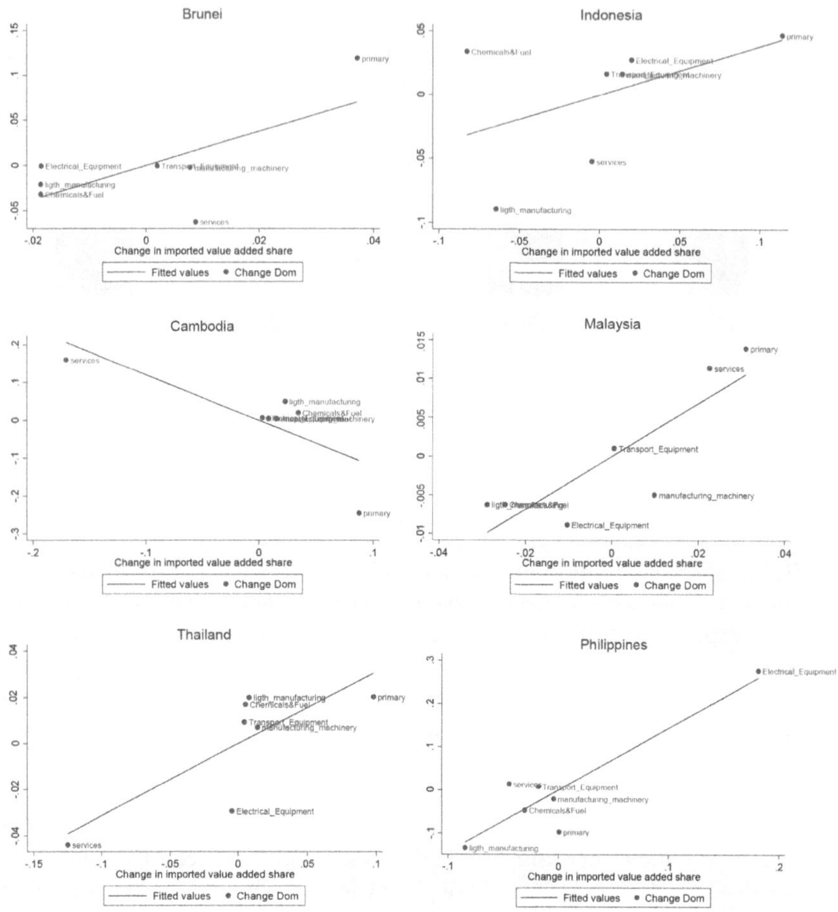

Figure 2.8 Change in domestic and foreign value added in exports across sectors, 1995–2009

Note: Horizontal axis shows change in the share of foreign value added used in a sector relative to total foreign value added used. Vertical axis shows the same but for domestic value added use.

Source: Authors' calculations based on OECD–WTO TiVA database.

relationships between a number of measures of GVC engagement and outcomes associated with such engagement. The first investigated GVC outcome measure was the overall per capita domestic value added embodied in a country's exports – a value-added measure of productivity changes in the production of exports or loosely a measure of process upgrading. The second was a measure of export sophistication as defined by Hausmann et al. (2007),[10] which can be used to capture product upgrading. The third measure captured the diversification of exported products, which is considered an important indicator of competitiveness and quality of integration with international markets (e.g., Cadot et al. 2011) and also broadly encapsulated functional or chain upgrading.[11]

Changes in per capita domestic value added embodied in exports provide a comprehensive indicator of GVC performance, since they capture the gains associated with exporting which accrue to domestic labour and capital.[12] Regressing this measure against changes in the use of foreign value added in exports (the value of the backward linkage) reveals the complementary nature between domestic value addition and the use of foreign value added in the form of imported intermediate inputs (Table 2.1, column 1).[13] It was also found that positive changes in the measure of the sophistication of imported manufacturing and primary intermediates have a positive effect on productivity along the value chain, suggesting that access to sophisticated inputs can help increase productivity. Positive changes in per capita gross domestic product (GDP) also translate into positive changes, but a growing distance from economic activity (in terms of being far away from growing poles of economic activity) is seen to have a negative impact.

The measure of sophistication of export bundles based on the methodology of Hausmann et al. (2007) is a proxy for product upgrading (Table 2.1, column 2). Positive changes in this variable imply growing sophistication of the products being exported, and it was found that a growing backward participation (in terms of the share of foreign value added in exports) is associated positively with exporting more sophisticated export bundles. Additionally, the use of more sophisticated inputs and higher per capita GDP is associated with product upgrading while positive changes in foreign direct investment (FDI) inflows are not.

Finally, other forms of gains from value chain participation such as, for example, functional upgrading through a measure of the diversification of exports were investigated (Table 2.1, column 3).[14] Both positive changes in backward participation and in the use of more sophisticated non-primary imported intermediates were associated with diversification, as was a growing distance from economic poles of activity.

Overall, the results suggest that engaging more widely in GVCs, whether by using more foreign value added embodied in imported intermediates or importing more sophisticated intermediates, correlates with positive outcomes. This in turn suggests that there might be a case to support further GVC engagement. What, then, is the role for policy in promoting it?

4. What role can policy play?

There are two practical reasons for looking at the determinants of GVC participation. The first is to assess whether countries are performing above or below their potential; the second is to disentangle the role that different structural and policy factors play in determining participation.

Table 2.1 Determinants of domestic content of exports across income groups

	(1) *log of per capita domestic value added in exports* *0.0124*'' (0.00568)*	*(2)* *log of export sophistication*	*(3)* *normalized trade concentration indicator*
Variables	0.0124** (0.00568)		
Backward log of value (lag)		0.192*** (0.0914)	-0.232*** (0.0488)
Backward (ratio)	9.427** (4.288)	6.852** (3.222)	-4.032** (1.720)
Sophistication of manufactured intermediates (square of log)	-0.502** (0.224)	-0.364*** (0.169)	0.211*** (0.0900)
Log of sophistication of primary intermediates	0.0310 (0.0250)	-0.0663*** (0.0185)	-0.00442 (0.00990)
FDI inflows (log)	0.000522 (0.000458)	-0.000723** (0.000342)	0.000125 (0.000183)
Imports covered by FTA (share)	0.000755 (0.0351)	-0.0170 (0.0261)	0.0190 (0.0140)
Per capita GDP at constant prices(log)	0.9333*** 0.0384	0.205*** (0.0286)	0.0542*** (0.0153)
Distance to economic activity (log)	-2.221*** (0.355)	-0.354 (0.264)	0.347*** (0.141)
Constant	-31.63 (21.24)	-21.05 (15.94)	15.96* (8.513)
Observations	2,050 0.814	2.064 0.374	2,064 0.037
R-squared	Y	Y	Y
Year FE	Y	Y	Y
Reporter FE			

Understanding how a country is integrated in production networks requires more than just looking at their relative backward participation rate. That is to say, it is not because Japan has a rate of 12 per cent and China one of 33 per cent that it can be said that Japan is less integrated than China. Indeed, larger countries tend to have lower rates of participation, and this is often attributed to the fact that they have larger domestic markets from which to draw their intermediates. To understand how countries are performing, it is important to engage in a benchmarking exercise that determines whether countries are performing above or below what would be expected on the basis of their characteristics.

This can be done by decomposing the policy and non-policy determinants of backward participation (Figure 2.9) by regressing the participation indicators

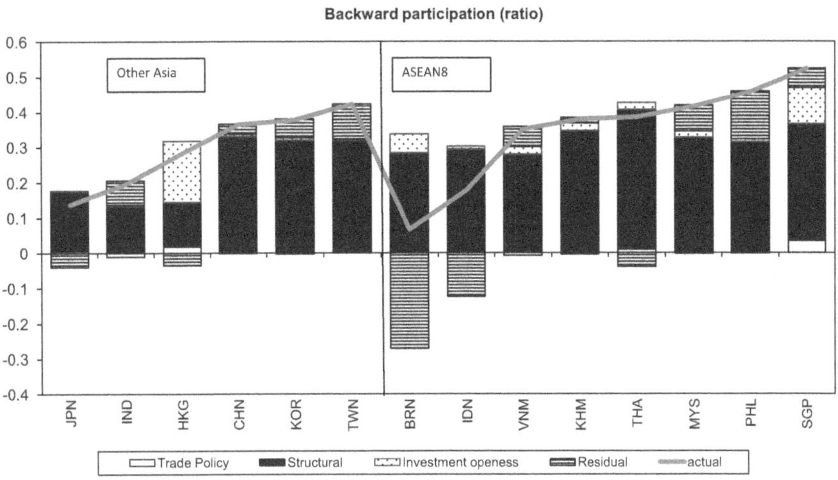

Figure 2.9 Determinants of participation

Note: The bars show how the predicted backward linkage (from the model) decomposes according to these structural and policy elements with the line then showing the actual observed degree of backward participation. Data shown for year 2005. See OECD (2015) for a more detailed description of the variables and estimation methods.

Source: Adapted from OECD (2015).

against a set of structural parameters (such as economic size, share of manufacturing in GDP, and distance to economic centres of activity) and policy variables (such as tariffs faced or charged, presence of free trade agreements [FTAs]), and openness to foreign direct investment (FDI) as in OECD (2015).

The first message is that the structural characteristics of countries are the main determinants of participation. That is to say that the size or geographical location of countries, as well as their manufacturing share in GDP, explains most of the variation in participation rates. Nevertheless, there is also an important role to be played by policy. For example, Singapore and Hong Kong exemplify how an open trade and investment policy is conducive to high levels of participation. Where trade policy alone is concerned, its contribution is generally modest, although it is important to note that this variable controls for both push and pull factors in that it aggregates the tariffs that industries apply, as well as those that are applied on exports. Hence, for example, the negative role of trade policy in Viet Nam is largely driven by high average tariffs on the products that Viet Nam exports.

To understand how well a country is 'performing' in terms of its integration into GVCs, one can look at the difference between predicted and actual participation. For example, the results suggest that Indonesia and Brunei may be 'underperforming' (i.e., the predicted participation rate is below the observed rate). This is due to unobserved (in our model) characteristics of countries, which,

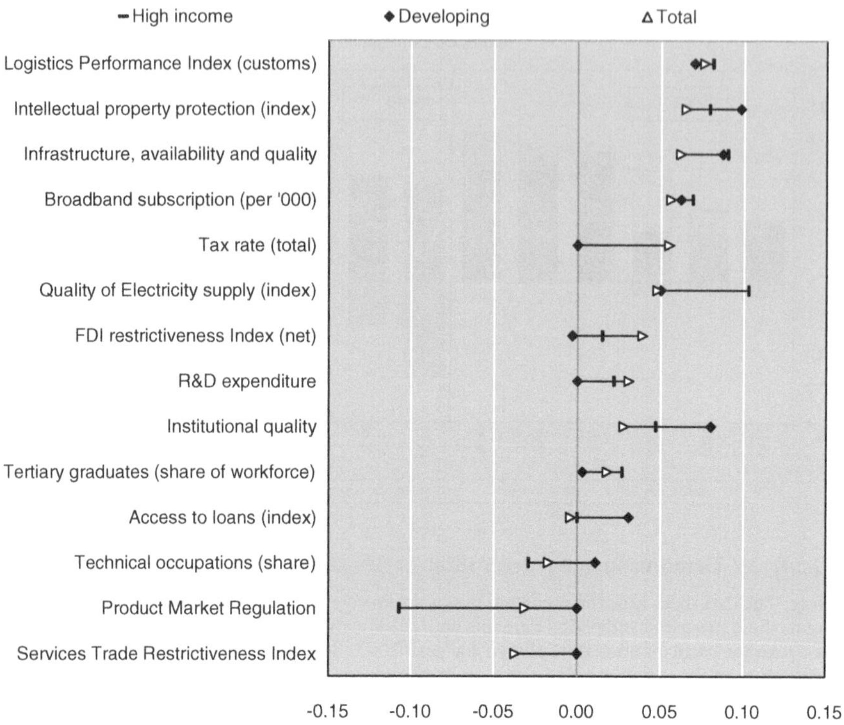

Figure 2.10 What other policies can influence participation?
Source: OECD (2015).

although hard to pin down, suggest that there is room for improvement where participation is concerned. For developing countries, this seems to be the case particularly in areas such as logistics performance, intellectual property (IP) protection, infrastructure, and the quality of institutions (Figure 2.10).

5. What can ASEAN policy makers do?

Important changes are afoot in a region that is increasingly looking inward for sources of intermediates. No longer do countries just specialize in assembly activities involving exports of final goods, they are now increasingly supplying intermediates to extra-regional destinations.

This general point masks what is otherwise relatively heterogeneous value chain participation across countries in the region. Whilst Brunei's domestic value added in exports is overwhelmingly specialized in the primary sector, Singapore is highly specialized in services. The Philippines, on the other hand, has significantly increased its domestic value added in exports from the electrical equipment sector, where it has also witnessed an important surge in the use of foreign value added.

Complementarities between domestic and foreign value added in exports are emerging and supported by the fact that changes in the domestic and imported elements of value added in exports appear to show a strong positive correlation (for all countries except Malaysia). This suggests that the use of foreign value added may indeed help develop, rather than substitute for, domestic value added creation.

These fast evolving patterns of specialization have arisen in the context of an evolving process of deepened economic and social integration with the near completion of the ASEAN Economic Community (by the end of 2015) and enhanced connectivity (through investment in hard and soft infrastructure underpinned in the ASEAN Blueprint on connectivity). Although these processes are linked to the positive GVC performance in the region, challenges still lie ahead, both internal and external. Competitive pressures are likely to grow as other countries increasingly look to joining GVCs, which reinforces the importance of continued efforts at further openness and domestic regulatory reform.

Looking inwards

Important differences remain in the levels of GVC participation within ASEAN, with the CLMV countries less advanced than the other six.[15] The ASEAN Economic Community (AEC) regional integration effort is likely to encourage greater convergence (with Pillar 3 giving particular consideration to the development aspect), but it is finalizing the internal market that is likely to be most conducive to this catch up. Anecdotally, it appears that the ASEAN–6 countries see their CLMV neighbours as important complements to their GVC strategies. They see them as offering new economic opportunities for offshoring parts of their production and therefore 'upgrading' within the value chain. These aims will be boosted by moves to further integration, first by eliminating intra-regional and most favoured nation (MFN) tariff barriers, and second by finalizing the internal market so that services and investment can move freely within the region (OECD 2014).

The ASEAN–6 countries have made good progress to date, but again there are lessons that can be learned. Countries like Indonesia, the Philippines, and Thailand, for example, can learn from Malaysia and Singapore. There are big differences between these in terms of logistics performance, infrastructure, and quality of institutions. The work to finalize the single market is likely to help convergence, and this, coupled with domestic reform aimed at increasing institutional quality and logistics performance, will also complement regional efforts.

Looking outwards

Recent OECD work (OECD 2015) shows that importing more intermediates from regional partners is associated with positive changes in comparative advantages. But it also shows that importing from non-regional partners can be important. Export competitiveness now depends on having access to the most competitive inputs, and sometimes these cannot be found in the region. This means that continued efforts to liberalize beyond the region, whether plurilaterally or, most effectively, multilaterally will likewise boost the benefits that ASEAN countries can derive from GVCs. In addition, efforts to streamline the number of trading arrangements would help to reduce costs for business.

In conjunction, access to new ideas and technologies embodied in foreign investments are also important. Openness to FDI in some ASEAN countries remains low (Indonesia, the Philippines, and Viet Nam are all below world averages in this respect), and this implies that there might be an opportunity cost in terms of foregone access to new technologies and evolving production techniques. Wider efforts to liberalize foreign investment should therefore also be pursued, particularly in light of the participation enhancing effects of FDI, as discussed.

Notes

1 This chapter draws heavily on a recent OECD report titled *Developing Countries' Participation in Global Value Chains: Implications for Trade and Trade Related Policies*. It was authored by Przemyslaw Kowalski, Javier López-González, Alexandros Ragoussis, and Cristian Ugarte. Research support was given by Pascal Achard and Charles Cadestin. This chapter has benefited from insightful comments and suggestions by Trudy Witbreuk. Views expressed in this chapter are the authors' alone, and do not necessarily represent those of the OECD or its members.
2 OECD, Trade and Agriculture Directorate, Development Division.
Corresponding author: Javier Lopez Gonzalez (Javier.lopezgonzalez@oecd.org).
3 Note that the numerator in each of these shares is the same as that of the backward linkage and that it is the denominator that changes.
4 Eight of 10 ASEAN countries are covered in TiVA – Brunei Darussalam, Cambodia, Indonesia, Malaysia, the Philippines, Singapore, Thailand, and Viet Nam. Those not covered are Lao PDR and Myanmar.
5 Since all foreign origins are captured, the column sum of the shares adds up to 100 per cent, and to improve readability darker shades identify more important sources of value added for each nation.
6 Note that although this analysis allows pinpointing the ultimate origin of value added, it does not allow us to distinguish whether this comes from domestic or foreign firms. That is to say, it is possible that Japanese firms are increasingly locating in ASEAN countries and that this is driving some of these source-switching changes. This is important because profits can ultimately be repatriated, and therefore value added may not necessarily be retained in the ASEAN country in question. Nevertheless, it is still the case that many of the workers within a country will be domestic and therefore value is retained in this form, irrespective of the ownership.
7 Note that this includes direct and indirect imports. The direct component captures value added which is directly sourced from service sectors, whereas the indirect element is the value added which is embedded in products used to produce exports. So a country can have large shares of imported value added in exports from the service sector solely by importing components which have a high service content.

8 Another example of this is found in the business model of the British firm ARM Holdings. The firm licences architectures for mobile devices and these are pervasive across the mobile device market, allowing the firm a very small share of a single mobile phone but a presence in most of the market.
9 This is especially true if an appropriate time frame is considered since some firms may make strategic choices and accept lower initial profitability to maximize profits in the long term.
10 Hausmann et al. (2007) posited that product sophistication can be proxied by per capita incomes of countries which typically export or supply them and that countries exporting more sophisticated products defined in this way tend to grow faster.
11 While the OECD TiVA is the preferred source of data, the EORA database is used as the source of trade-in-value-added data to maximize the coverage of countries. Non-value-added-based measures of upgrading and controls are calculated using the BACI data set and the World Bank Development Indicators. See the appendix for a discussion of these.
12 Since interest lies in the dynamic nature of GVC specialization and performance, the estimations isolate the impact of *changes* in explanatory variables on *changes* in the performance variables through the use of country-level fixed effects, which serve to restrict the variance of the variables to a temporal dimension only. This also means that time-invariant, country-specific characteristics such as geographical location or absolute differences in aggregate technology, which are likely to also affect gains from GVC participation, are controlled for.
13 This variable is lagged by one year to avoid mechanical changes imposed by the I/O structure of the system. There is an implicit linear relationship between domestic and foreign value added in exports (which is determined by the Leontief technology). For example, if the value added in exports is 70 per cent domestic and 30 per cent foreign, then one would expect that a 1-unit increase in exports will mechanically lead to a 0.7-unit increase domestic value added in exports and a 0.3-unit increase in foreign value added in exports. To avoid this mechanical association, lags of the independent variable are used.
14 This would be the case especially under the proviso that a lower degree of export concentration should be correlated with a more diversified exporting structure and therefore new export functions within the economy. We use a normalized trade concentration indicator (nTCI) and regress this on the same variables as above. A positive change in the nTCI implies a concentration of exporting activities, and therefore we are looking at whether our GVC indicators have a negative coefficient (i.e., whether they are correlated with diversification).
15 The CLMV countries are Cambodia, Lao PDR, Myanmar, and Viet Nam.
16 This indicator of value chain activity is the one preferred by Timmer et al. (2012).
17 Final demand incorporates household and government consumption as well as gross fixed capital formation (GFCF).
18 This is a similar metric to what was presented as the ASEAN-8 row in Figure 2.5, but for the production of exports rather than to satisfy final demand.

References

Baldwin, R and López-González, J 2013, 'Supply-chain trade: a portrait of global patterns and several testable hypotheses', *CEPR Discussion Paper Series* No. 9421, CEPR, Washington, DC.

Cadot, O, Carrère, C, and Strauss-Kahn, V 2011, 'Export diversification: what's behind the hump?', *The Review of Economics and Statistics*, vol. 93, no. 2, pp. 590–605.

Gereffi, G and Fernandez-Stark, K 2011, *Global Value Chain Analysis: A Primer*, Center on Globalization, Governance & Competitiveness (CGGC), Duke University, Durham, NC.

Hausmann, R, Hwang, J, and Rodrik, D 2007, 'What you export matters', *Journal of Economic Growth*, vol. 12, no. 1, pp. 1–25.

Hummels, D, Ishii, J, and Yi, K-M 2001, 'The nature and growth of vertical specialization in world trade', *Journal of International Economics*, vol. 54, no. 1, pp. 75–96.

Humphrey, J and Schmitz, H 2002, 'How does insertion in global value chains affect upgrading in industrial clusters?', *Regional Studies*, vol. 36, no. 9, pp. 1017–1027.

Koopman, R, Powers, W, Wang, Z, and Wei, S-J 2011, 'Give credit where credit is due: tracing value added in global production chains', *NBER Working Paper* No. 16426, NBER, Cambridge, MA.

Kraemer, KL, Linden, G, and Dedrick, J 2011, *Capturing Value in Global Networks: Apple's iPad and iPhone*. Available from: http://pcic.merage.uci.edu/papers/2011/Value_iPad_iPhone.pdf [19 June 2015].

Milberg, W and Winkler, D 2010, 'Trade crisis and recovery: restructuring of global value chains', *Policy Research Working Paper Series* No. 5294, The World Bank, Washington, DC.

OECD 2014, *Southeast Asia Investment Policy Perspectives*, OECD, Paris.

OECD 2015, 'The participation of developing countries in global value chains: implications for trade and trade-related policies', *OECD Working Paper* No. 179, OECD, Paris.

Timmer, M, Erumban, AA, Gouma, R, Los, B, Temurshoev, U, de Vries, GJ, Arto, I, Genty, VAA, Neuwahl, F, Rueda-Cantuche, JM, Villanueva, A, Francois, J, Pindyuk, O, Poschl, J, Stehrer, R, and Streicher, G 2012, *The World Input–Output Database (WIOD): Contents, Sources and Methods*. WIOD Background Document. Available from: www.wiod.org.

Xing, Y and Detert, N 2010, 'How the iPhone widens the United States trade deficit with the People's Republic of China', *ADBI Working Paper Series* No. 257, ADBI, Tokyo.

Annex

	France	*Germany*	*UK*	*RoEU*	*Turkey*	*South Africa*	*Russia*	*India*	*China*	*Korea*	*Japan*	*RoAsia*	*Australia*	*New Zealand*	*South America*	*Mexico*	*USA*	*Canada*	*RoW*
France			−1%	−1%															−1%
Germany	1%		−1%	−1%	1%				1%	1%									
UK																			0%
RoEU	2%	3%	−1%	1%	2%		−1%	1%	2%	1%	1%					1%			
Turkey																			
South Africa																			
Russia	1%	1%		1%	2%					1%									
India									1%			1%							
China	1%	1%		1%	1%	1%		2%		4%	1%	2%	1%	1%	1%	3%	1%	1%	
Korea									2%							1%			
Japan			−1%						2%	−1%		−4%					−1%	−1%	
RoAsia								2%	4%	3%	1%	1%	2%	1%		2%			
Australia								1%	1%	1%	1%								
New Zealand																			
South America									1%	1%									
Mexico																			
USA	1%	1%	−1%				−1%	1%	2%			−1%	−1%	−2%		−5%		−6%	
Canada																			
RoW	2%	2%	1%	2%	2%	3%	−1%	5%	4%	7%	2%	3%	1%	1%	1%	1%	1%	2%	1%
Imported	7%	8%	−3%	3%	11%	5%	−4%	12%	21%	17%	8%	1%	1%	1%	1%	4%	3%	−4%	0%

Annex Figure 2.1 Changes in value added in exports, 2009

Note: Entries below 1.5 per cent are omitted to improve readability.

Source: Authors' calculations based on OECD TiVA database.

Domestic Value Added embodied in foreign final demand	*France*	*Germany*	*United Kingdom*	*Rest of EU*	*Turkey*	*South Africa*	*Russian Federation*	*India*	*China*
France				1%					
Germany	1%		1%	3%					
United Kingdom				1%					
Rest of EU	2%	2%	1%	5%			1%		1%
Turkey									
South Africa									
Russian Federation				1%					
India									
China				1%					
Korea									
Japan									1%
Rest of Asia				1%					1%
Australia									
New Zealand									
South America									
Mexico									
United States		1%	1%	2%					1%
Canada									
RoW	1%	1%	1%	2%				1%	1%
Total	5%	6%	5%	19%	1%	1%	2%	3%	6%

Annex Figure 2.2 Global matrix of value-added trade, 2009

Note: numbers show the value of column nation sales to row nation as a share of globally traded value added in final demand. Values below 0.5 per cent are zeroed out.

Source: Adaptation of Baldwin and López-González (2013) using TiVA.

Global matrix of trade in value added

Annex Figure 2.2 shows how global production is fragmented across countries in 2009 by mapping the ***domestic value added embodied in foreign final demand*** (as extracted from the OECD–WTO TiVA database).[16] This measure captures the extent to which countries rely on different foreign sources of value added for their consumption.[17] The matrix shows column nation sales of value added used by row nations as a share of global trade in value added. The entries therefore identify the global importance of the bilateral production links that tie countries together (only values above 0.5 per cent are highlighted). For example, where the column entry for France meets the row entry for Germany, the 1 per cent represents the share of French sales of value added to Germany over global traded value added in foreign final demand. This global matrix allows us to pinpoint the focal points of value creation across the world.

	Korea	*Japan*	*Rest of Asia*	*Australia*	*New Zealand*	*South America*	*Mexico*	*United States*	*Canada*	*RoW*	*Total*
										1%	4%
								1%		1%	8%
								1%		1%	4%
								2%		3%	20%
											1%
											1%
										1%	3%
										1%	2%
		1%	1%					2%		1%	8%
											2%
			1%					1%		1%	5%
		1%	1%					1%		1%	7%
											2%
											0%
										1%	3%
								1%			2%
		1%	1%				1%		1%	2%	12%
								2%			3%
		1%	1%					3%		1%	13%
	2%	5%	5%	2%	0%	2%	2%	16%	3%	15%	100%

The message is relatively straightforward: global production appears to be organized along three distinct 'factories' – Factory Europe, Factory Asia, and Factory North America (see Baldwin and López-González 2013). Factory Europe appears to be the most integrated (counting intra-EU flows), but Factory Asia and Factory North America follow closely. The off-diagonal elements of the matrix, which capture non-regional or between-factory activity, are also important and have been growing since 1995. They involve large 'headquarter' economies such as the United States, Germany, Japan, and, to a lesser albeit growing extent, China.

Factory Asia absorbs 18 per cent of traded value added (column sums) and importantly its sales occupy 23 per cent (row sums), giving rise to a large trade in value added surplus. However, the asymmetries within the region are apparent. Japan and China are dominant with the RoAsia, which includes 8 of the 10 ASEAN countries amongst others (Taiwan and Hong Kong, China), being cumulatively as big as China.

Regional decomposition of GVC participation

The regional element of GVC participation is presented in Annex Figure 2.3. Here regional 'buying' is the use of ASEAN-8 value added to produce exports as a share

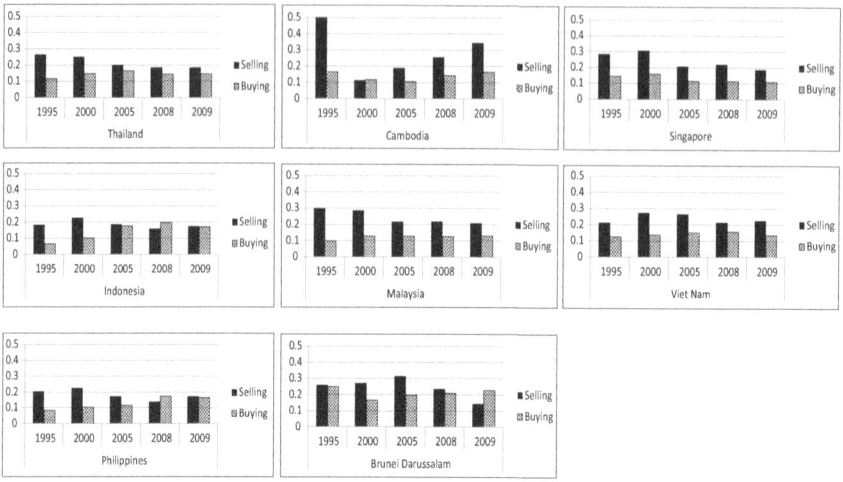

Annex Figure 2.3 Intra-ASEAN buying and selling of value added in exports, 1995 to 2009

Note: Pattern bars show the value of sales of value added that are sold to other ASEAN-8 countries for these to produce exports as a share of total foreign sales of value added. Black bars show the value of foreign value added that originates from other ASEAN-8 countries as a share of total foreign value added used.

Source: Authors' calculations based on OECD–WTO TiVA database.

of all foreign value added used for the reference country.[18] The regional 'selling' element is the value of sales of value added to other ASEAN-8 countries for these to produce exports as a share of total sales of foreign value added in exports. The shares essentially show us the regional orientation of the backward and forward participation measures. For example, in the case of Thailand in 2009, around 20 per cent of the value added that Thailand sold abroad for the production of exports went to its ASEAN-8 partners (recalling from above that this represents 19 per cent of gross exports). In contrast, the ASEAN-8 share of buying was just over 10 per cent (out of the 33 per cent that the backward linkage represents).

Differences in the absolute level of these buying and selling elements give us an indication of the nature of the regional linkages. In absolute terms, the regional selling element tends to be larger than the regional buying element. This suggests that, proportionately, there is a highly developed market for intermediate value added sales in the region, but that intermediate purchases, as a share of all purchases, are relatively larger with respect to extra-regional partners. The intuition is that ASEAN countries are increasingly sourcing their intermediates from regional sources (rising buying element), but that they are also increasingly selling their intermediate value added into other non-regional value chains (falling selling element). To a certain extent, this may be evidence of moving away from simple assembly activities towards other upstream segments of value chain activity involving the production of intermediates rather than final goods.

ORG_COUNTRY	*EU*	*IND*	*ASEAN-8*	*CHN*	*TWN*	*HKG*	*KOR*	*JPN*	*AUS-NZ*	*NAFTA*	*ROW*	*Share by sector*
Primary	5%	1%	18%	4%	0%	0%	1%	1%	7%	11%	52%	9%
Chemicals and fuel	19%	1%	13%	3%	6%	0%	6%	23%	1%	18%	9%	11%
Electrical equipment	14%	0%	12%	1%	5%	1%	6%	31%	0%	28%	2%	20%
Transport equipment	19%	0%	8%	2%	3%	0%	2%	47%	2%	12%	4%	1%
Light manufacturing	19%	1%	8%	4%	5%	0%	8%	24%	5%	16%	11%	13%
Manufacturing and machinery	27%	0%	8%	3%	2%	0%	3%	25%	1%	21%	8%	3%
Services	29%	0%	10%	2%	3%	3%	3%	20%	3%	19%	8%	43%
2009												

ORG_COUNTRY	*EU*	*IND*	*ASEAN-8*	*CHN*	*TWN*	*HKG*	*KOR*	*JPN*	*AUS-NZ*	*NAFTA*	*ROW*	*Share by sector*
Primary	2%	2%	15%	6%	0%	0%	0%	0%	5%	5%	65%	15%
Chemicals and fuel	13%	2%	19%	11%	4%	0%	5%	12%	1%	14%	18%	10%
Electrical equipment	8%	1%	18%	14%	10%	0%	9%	12%	0%	22%	5%	14%
Transport equipment	13%	1%	10%	9%	1%	1%	6%	29%	1%	26%	4%	1%
Light manufacturing	14%	2%	10%	17%	3%	0%	7%	16%	4%	13%	13%	10%
Manufacturing and machinery	26%	2%	11%	12%	1%	0%	4%	16%	1%	19%	7%	4%
Services	28%	3%	8%	6%	3%	2%	4%	11%	3%	19%	12%	45%

Annex Figure 2.4 Foreign value added embodied in ASEAN-8 gross exports by source sector and country, 1995 to 2009

Note: Table shows origin of foreign value added used by ASEAN-8 countries to produce exports. For example, first entry on top panel shows that 5 per cent of the value added that ASEAN-8 uses to produce gross exports comes from the European Union. Across the rows the entries sum to 100 per cent. The final column shows the share that each sector occupies in total foreign use of value added to produce exports.

Source: Authors' calculations based on OECD–WTO TiVA database.

3 The impact of China's trade on ASEAN's trade

Miaojie Yu and Xiaomin Cui[1]

1. Introduction

Based on gross domestic product (GDP), China became the second-largest global economy in 2013, with the first and the third economies being the United States (US) and Japan, respectively. In 2013, total GDP of EAP was US$20.5 trillion and that of developing East Asia Pacific (EAP) countries accounted for 55.7 per cent of this total. Export-led growth patterns have played a crucial role in the development of China and most Association of Southeast Asian Nations (ASEAN) countries. By exporting goods and commodities in which they have a comparative advantage, China and ASEAN countries have realized rapid industrialization and development. Since its economic reforms and opening up in 1978, China's annual growth rate of exports has been about 13.7 per cent. Its ratio of exports to GDP has continued to increase, from 7.8 per cent in 1982 to 35.7 per cent in 2006, although decreasing subsequently to 22.6 per cent in 2014. The average ratio of exports to GDP of the six relatively developed ASEAN countries – Brunei Darussalam, Indonesia, Malaysia, the Philippines, Singapore, and Thailand – was 79.7 per cent from 1990 to 2014, whereas the export ratio of Singapore is close to 200 per cent. The ratio of exports to GDP of the three less-developed ASEAN countries – Cambodia, Lao PDR, and Viet Nam – was 46.2 per cent from 1993 to 2014. The average annual export growth rate of all ASEAN countries, excluding Myanmar, was 8.0 per cent from 1999 to 2013.

Bilateral trade in goods between China and ASEAN has increased dramatically, especially after the launch of the ASEAN–China Free Trade Agreement (FTA) in 2010. Based on 2014 data, China is ASEAN's largest trade partner and ASEAN is China's third-largest trade partner, after the European Union (EU) and the United States, respectively. As shown in Figure 3.1, bilateral trade increased dramatically from 2000 to 2011, but ASEAN's exports to China weakened in 2012 and recovered in 2013. In 2000, China's total exports were $249.2 billion, and only 6.96 per cent of these exports went to ASEAN countries. In 2000, ASEAN's total exports (excluding services) were $330.0 billion, with exports to China comprising only 4.96 per cent. In 2013, China's total exports were $2,209.0 billion, with the share of exports going to ASEAN countries increasing to 11.0 per cent of the total. In 2013, ASEAN's total exports were $941.4 billion, with exports to China accounting for 16.2 per cent.

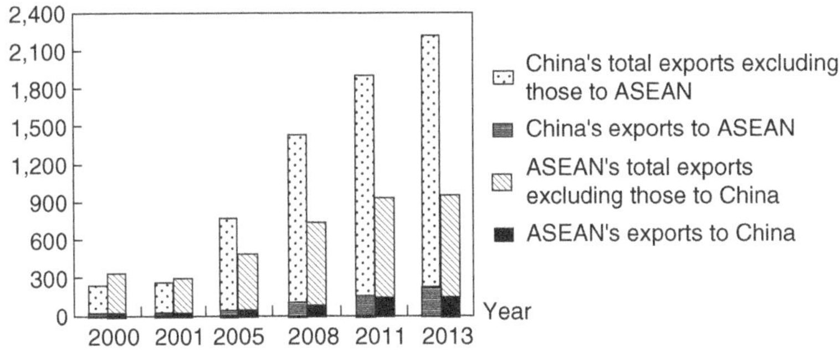

Figure 3.1 ASEAN's and China's bilateral and total exports (in goods, US$ bn)

Source: Data for Viet Nam's exports to China in 2013, Viet Nam's and Brunei's total exports, excluding those to other ASEAN countries in 2012 and 2013, are from the UN Comtrade database. All data come from the CEIC database, if not otherwise specified.

Bilateral imports in goods between China and ASEAN countries also increased steadily. China's imports from ASEAN totalled $200 billion in 2013, accounting for 10.2 per cent of total imports. At the same time, ASEAN, with the exception of Myanmar and Lao PDR, imported $194 billion in commodities from China, or 15.8 per cent of its total imports. Many studies have found that the industrial structures of China and ASEAN are relatively complementary. As imports are largely used as inputs for exports, production networks shared across China and ASEAN countries make it possible to manufacture certain products efficiently and effectively.

Nevertheless, China's export performance has been anaemic since 2007 and has weakened further in recent years. On one hand, its ratio of exports of goods and services to GDP decreased from 34.9 per cent in 2007 to 22.6 per cent in 2014. On the other hand, its annual growth rate of exports of goods and services slowed rapidly – from 35.1 per cent in 2006 to 4.0 per cent in 2014. As China's growth pattern changes as it moves up the income ladder to one that relies more on consumption and less on investment and exports, and more on services and less on industry, opportunities for Southeast Asian countries to expand their economies through trade and investment will emerge. Several questions arise, such as how China's trade impacts Southeast Asia countries. How can Southeast Asia improve its position in international trade by tapping into the opportunities emerging from the rebalancing of growth in China? And how should Southeast Asian countries increase the value added of their products and improve trade competitiveness?

This chapter tries to answer these questions in three ways. First, we study the influence of China's exports on ASEAN countries' exports using econometric analysis, distinguishing the effect of China's exports to ASEAN from its exports to the rest of the world, and considering industrial heterogeneity. Second, we calculate the revealed comparative advantages of China and ASEAN countries, comparing the relative competitiveness of each industry in ASEAN countries with those industries in China. Third, we calculate the export value-added ratios of China and ASEAN countries, allowing us to study how Southeast Asia can improve value added in exports and its trade competitiveness.

The remainder of this chapter is organized as follows: Section 2 introduces the data sets used in the empirical analysis and describes the measures of key variables; Section 3 presents the empirical evidence for the impact of China's trade on ASEAN's trade; Section 4 provides some robustness checks; and Section 5 concludes.

2. Data and measures

All industry-level trade data for ASEAN and China are from the United Nations (UN) Comtrade database. The UN Comtrade database is a widely used database of trade in various goods and commodities, and provides disaggregated commodity trade data for each country around the world. This database provides trade data based on three international standards – the International Convention for Harmonized Commodity Description and Coding System (HS), the Standard International Trade Classification (SITC), and the Classification by Broad Economic Categories (BEC). This chapter makes most use of data at the SITC one-digit level to identify the industrial heterogeneous effect of China's trade on ASEAN's trade. However, country-level trade data are from the global database,[2] a subset of the CEIC database. The CEIC database also has industry-level trade data, but these data are incomplete and the CEIC does not have trade data at the HS or SITC levels.

In addition, data on GDP, value added of manufacturing industries, total labour forces, urban populations, simple average applied tariffs, and ASEAN countries' exchange rates to the US dollar are all drawn from the CEIC database. Labour productivity of the manufacturing sector is equal to the ratio of value added to total labour force of the manufacturing sector. As data for Myanmar's GDP in the CEIC database are collected at the end of March rather than at the end of December, we obtain these data collected at the end of December from the IMF global database. It is worth pointing out that all exports from UN Comtrade and CEIC database only include exports of goods and commodities, excluding services. Henceforth, by 'exports' we mean exports of goods and commodities.[3]

2.1 Export values

After China's accession to the WTO in 2001, ASEAN countries' exports to China increased dramatically. The annual growth rate of ASEAN's exports to China peaked in 2003, at 41.8 per cent; they subsequently decreased gradually until finally turning negative during the global financial crisis, then rebounded significantly in 2010–2011, but steadily became anaemic from 2012 to 2013. According to Table 3.1, ASEAN's exports to China in 2011 were 8.7 times larger than in 2000. In addition, ASEAN's exports to China increased when the Chinese yuan appreciated against the US dollar in 2005. However, during the global financial

Table 3.1 ASEAN's exports to China

Affairs	*Year*	*All 10 countries of ASEAN*			*The six relatively developed countries* ***(Brunei, Indonesia, Malaysia, Philippines, Singapore, Thailand)***			*The four less-developed countries* ***(Cambodia, Lao PDR, Myanmar, Viet Nam)***		
		Total exports to China (US$ m)	*Exports to China percentage to total exports (%)*	*Exports to China percentage to its GDP (%)*	*Total exports to China (US$ m)*	*Exports to China percentage to total exports (%)*	*Exports to China percentage to its GDP (%)*	*Total exports to China, (US$ m)*	*Exports to China percentage to total exports (%)*	*Exports to China percentage to its GDP (%)*
	2000	16,376	4.96	2.68	14,696	4.66	2.60	1,679	11.22	3.59
China access to WTO	2001	16,719	5.57	2.85	15,157	5.33	2.81	1,563	9.75	3.40
CNY appreciates against US$	2005	52,680	10.78	5.69	49,146	10.76	5.81	3,533	11.15	4.38
Global financial crisis	2008	88,997	12.09	5.77	83,412	12.32	5.98	5,585	9.45	3.74
	2011	142,243	15.32	6.44	128,207	15.36	6.42	14,035	14.93	6.60
	2013	152,373	16.19	6.20	135,385	16.63	6.15	16,988	13.34	6.63

Notes: The word 'Exports' in this table only includes exports of goods and commodities, excluding services. If the original data are measured with the national currency, we used the period-average exchange rate to calculate the data measure in US dollars. Total exports of ASEAN, the six relatively developed countries and the four less-developed countries, do not include the trade volume within ASEAN countries.

Source: Data on Myanmar's GDP are from the IMF global database. Data for China's and ASEAN's GDP in 2013 are from the World Bank database. All data come from CEIC database, if not otherwise specified.

crisis (2008–2009), ASEAN's exports to China contracted by 7.1 per cent. In addition, with the exception of 2009, ASEAN's exports to China increased far faster than its total GDP and total exports. As shown in Table 3.1, the ratio of ASEAN's exports to China as a share of its total exports, which excludes exports within ASEAN countries, was 15.3 per cent in 2011. Meanwhile, the ratio of ASEAN's exports to China accounted for 6.4 per cent of its total GDP.

We divide the 10 ASEAN countries into two groups – the six relatively developed countries and the four less-developed countries. From 2000 to 2011, the six relatively developed countries – Brunei Darussalam, Indonesia, Malaysia, the Philippines, Singapore, and Thailand – all posted consistent increases in exports of goods and services to China. The ratios of these countries' exports to China as a share of total exports and GDP were close to those of all ASEAN countries, and have continued to increase in 2000–2011 except during the global financial crisis. However, the six relatively developed countries' exports to China and the share of GDP weakened in 2012 and recovered slightly in 2013. The six relatively developed countries contributed to about 92 per cent of ASEAN's total exports to China. Although the four less-developed ASEAN countries – Cambodia, Lao PDR, Myanmar, and Viet Nam – only accounted for a small share of total exports, the nominal value of their exports to China increased rapidly from 2007 to 2011, even during the financial crisis. However, the growth rate of the nominal value of their exports slowed in 2012 and rebounded to 17 per cent in 2013. The exports of the four less-developed ASEAN countries to China decreased slightly when China joined the World Trade Organization (WTO) in 2001, but their exports to China as a share of total exports and GDP decreased when the Chinese yuan appreciated against the US dollar. The share of total exports recovered to 11.6 per cent in 2009 and increased steadily until 2011, before declining somewhat in 2012–2013.

2.2 Revealed comparative advantage

According to traditional trade theory, especially the Ricardian model, trade patterns are closely related to each country's comparative advantage. There are many measures of a country's comparative advantage. One of them is the 'revealed comparative advantage', which was initially proposed by Balassa (1965). This defines a country's comparative advantage based on relative exports. The revealed comparative advantage (RCA) index of sector *k* in country *c* at time *t* is:

$$RCA_{kt}^{c} = \frac{EX_{kt}^{c} / \sum_{k} EX_{kt}^{c}}{\sum_{k} EX_{kt}^{c} / \sum_{c}\sum_{k} EX_{kt}^{c}}$$

where, EX_{kt}^{c} is the total exports of sector *k* in country *c* at time *t*. The RCA index measures the relative exports and trade competitiveness of sector *k* in country *c*, compared with the average relative exports of sector *k* in the world. If RCA_{kt}^{c} is larger than 1, then country *c* has a comparative advantage in industry *k* at time *t*. In the case where RCA_{kt}^{c} is less than 1, country *c* has a comparative disadvantage in this industry.

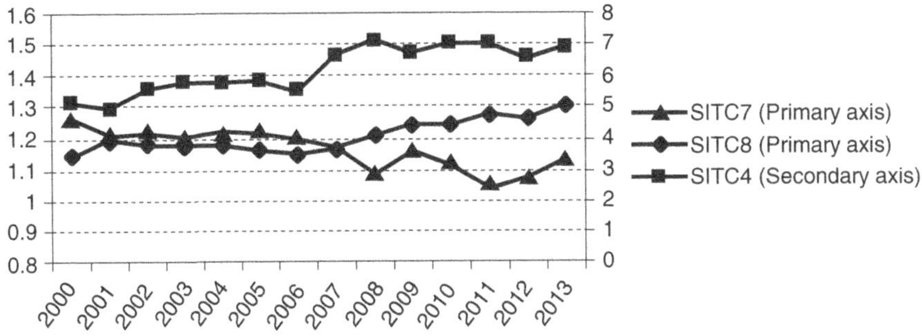

Figure 3.2 The RCA indices of ASEAN's three representative industries from 2000 to 2013

Notes: Lines of SITC7 and SITC8 correspond to the primary axis, while line of SITC4 corresponds to the secondary axis.

Source: UN Comtrade Database.

Because different countries have different comparative advantages, the impact of China's trade on ASEAN's trade should show industrial heterogeneity. Therefore, we measure the RCA of China and ASEAN at the SITC one-digit industry level[4] from 2000–2013. As can be seen in Figure 3.2 and Table 3.2, ASEAN had the strongest RCA in animal and vegetable oils, fats, and waxes, where the RCA index was about 6.2. ASEAN had an increasing comparative advantage in miscellaneous manufactured articles from 2000–2013 and also suffered a gradual loss in comparative advantage in manufacturing machinery and transport equipment from 2000 to 2011, even though the RCA index recovered slightly in 2012–2013. However, ASEAN includes 10 countries and each country has its own comparative advantage. We therefore analyze the comparative advantage of each ASEAN country separately.

Overall, from 2000 to 2013, China did not have comparative advantages in food and live animals, beverages and tobacco, crude materials, mineral fuels and lubricants, animal and vegetable oils, and chemicals, given that the RCA indices of these sectors are all less than 1. In particular, China had the weakest RCA in animal and vegetable oils, fats, and waxes, with an average RCA index of 0.08. Conversely, ASEAN had the strongest RCA in this sector. This reflects the structural complementarity of exports between China and ASEAN. However, China remained competitive in manufacturing, with the RCA index for manufacturing machinery and transport equipment increasing steadily from 0.8 in 2000 to 1.45 in 2011. Until 2013, ASEAN also had weak RCA in manufacturing, where China and ASEAN may fiercely compete.

2.2.1 RCA indices of ASEAN countries before China joined the WTO

As shown in Table 3.3, Cambodia had a strong comparative advantage in miscellaneous manufactured articles in 2000, with an RCA index of 7.5. The reason for this might be that Cambodia had weak comparative advantage in other sectors.

Table 3.2 Revealed comparative advantage of China and ASEAN

Country	*Year*	*SITC0*	*SITC1*	*SITC2*	*SITC3*	*SITC4*	*SITC5*	*SITC6*	*SITC7*	*SITC8*	*SITC9*
China	2000	0.94	0.34	0.59	0.32	0.15	0.54	1.25	0.80	2.81	0.05
China	2005	0.58	0.19	0.31	0.19	0.09	0.44	1.22	1.21	2.20	0.06
China	2008	0.44	0.14	0.23	0.14	0.07	0.53	1.34	1.37	2.26	0.03
China	2011	0.46	0.16	0.18	0.11	0.05	0.56	1.28	1.45	2.25	0.02
China	2013	0.42	0.15	0.16	0.09	0.05	0.50	1.33	1.41	2.32	0.02
ASEAN	2000	1.04	0.27	0.93	0.92	5.14	0.46	0.59	1.27	1.14	0.50
ASEAN	2005	1.04	0.25	1.07	0.97	5.85	0.62	0.57	1.22	1.16	0.66
ASEAN	2008	1.21	0.30	1.01	0.94	7.13	0.56	0.60	1.09	1.21	1.40
ASEAN	2011	1.13	0.38	1.00	0.95	7.03	0.69	0.63	1.04	1.27	1.16
ASEAN	2013	1.06	0.41	0.80	0.88	6.89	0.62	0.67	1.13	1.30	1.00

Notes: ASEAN only includes countries for which we can find data in the UN Comtrade Database. All SITC categories are based on *Standard International Trade Classification, Revision 3.*

Source: UN Comtrade Database.

Table 3.3 Revealed comparative advantage of ASEAN countries in 2000

2000	*SITC0*	*SITC1*	*SITC2*	*SITC3*	*SITC4*	*SITC5*	*SITC6*	*SITC7*	*SITC8*	*SITC9*
Cambodia	0.14	0.22	0.96	0.00		0.00	0.23	0.02	7.49	0.12
China	0.94	0.34	0.59	0.32	0.15	0.54	1.25	0.80	2.81	0.05
Indonesia	1.08	0.43	2.29	2.54	9.40	0.57	1.45	0.42	1.31	0.15
Malaysia	0.33	0.37	0.91	0.97	11.42	0.43	0.50	1.51	0.65	0.20
Philippines	0.65	0.15	0.44	0.13	4.12	0.10	0.27	1.84	0.94	0.10
Singapore	0.24	0.91	0.23	0.74	0.52	0.78	0.28	1.63	0.66	0.81
Thailand	2.69	0.28	1.30	0.32	0.37	0.66	0.85	1.05	1.18	0.67
Viet Nam	4.67	0.14	0.86	2.66	1.41	0.11	0.39	0.21	2.26	0.77

Notes: ASEAN only includes countries for which we can find data in the UN Comtrade Database. All SITC categories are based on *Standard International Trade Classification, Revision 3.*

Source: UN Comtrade Database.

Table 3.4 Revealed comparative advantage of ASEAN countries in 2005

2005	*SITC0*	*SITC1*	*SITC2*	*SITC3*	*SITC4*	*SITC5*	*SITC6*	*SITC7*	*SITC8*	*SITC9*
Cambodia	0.10	0.36	0.53	0.00	0.07	0.01	0.09	0.01	8.24	0.13
China	0.58	0.19	0.31	0.19	0.09	0.44	1.22	1.21	2.20	0.06
Indonesia	1.04	0.50	3.31	2.25	15.82	0.49	1.21	0.41	1.04	0.07
Malaysia	0.39	0.40	0.83	1.09	12.41	0.55	0.52	1.41	0.73	0.39
Philippines	0.77	0.57	0.52	0.15	4.54	0.13	0.31	1.94	0.88	0.04
Singapore	0.20	0.66	0.18	0.99	0.30	1.07	0.33	1.53	0.59	1.08
Thailand	2.19	0.29	1.61	0.35	0.42	0.76	0.89	1.17	1.05	0.43
Viet Nam	3.81	0.57	1.19	2.10	0.15	0.15	0.48	0.25	2.77	0.12

Notes: ASEAN only includes countries for which we can find data in the UN Comtrade Database. All SITC categories are based on *Standard International Trade Classification, Revision 3.*

Source: UN Comtrade Database.

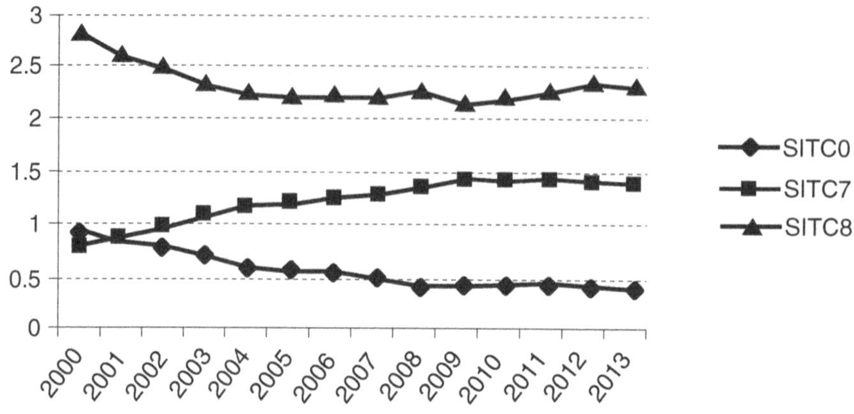

Figure 3.3 RCA indices of China's three representative industries from 2000 to 2013

Notes: More details about China's revealed comparative advantage are listed in Tables 3.2–3.6. Here only three representative sectors are listed.

Source: UN Comtrade Database

Compared with other countries, Indonesia and Malaysia had extremely strong RCAs in animal and vegetable oils, fats, and waxes, with an average RCA index of 10.41. The Philippines also had a strong RCA in this sector. Indonesia had an RCA in crude materials and mineral fuels in 2000. In addition, Malaysia, the Philippines, Singapore, and Thailand all had weak comparative advantages in manufacturing machinery and transport equipment. Thailand and Viet Nam had strong RCAs in food and live animals.

2.2.2 RCA indices of ASEAN countries when the Chinese yuan appreciated

Compared with 2000, the RCA indices of Cambodia and Singapore remained almost the same in 2005, except for a slight increase in Cambodia's RCA in miscellaneous manufactured articles. China had a comparative advantage in manufacturing machinery and transport equipment in 2005, although its RCA index had been less than 1 in 2000. With a slight decrease in its RCA in mineral fuels and lubricants, Indonesia improved its competitiveness in crude materials and in animal and vegetable oils. Malaysia and the Philippines also strengthened their exports in this sector. Thailand and Viet Nam improved their comparative advantage in crude materials, whereas RCA indices in the food and live animals industries decreased. However, Viet Nam saw a significant fall in competitiveness in animal and vegetable oils, fats, and waxes, with an 89 per cent decline in its RCA index.

2.2.3 RCA indices of ASEAN countries during the global financial crisis

In 2008, Cambodia improved its comparative advantage in manufacturing miscellaneous manufactured articles, but its comparative advantage in crude materials became weak. Since 2005, Indonesia and Malaysia have continued to strengthen their comparative advantage in animal and vegetable oils, fats, and waxes, although their RCA indices decreased somewhat during the global financial crisis. The RCA indices of the Philippines, Singapore, and Thailand changed little in 2008 compared with those in 2005. However, from 2000 to 2008, Viet Nam steadily lost its strong comparative advantage in mineral fuels and lubricants, but improved it in miscellaneous manufactured articles. This is consistent with the rapid expansion of Viet Nam's manufacturing industry in recent years supported by the country's relatively low labour cost.

2.2.4 RCA indices of ASEAN countries from 2000 to 2013

Since 2009, Cambodia has lost its comparative advantage in crude materials. In 2013, the Philippines's comparative advantage in food, live animals, animal and vegetable oils, and fats and waxes strengthened somewhat, compared with 2008. Viet Nam's comparative advantage in mineral fuels weakened significantly in 2010–2013 with its RCA index decreasing to 0.45, although its RCA index was around 2.7 in 2000. From 2000 to 2011, Indonesia's comparative advantage in manufacturing gradually declined despite a mild increase in 2012–2013. Malaysia retained its strong competitiveness in producing animal and vegetable oils, fats, and waxes. Malaysia and Singapore maintained their weak comparative advantage in mineral fuels and machinery and transport equipment. Thailand's RCA index changed little from 2000 to 2013, with the exception that the country became less competitive in producing miscellaneous manufactured articles.

2.3 The value added of exports

Given the increasing prevalence of global production networks and vertical FDI, an increasing number of countries are engaged, both directly and indirectly, in producing final products, with an increasing number of intermediate inputs being transported between countries. Therefore, it is value added that matters most, rather than gross value. In this section, we focus on export value added ratios of China and ASEAN countries. Most papers use input–output tables[5] to calculate export value-added ratios. Koopman et al. (2014) calculated the domestic value-added ratios of 41 countries at the industry level. These include Indonesia and China but exclude other ASEAN countries, as input–output data for these countries are lacking in the WIOD database. Therefore, we use the net export ratio as a proxy variable of the export value added ratio.

$$Value\ Added\ Ratio_{ct} = \frac{Export - Import}{Export}$$

We then divided ASEAN countries into three groups – the developed, less developed, and least developed – based on their economic strength, and calculated

Table 3.5 Revealed comparative advantage of ASEAN countries in 2008

2008	*SITC0*	*SITC1*	*SITC2*	*SITC3*	*SITC4*	*SITC5*	*SITC6*	*SITC7*	*SITC8*	*SITC9*
Cambodia	0.04	0.29	1.04	0.00	0.37	0.01	0.05	0.07	8.87	0.07
China	0.44	0.14	0.23	0.14	0.07	0.53	1.34	1.37	2.26	0.03
Indonesia	1.10	0.54	3.09	1.78	20.45	0.48	1.09	0.37	0.90	0.13
Malaysia	0.51	0.49	0.76	1.12	15.40	0.57	0.63	0.96	0.76	2.54
Philippines	0.91	0.65	0.70	0.20	3.93	0.22	0.59	1.95	0.80	0.18
Singapore	0.20	0.83	0.18	1.13	0.34	0.90	0.33	1.47	0.60	1.74
Thailand	2.40	0.38	1.51	0.39	0.68	0.75	0.91	1.23	1.01	0.42
Viet Nam	3.68	0.41	1.12	1.23	0.30	0.22	0.74	0.34	2.95	0.27

Notes: ASEAN only includes countries for which we can find data in the UN Comtrade Database. All SITC categories are based on *Standard International Trade Classification, Revision 3.*

Source: UN Comtrade Database.

Table 3.6 Revealed comparative advantage of ASEAN countries in 2013

2013	*SITC0*	*SITC1*	*SITC2*	*SITC3*	*SITC4*	*SITC5*	*SITC6*	*SITC7*	*SITC8*	*SITC9*
Brunei	0.02	0.02	0.03	5.96	0.00	0.07	0.06	0.03	0.04	0.03
Cambodia	0.62	0.47	0.70	0.00	0.35	0.02	0.12	0.22	7.40	0.01
China	0.42	0.15	0.16	0.09	0.05	0.50	1.33	1.41	2.32	0.02
Indonesia	0.99	0.70	2.63	1.94	20.25	0.56	0.98	0.36	0.83	0.22
Malaysia	0.51	0.68	0.67	1.38	13.68	0.61	0.76	1.14	0.82	0.14
Philippines	1.28	0.83	1.44	0.24	4.51	0.42	0.85	1.64	0.81	0.12
Singapore	0.23	1.13	0.17	1.07	0.13	1.13	0.31	1.39	0.76	1.84
Thailand	1.97	0.70	1.33	0.39	0.66	0.99	1.05	1.25	0.77	0.32
Viet Nam	2.28	0.51	0.89	0.45	0.36	0.27	0.85	0.98	2.50	0.10

Notes: ASEAN only includes countries for which we can find data in UN Comtrade Database. All SITC categories are based on *Standard International Trade Classification, Revision 3.*

Source: UN Comtrade Database.

their export value-added ratios. According to Figure 3.4, the value-added ratio of ASEAN declined dramatically from 2000 to 2013, even though China's value-added ratio has increased significantly in this period. Specifically, the value-added ratio of the developed group, namely Brunei and Singapore, has decreased remarkably in the period 2000–2013. Analogously, the value-added ratios of the less developed group, namely Indonesia, Malaysia, the Philippines, and Thailand, also decreased steadily from 2000 to 2013. Conversely, the value-added ratios of the least developed group, namely Cambodia, Lao PDR, Myanmar, and Viet Nam, were positive and rose significantly in this period, meaning that the growth rate of the nominal value of their exports was higher than that of their imports.

3. Estimation results

The literature states that both competition and complementarity effects of China's trade on ASEAN's trade exist, as these countries have similar cultural and export structures. Competition effects would tend to increase the threat of China's expansion into ASEAN countries. However, similar cultural backgrounds promote regional cooperation between China and ASEAN countries. In addition, China's economic growth has weakened since 2012, which may offer opportunities for Southeast Asia countries. We provide empirical evidence of the relationships between China and ASEAN, and distinguish between the competition and complementarity effects of Chinese expansion. The impact of China's trade on ASEAN countries should be heterogeneous among industries, as each country has its own comparative advantages.

In view of this, we study the heterogeneous effect of China's trade on ASEAN countries empirically at the industry level in three respects. First, we study the influence of China's exports on ASEAN countries' exports, distinguishing China's exports to ASEAN from its exports to the rest of the world. Second, we study the impact of China's exports on ASEAN countries' export value-added ratios. Third, we study how ASEAN countries might seize the opportunity and improve the value-added ratios of their exports and trade competitiveness using a gravity model.

3.1 The impact of China's exports on ASEAN's exports

To separate the competition effect of China's exports from the complementarity effect, we consider two indicators: China's exports to ASEAN countries, and China's exports to the rest of the world. If ASEAN countries import from China, they can choose more intermediate inputs and learn from the technology embedded in the imported goods, which should help them to improve their own productivity and the quality of their finished goods. In this case, China's exports to ASEAN will promote ASEAN's exports. We call this the complementarity effect, reflecting the complementarity between China's and ASEAN's exports. Conversely, ASEAN competes with China in the world market, such

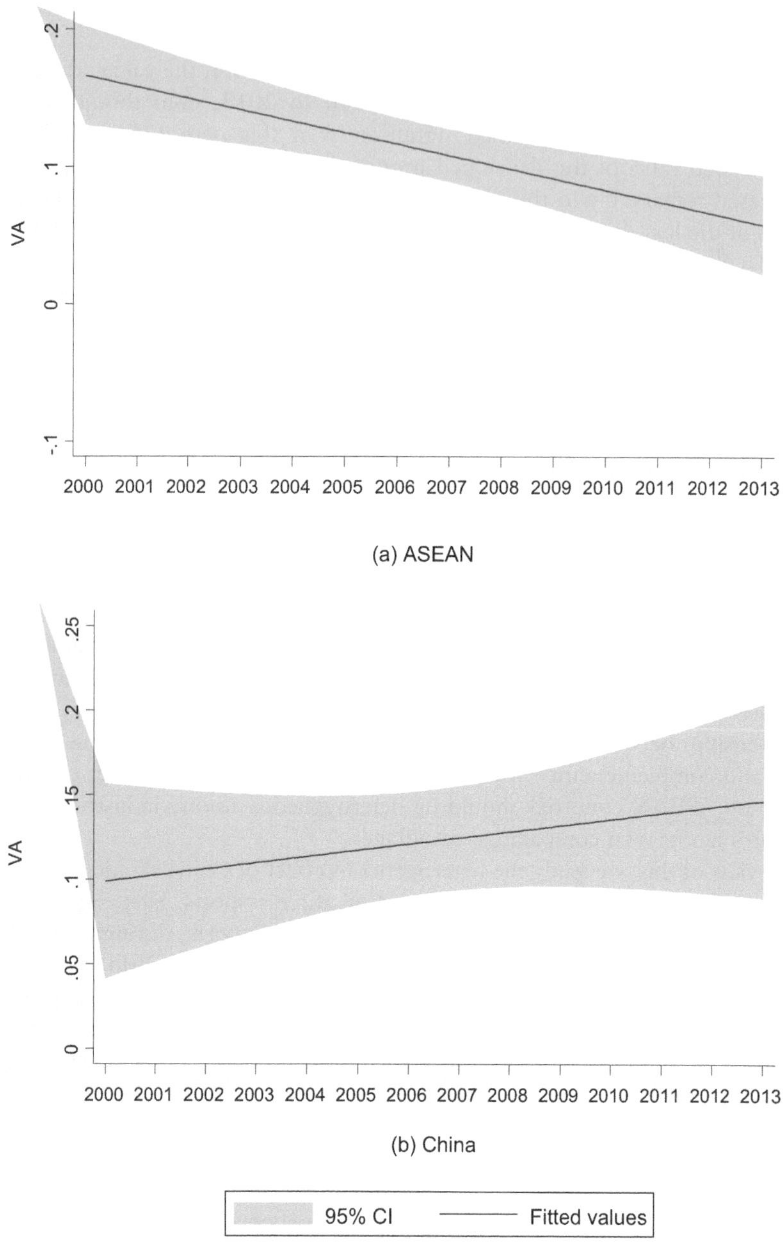

Figure 3.4 Value-added ratios of ASEAN and China from 2000 to 2013

Notes: In the UN Comtrade Database, there are not data for Lao PDR and Myanmar in the period 2000–2013, with the exception of data for Myanmar in 2010. In addition, exports within ASEAN countries are excluded when calculating the value-added ratio of ASEAN.

Source: UN Comtrade Database.

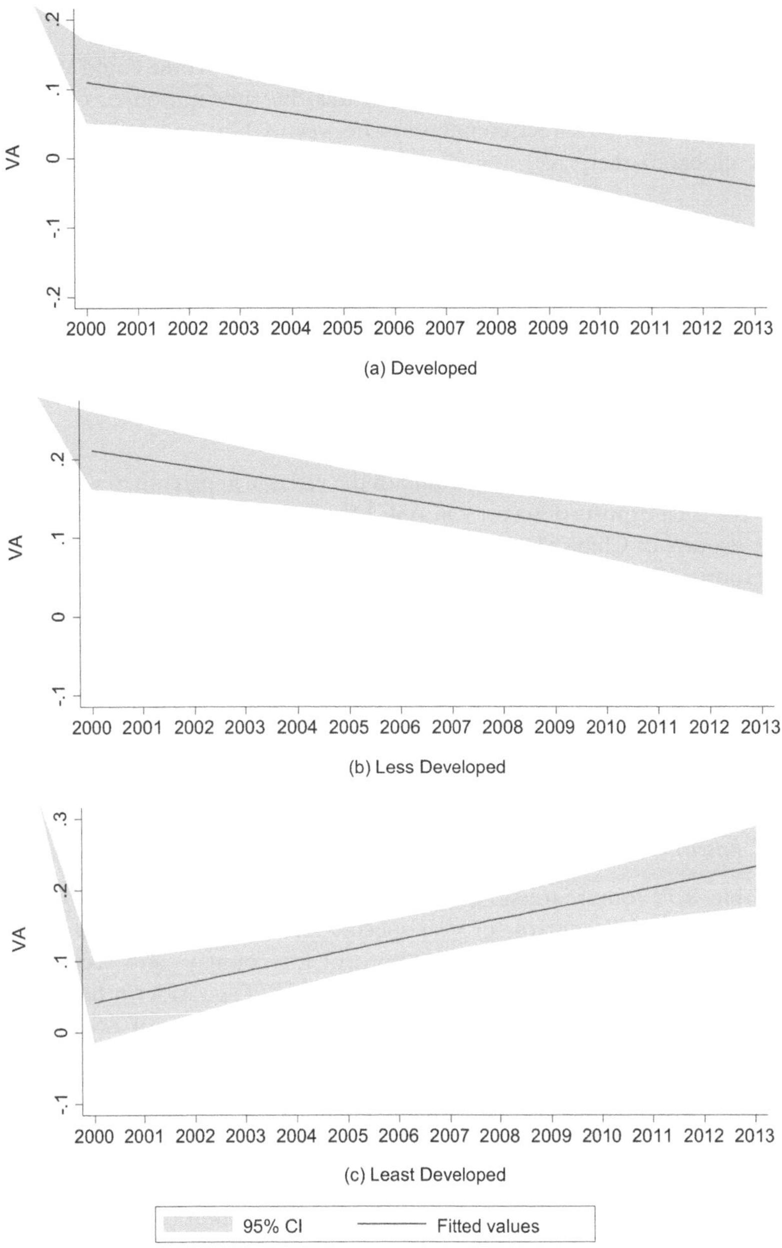

Figure 3.5 Value-added ratios of ASEAN countries from 2000 to 2013

Notes: We divide ASEAN countries into three groups according to their economic power. The developed ASEAN countries include Brunei Darussalam and Singapore. The less developed ASEAN countries include Indonesia, Malaysia, the Philippines, and Thailand. The least developed ASEAN countries include Cambodia, Lao PDR, Myanmar, and Viet Nam. However, in the UN Comtrade Database, there are not data for Lao PDR and Myanmar in the period 2000–2013. In addition, exports within each group are excluded when calculating the value-added ratios.

Source: UN Comtrade Database.

that an increase in China's exports to the rest of the world could crowd out exports from ASEAN countries. We call this the substitution or competition effect, reflecting the direct competition between China and ASEAN countries in the global marketplace.

3.1.1 Empirical specification

To distinguish the complementarity effect from the substitution/competition effect, we make use of the following empirical framework:

$$\ln(EX_{kct}) = \beta_1 \ln(CEX\_ASEAN_{kct}) + \beta_2 \ln(CEX\_ROW_{kt}) + \beta_3 X_{kct} + \alpha_k + \gamma_c + \epsilon_t + \omega_{kct} \quad (1)$$

where ln(·) means taking the logarithm of the variable in parentheses. EX_{kct} represents the total exports of sector k in ASEAN country c at time t. CEX_ASEAN_{kct} is the exports of China's sector k to ASEAN country c at time t. CEX_ROW_{kt} is the exports of China's sector k to the rest of the world at time t. X_{kct} are covariates, such as the simple average tariff of country c and the exchange rate. α_k, γ_c, and ϵ_t represent the sector-, country-, and year-specific fixed effects, respectively. ω_{kct} represents the heterogeneous shock. β_1 reflects the complementarity effect of China's exports on ASEAN's exports, so its estimator should be positive. Conversely, β_2 reflects the substitution/competition effect of China's exports on ASEAN's export, so its estimator should be negative.

3.1.2 Baseline results

In Table 3.7, four regression specifications are considered. All the regressions in Table 3.7 control for year-specific fixed effects. However, only the regressions in columns (2) and (4) control for country- and sector-specific fixed effects. The estimations in columns (3) and (4) consider other variables, including the simple average tariff and the exchange rate of ASEAN countries' currencies against the US dollar. As shown in column (1), an increase in China's exports to ASEAN leads to an increase in ASEAN countries' exports to the rest of the world, consistent with the complimentary effect. However, an increase in China's exports to the rest of the world results in a decrease in ASEAN countries' exports, proving the substitution/competition effect. In addition, the coefficients of year dummies in column (1) are significantly positive and increase steadily, which is consistent with the gradual increase of ASEAN's exports.

After controlling for country- and industry-specific fixed effects in column (2), the coefficient of the logarithm of China's exports to ASEAN decreases significantly by 68.4 per cent. However, the magnitude of the estimator of β_2 increases slightly. Compared with column (1), the regression in column (3) controls for the effect of simple average tariffs and exchange rates. The magnitudes of the two key regression coefficients both descend slightly. The regression in column (4) considers all these factors. Compared with column (2), the complementarity effect weakens slightly,

Table 3.7 The impact of China's exports on ASEAN's exports

Dependent variable: ln(EX_{kt}^{c})	*(1)*	*(2)*	*(3)*	*(4)*
ln(China's exports to ASEAN)	0.788*** (0.059)	0.249*** (0.072)	0.686*** (0.068)	0.217** (0.090)
ln(China's exports to the rest of the world)	−0.335*** (0.065)	−0.427* (0.234)	−0.215*** (0.073)	−0.450* (0.273)
Tariff simple average			−0.128*** (0.024)	−0.043 (0.060)
National currency/US$			0.000 (0.000)	0.000 (0.000)
Constant	14.62*** (0.813)	25.75*** (5.027)	15.380*** (0.886)	26.895*** (6.109)
Year-specific fixed effect	Yes	Yes	Yes	Yes
Country-specific fixed effect	No	Yes	No	Yes
Industry-specific fixed effect	No	Yes	No	Yes
Adjusted R^2	0.456	0.750	0.502	0.747
Observations	877	877	664	664

Notes: EX_{kt}^{c} represents the total export of sector k in country c in year t; All standard errors of linear regressions in the brackets are heterogeneous robust. *, **, *** represent 10 per cent, 5 per cent, 1 per cent significance, respectively.

Source: UN Comtrade and CEIC Database.

but the substitution/competition effect strengthens. Both of the two coefficients are statistically significant. Specifically, a 1 per cent increase in China's exports to ASEAN countries leads to ASEAN countries' exports increasing by 0.22 per cent on average. Meanwhile, a 1 per cent increase in China's exports to the rest of the world crowds out 0.45 per cent of ASEAN countries' exports.

3.2 The impact of China's exports on ASEAN's value added of exports

3.2.1 Empirical specification

To continue studying the effect of China's exports on ASEAN countries' value added of exports, the following specification is considered:

$$(VA_{kct}) = \theta_1 \ln(CEX\_ASEAN_{kct}) + \theta_2 \ln(CEX\_ROW_{kt}) + \theta_3 X_{kct} + \alpha_k + \gamma_c + \epsilon_t + \omega_{kct} \quad (2)$$

where VA_{kct} is the export value-added ratio of industry *k* in ASEAN country *c* in year *t*. The definitions of other variables are the same as those in equation (1).

Given their participation in global supply chains, an increase in exports from China to ASEAN will promote exports from ASEAN countries. However, the impact of an increase in exports from China to ASEAN on the export value-added ratios of ASEAN countries should be the opposite. The value added of exporters who are responsible for one or several production stages should be lower than those controlling the whole production process. Therefore, if ASEAN countries import more materials, especially intermediate inputs, from China, their value-added ratios should be lower on average. So we would expect the sign of θ_1 to be negative. China has a strong comparative advantage in labour-intensive industries, so it should be able to export more goods at lower prices. Only the exporters with higher productivity efficiency and value added can compete with China in the world market, and firms with lower productivity and value added will be forced to exit over time. So in contrast to β_2, θ_2 would be expected to be positive.

3.2.2 Baseline results

Similar to Table 3.7, four regression specifications are considered in Table 3.8. The regression in column (1) only considers year-specific fixed effects, finding that the coefficient of the logarithm of China's exports to ASEAN countries is significantly negative, and that the estimator of θ_2 is significantly positive. When controlling for the country- and sector-specific fixed effects in column (2) and considering other covariates in column (3), the results are consistent with those in column (1), except that the magnitudes of these two coefficients increase significantly. The regression in column (4) considers all these factors. In column (4), an increase in exports from China to ASEAN countries would lead to a significant decrease in ASEAN countries' export value-added ratios. Conversely, China's exports to the rest of the world are significantly and negatively correlated with ASEAN countries' export value-added ratios.

It is worth mentioning that the value-added ratios in these estimations could be lower than 0 and larger than 1, as we use the net export ratio as a proxy for the export value-added ratio. Therefore, it is reasonable that the coefficients of the two key variables – China's exports to ASEAN countries and China's exports to the rest of the world – are much larger than 1. However, the magnitudes are less important than the signs in our study. It would require additional data and deepening studies to calculate the true export value-added ratios of ASEAN countries, given that input–output tables do not exist for all ASEAN countries for the period 2000–2011.

3.3 How to improve the value added of ASEAN's exports?

Based on the literature and the empirical analysis as cited earlier, there exist both complementarity and substitution/competition effects from China's exports on ASEAN's exports and value added. China's economic growth rate has been slowing since 2012, creating challenges but also offering opportunities to ASEAN

Table 3.8 China's exports and ASEAN's value added

Dependent variable: VA_{kt}^{c}	*(1)*	*(2)*	*(3)*	*(4)*
ln(China's exports to ASEAN)	-8.130**	-14.22**	-12.42**	-19.94**
	(3.838)	(6.910)	(5.558)	(9.408)
ln(China's exports to the rest of the world)	5.940	30.12**	9.636*	37.48*
	(3.698)	(14.99)	(5.276)	(19.25)
Tariff simple average			-2.799**	-4.766
			(1.198)	(3.548)
National currency/US$			0.001**	-0.010
			(0.001)	(0.009)
Constant	8.768	-444.1*	33.00	-296.0
	(30.02)	(245.3)	(34.11)	(234.3)
Year-specific fixed effect	Yes	Yes	Yes	Yes
Country-specific fixed effect	No	Yes	No	Yes
Industry-specific fixed effect	No	Yes	No	Yes
Adjusted R^2	0.028	0.061	0.048	0.071
Observations	876	876	664	664

Notes: VA_{kt}^{c} is the gross value-added ratio in export of sector k in country c in year t. All standard errors of linear regressions in the brackets are heterogeneous robust. *, **, *** represent 10 per cent, 5 per cent, 1 per cent significance, respectively.

Source: UN Comtrade and CEIC Database.

countries. In view of this, we need to ask how Southeast Asian countries can best use these opportunities to improve their trade competitiveness. With country-level data from the CEIC database,[6] we study this question using a gravity model,[7] which helps to explain international trade flows (Feenstra et al. 2001). Using the gravity model, we add the GDPs of each ASEAN country and global GDP excluding China's exports to the rest of the world, to the regressions. To study how ASEAN countries can improve their trade competitiveness, we control for manufacturing labour productivity and the logarithm of the labour force.

Four empirical specifications are considered in Table 3.9. All regressions control the year-specific fixed effects, but only regressions (2) and (4) consider the country-specific fixed effects. In addition, regressions in columns (1) and (2) focus on the impact of manufacturing labour productivity on exports. But regressions in columns (3) and (4) study its influence on the value added of exports. As shown in columns (1) and (2), the coefficient of the logarithm of China's exports to ASEAN is similar to that in Table 3.7. The coefficient of the logarithm of China's exports to the rest of the world is still negative but insignificant. According to regressions in columns (3) and (4), the impact of China's exports to ASEAN on ASEAN's value-added ratio of exports is still negative, but its magnitude is

Table 3.9 Export value, value-added ratio, and labour productivity

Dependent variable	$ln(EX_{kt}^{c})$		VA_{kt}^{c}	
	(1)	*(2)*	*(3)*	*(4)*
ln(China's exports to ASEAN)	0.412***	0.305***	−0.183***	−0.095
	(0.021)	(0.075)	(0.024)	(0.066)
ln(China's exports to the rest of the world)	-0.647	-0.969	1.615	1.553
	(1.439)	(0.925)	(1.364)	(1.246)
ln(GDP)	0.178	0.066	−0.308*	−0.035
	(0.125)	(0.229)	(0.180)	(0.279)
ln(World GDP–China's exports to the rest of the world)	0.971	0.984	-4.128	-4.149
	(3.928)	(2.564)	(3.708)	(3.240)
ln(Manufacturing labour productivity)	0.600***	0.900***	0.535***	0.062
	(0.097)	(0.164)	(0.140)	(0.200)
ln(Labour force)	0.248**	3.400***	0.429***	0.553
	(0.096)	(0.553)	(0.142)	(0.798)
Constant	-11.60	-62.40*	45.52	43.48
	(50.10)	(33.81)	(47.40)	(49.64)
Year-specific fixed effect	Yes	Yes	Yes	Yes
Country-specific fixed effect	No	Yes	No	Yes
Adjusted R^2	0.990	0.996	0.648	0.874
Observations	112	112	112	112

Notes: All standard errors in the brackets are heterogeneous robust. *, **, *** represent 10 per cent, 5 per cent, 1 per cent significance, respectively.

Source: CEIC database.

much smaller. In addition, the coefficient of the logarithm of China's exports to the rest world is not statistically significant.

Consistent with the gravity model, the coefficient of the logarithm of GDP is positive in columns (1) and (2), but negative in columns (3) and (4). None of the coefficients of the logarithm of world GDP excluding China's exports to the rest of the world are significant. When controlling for the labour force, the coefficients of the logarithm of manufacturing labour productivity in the four regressions are all positive. Specifically, a 1 per cent increase in ASEAN countries' manufacturing labour productivity will lead to a 0.9 per cent increase of their exports. However, the coefficient of the logarithm of manufacturing labour productivity is insignificant when controlling country-specific fixed effect in column (4). The coefficients of the logarithm of the labour force are also positive. In summary, ASEAN countries could improve their exports and the value added of their exports by improving their labour productivity, especially manufacturing labour productivity.

3.4 How to improve ASEAN's labour productivity?

If ASEAN countries can increase their exports and the value added of their exports by increasing their manufacturing labour productivity, we need to ask how ASEAN countries can improve their labour productivity. There are two possible approaches – trade liberalization and internal improvement. Trade liberalization, such as tariff reductions and free trade agreements, forces domestic firms to face tougher competition from imports. Only firms with high levels of productivity will survive the increased competition, whereas firms with low productivity will be forced to exit over time. As a result of these low-productivity firms' exits, the average productivity of domestic firms will increase and their trade competitiveness will also improve. The other approach is internal improvement, whereby ASEAN countries improve their trade competitiveness through their own efforts, including increasing investment in R&D, and the hiring or training of more highly skilled labour.

In Table 3.10, we use a simple average tariff and urban population[8] to identify these two channels. All the regressions in Table 3.10 control for year- and country-specific fixed effects. Regressions in columns (1) and (2) study the effect of trade liberalization. The regression in column (1) controls for the simple average tariff and finds that a decrease in tariffs will lead to an increase of ASEAN countries' export value-added ratios. However, the estimator is not statistically significant. Compared with column (1), the regression in column (2) controls for the interaction of tariffs and manufacturing labour productivity. However, the coefficient of the interaction term is not consistent with expectation, and its magnitude is quite small.

Regressions in columns (3) and (4) focus on the internal improvement channel. The regression in column (3) controls for the logarithm of urban population, which has a significantly positive effect on ASEAN countries' value added of exports. Furthermore, the regression in column (4) brings in the interaction of manufacturing labour productivity and urban population. The coefficient of the interaction is positive, but not significant. In addition, the coefficient of the logarithm of labour force is both statistically and economically significant. However, the coefficient of the logarithm of manufacturing labour productivity turns negative but insignificant.

4. Robustness checks

4.1 Revealed comparative advantage

In studying the effect of China's exports on ASEAN countries' revealed comparative advantage we consider four specifications in Table 3.11. As in Tables 3.7 and 3.8, only regressions in columns (2) and (4) control for the sector- and country-specific fixed effects. Furthermore, only regressions in columns (3) and (4) consider other variables, including the sample average tariff and the exchange rate of the domestic currency against the US dollar. The industry-level data provide the

Table 3.10 Tariff, urban population, and labour productivity

Dependent variable: VA_{kt}^{c}	*(1)*	*(2)*	*(3)*	*(4)*
ln(China's exports to ASEAN)	−0.016 (0.076)	−0.052 (0.077)	−0.130** (0.058)	−0.121** (0.055)
ln(China's exports to the rest of the world)	1.091 (0.996)	1.291 (1.037)	1.454 (1.091)	1.499 (1.105)
ln(GDP)	−0.182 (0.234)	−0.100 (0.240)	−0.196 (0.232)	−0.161 (0.239)
ln(World GDP– China's exports to the rest of the world)	−3.115 (2.739)	−3.892 (2.867)	−3.646 (2.962)	−3.977 (3.074)
ln(Manufacturing labour productivity)	0.086 (0.200)	0.074 (0.206)	−0.069 (0.141)	−0.053 (0.147)
ln(Labour force)	0.259 (0.804)	0.658 0.954		
Tariff simple average	−0.001 (0.013)	−0.007 (0.015)		
ln(Manufacturing labour productivity) × Tariff simple average		0.00002* (0.000)		
ln (Urban population)			1.132*** (0.408)	1.218*** (0.442)
ln(Manufacturing labour productivity) × ln (Urban population)				0.000002 (0.000)
Constant	37.37 (41.88)	40.86 (43.72)	29.32 (38.60)	32.54 (39.73)
Year-specific fixed effect	Yes	Yes	Yes	Yes
Country-specific fixed effect	Yes	Yes	Yes	Yes
Adjusted R^2	0.923	0.926	0.882	0.885
Observations	82	82	112	112

Notes: All standard errors in the brackets are heterogeneous robust. *, **, *** represent 10 per cent, 5 per cent, 1 per cent significance, respectively.

Source: CEIC database.

comparative advantage index and China's exports are used in the regressions. In addition, we control for the relative export scale of an industry, which is equal to the ratio between the sector and total exports.[9]

Similar to the impact of China's exports on ASEAN countries' exports, an increase in exports from China to ASEAN will improve ASEAN's exports and RCA. Meanwhile, China's expansion into the world market will tend to crowd

Table 3.11 China's exports and ASEAN's revealed comparative advantage

Dependent variable: RCA_{kt}^{c}	*(1)*	*(2)*	*(3)*	*(4)*
ln(China's exports to ASEAN)	0.033 (0.054)	0.161 (0.101)	0.018 (0.078)	0.082 (0.124)
ln(China's exports to the rest of the world)	-0.330*** (0.125)	-0.207 (0.403)	-0.230 (0.475)	-0.277 (0.461)
Relative sector size	1.549* 0.910	-6.582 (4.798)	-9.333 (6.076)	-9.620 (6.200)
Tariff simple average			-0.013 (0.023)	0.028 (0.094)
National currency/US$			0.001** (0.000)	-0.000 (0.000)
Constant	7.956*** (2.140)	3.062 (9.180)	6.530 (11.96)	6.837 (12.12)
Year-specific fixed effect	Yes	Yes	Yes	Yes
Industry-specific fixed effect	No	Yes	Yes	Yes
Country-specific fixed effect	No	Yes	No	Yes
Adjusted R^2	0.03	0.274	0.241	0.284
Observations	877	877	664	664

Notes: RCA_{kt}^{c} is the revealed comparative advantage of sector k in country c in time t. Relative sector size is equal to the ratio of $\sum_{k} EX_{kt}^{c} / \sum_{c}\sum_{k} EX_{kt}^{c}$. All standard errors of linear regressions in the brackets are heterogeneous robust. *, **, *** represent 10 per cent, 5 per cent, 1 per cent significance, respectively.

Source: UN Comtrade and CEIC Database.

out ASEAN's exports and ASEAN's RCA index will also decrease. As shown in Table 3.11, the coefficient of the logarithm of China's exports to ASEAN countries is positive, albeit statistically insignificant. In addition, the coefficient of the logarithm of China's exports to the rest of the world is negative, and also insignificant.

4.2 The trade data at the SITC two-digit industry level

Overall, ASEAN countries have relatively strong comparative advantages in animal and vegetable oils, fats, and waxes. However, China has a relatively strong comparative advantage in the manufacture industry. In this section, we only use trade data for the animal and vegetable oils, fats, and waxes industry, the machinery and transport equipment industry, and the miscellaneous manufactured articles industry to test the robustness of the complementarity and substitution/competition effects. All trade data are at the SITC two-digit level. We study the influence of China's exports to ASEAN and to the rest of the world on ASEAN countries' exports, the value added of exports, and RCA.

All regressions in Table 3.12 consider the year-, industry- and country-specific fixed effects. According to columns (1) and (3) in Table 3.12, SITC two-digit trade data confirm the complementarity effect. The coefficients of exports from China to ASEAN in columns (1) and (3) are strongly positive. However, there is no clear evidence of a competition effect in Table 3.12, as the coefficient of the logarithm of China's exports to the rest of the world is positive and insignificant in column (1). In addition, the effects of trade liberalization and exchange rates on exports in column (1) are all consistent with expectations. However, all coefficients in regression (2) are insignificant. One possible reason for this may be that the net export ratio is not a good proxy of the export value-added ratio. Therefore, we consider other measures of the export value-added ratio in the following robustness checks.

Table 3.12 The impact of China's trade on ASEAN's trade with trade data at the SITC two-digit level

Dependent variable	*(1)*	*(2)*	*(3)*
	$ln(EX_{kt}^{c})$	VA_{kt}^{c}	RCA_{kt}^{c}
ln(China's exports to ASEAN)	0.179***	96.16	0.298***
	(0.047)	(158.8)	(0.095)
ln(China's exports to the rest of the world)	0.022	-855.1	-0.520
	(0.189)	(708.9)	(0.437)
Tariff simple average	-0.087*	-143.6	0.016
	(0.046)	(114.0)	(0.092)
National currency/US$	0.0002*	-0.169	-0.00004
	(0.0001)	(0.135)	(0.0002)
Constant	9.155**	17050	6.092
	(3.647)	(12454)	(8.478)
Year-specific fixed effect	Yes	Yes	Yes
Country-specific fixed effect	Yes	Yes	Yes
Industry-specific fixed effect	Yes	Yes	Yes
Adjusted R^2	0.790	0.045	0.271
Observations	1315	1314	1315

Notes: All standard errors of linear regressions in the brackets are heterogeneous robust. *, **, *** represent 10 per cent, 5 per cent, 1 per cent significance, respectively.

Source: UN Comtrade and CEIC Database.

4.3 Another measure of the export value-added ratio

Normal export firms may use imported intermediate inputs in the production of goods sold onto the domestic market. Therefore, the net export ratio may not be a good proxy variable of the export value-added ratio. In this section, we use the domestic value-added ratio as calculated by Wang et al. (2013) and Koopman et al. (2014) to study the impact of China's trade on ASEAN's trade. Based on the world input–output database (WIOD), Wang et al. (2013) estimate the domestic value-added ratios for 35 sectors across 39 countries, excluding ASEAN countries, with the exception of Indonesia. Due to different classification methods, we cannot merge these data with the trade data from the UN Comtrade database. Therefore, we use the export data from the WIOD, estimated data

Table 3.13 Another measure of the export value-added ratio

Dependent variable	*(1)*	*(2)*	*(3)*
	$ln(EX_{kt}^{c})$	VA_{kt}^{c}	VA_{kt}^{c}
ln(China's exports to Indonesia)	0.019	-0.004*	-0.004**
	(0.041)	(0.002)	(0.002)
ln(China's exports to the rest of the world)	0.434***	-0.011	-0.012
	(0.165)	(0.012)	(0.012)
Tariff simple average	-0.304**	-0.023	
	(0.127)	(0.015)	
National currency/ US$	-0.0004	0.00002	
	(0.0003)	(0.00003)	
ln(GDP)			-1.221***
			(0.242)
ln(World GDP– China's exports to the rest of the world)			-0.774**
			(0.324)
ln(Manufacturing labour productivity)			2.005***
			(0.443)
Constant	10.48**	1.112**	44.34***
	(5.221)	(0.467)	(12.06)
Year-specific fixed effect	Yes	Yes	Yes
Industry-specific fixed effect	Yes	Yes	Yes
Adjusted	0.947	0.920	0.923
Observations	303	303	331

Notes: All standard errors of linear regressions in the brackets are heterogeneous robust. *, **, *** represent 10 per cent, 5 per cent, 1 per cent significance, respectively.

Source: Data of exports are from the world input–output database. Data of value-added come from Wang et al. (2013), and other data are from CEIC.

from Wang et al. (2013), and the country-level data from CEIC to study the impact of China's trade.

All regressions in Table 3.13 control for the year- and industry-specific fixed effects. The regression in column (1) focuses on the impact of China's exports on Indonesia's exports. According to column (1), the coefficient of the logarithm of China's exports to Indonesia is still positive, albeit insignificant. In addition, the coefficient of the logarithm of China's exports to the rest of the world is strongly positive. This may be because China's exports to the rest of the world include those to other ASEAN countries, which would promote Indonesia's exports.

Regressions in columns (2) and (3) study the effect of China's trade on Indonesia's value added of exports. The coefficients of China's exports to Indonesia in these two regressions are strongly negative, showing that the export value-added ratio would decrease the more Indonesia imports from China. However, all other coefficients in column (2) are insignificant. The regression in column (3) studies the impact of China's exports under a gravity model. Compared with regressions in Table 3.9, all coefficients are consistent, except the coefficient of the logarithm of China's exports to the rest of the world. In addition, the impact of manufacturing labour productivity is strongly positive. When controlling for the logarithm of labour force, the main results are largely the same.

5. Concluding remarks

To study the impact of China's trade on ASEAN countries' trade, this chapter calculates the RCA indices and the export value-added ratios of ASEAN countries, and then uses trade data at the SITC one-digit level to study this impact empirically. Our research finds that both complementarity and substitution/competition effects exist. This means that, given global supply chains, an increase in exports from China to ASEAN will promote ASEAN countries' exports. However, China also competes with ASEAN countries on the world market. Therefore, an increase in China's exports to the rest of the world will tend to crowd out ASEAN countries' own exports. However, the research on the impact of China's exports on ASEAN countries' export value-added ratio indicates the opposite effect. The average export value-added ratio of an industry in ASEAN countries is lower the more these countries import from China. Meanwhile, China's increasing exports to the world market will promote the improvement of ASEAN countries' export value-added ratios.

Furthermore, we study how ASEAN countries can improve their value-added ratio of exports and trade competitiveness under a gravity model. ASEAN countries can increase exports and value added by increasing their manufacturing labour productivity. In addition, trade liberalization and internal improvements will help ASEAN countries to improve their labour productivity. We study the impact of China's exports on ASEAN countries' RCA using trade data at the SITC two-digit level and another measure of the export value-added ratio to test the robustness of the baseline results. The results show that our conclusions are robust. However, considerable research remains to be undertaken, especially in calculating the true value added of exports.

Notes

1 China Center for Economic Research (CCER), National School of Development, Peking University, Beijing 100871, China. The email addresses of Miaojie Yu and Xiaomin Cui are mjyu@ccer.pku.edu.cn and sunnylizzie@163.com, respectively. We would like to thank Lili Yan Ing, Stefan Wesiak, Ms. Fadriani Trianingsih, Ms. Maria Rosario, Ms. Chrestella Budyanto, Mr. Rizqy Anandhika, Elisa Ayu Candra Rini, Prof. Fukunari Kimura, and workshop participants of Tiva project at the Economic Research Institute for ASEAN and East Asia (ERIA). Financial support from China's National Science Fund for Distinguished Young Scholars (71625007), China's National Natural Science Foundation Grant (71573006) and Key Project of Humanities and Social Science of Ministry of Education in 2015 (15JJD780001) is gratefully acknowledged. All errors are our own.

2 We obtain country-level trade data through summing up industry-level data, which should be the same as those from the CEIC database. In fact, CEIC collects these data from the IMF.

3 In fact, data of services trade at the industry level seem to be unavailable.

4 Commodity names of SITC 0–9: 0 — Food and live animals; 1 — Beverages and tobacco; 2 — Crude materials, inedible, except fuels; 3 — Mineral fuels, lubricants and related materials; 4 — Animal and vegetable oils, fats, and waxes; 5 — Chemicals and related products, n.e.s.; 6 — Manufactured goods classified chiefly by material; 7 — Machinery and transport equipment; 8 — Miscellaneous manufactured articles; 9 — Commodities and transactions not classified elsewhere in the SITC. More details about the SITC can be found on its website.

5 More details can be found in Koopman et al. (2014).

6 There is a lack of industry-level data about industrial characteristics.

7 We could also study the heterogeneous impact of China's exports under a gravity model. However, this would not affect the main results, as GDP and distance can be controlled for by the year- and country-specific fixed effects.

8 We use urban population as a proxy variable for the size of skilled labour, as we cannot find data for the true size of skilled labour for most ASEAN countries.

9 While we do not control for the relative size in these regressions, this would not affect the main results, as it can be controlled for in the year- and country-specific fixed effects.

References

Balassa, B 1965, 'Trade liberalisation and "revealed" comparative advantage', *The Manchester School*, vol. 33, no. 2, pp. 99–123.

Balassa, B 1969, 'Country size and trade patterns: comment', *American Economic Review*, vol. 59, issue 1, pp. 201–204.

Balassa, B 1979, 'The changing pattern of comparative advantage in manufactured goods', *Review of Economics and Statistics*, vol. 61, no. 2, pp. 259–266.

Feenstra, RC, Markusen, JR and Rose, AK 2001, 'Using the gravity equation to differentiate among alternative theories of trade', *Canadian Journal of Economics/Revue canadienne d'économique*, vol. 34, no. 2, pp. 430–447.

Koopman, R, Wang, Z and Wei, S 2014, 'Tracing value-added and double counting in gross exports', *American Economic Review*, vol. 104, no. 2, pp. 459–494.

Wang, Z, Wei, S-J, and Zhu, K 2013, *Quantifying International Production Sharing at the Bilateral and Sector Levels*, National Bureau of Economic Research, Cambridge, MA.

4 Imported inputs on Indonesia's product variety and quality

Lili Yan Ing[1] *and Chandra Putra*

1. Introduction

Trade evolves. Production is sliced. Much of production is based in production networks. Imports are largely used as inputs for exports. Many countries are engaged directly and indirectly in producing final products. The development of global production chains, with an increased use of imported inputs, caused a reduction of the domestic value-added content for each unit of manufacturing production and exports. However, at the same time, the increased use of imported inputs may increase overall value added (productivity).

A number of key studies analyse how trade affects overall productivity, starting from how exposure to trade leads to increasing inter-firm reallocations towards more productive firms (Melitz 2003), how trade liberalisation affects plant productivity (Fernandes 2007), how trade liberalisation fosters productivity growth within and across firms, and in aggregate by inducing firms to produce only marginally productive products and forcing the lowest-productivity firms to exit (Bernard et al. 2006), how tariff reductions on intermediate inputs affect productivity more than tariff reductions on final goods (Amiti and Konings 2007), to how tariff reductions on imported inputs affect product diversification and product quality (Goldberg et al. 2008).

The recent literature further explores that there are at least two mechanisms through which imported inputs could affect value added through product development. Imported inputs are substitutes or complementary for domestic inputs and thus may have effects on prices as well as quality of inputs. For a given quality of inputs, firms always try to find inputs at the lowest prices. First, imported inputs can act as a catalyst to expand product diversification (Iacovone and Javorcik 2010; Kee and Feenstra 2011). Second, imported inputs may increase product quality (Kugler and Verhoogen 2012; Amiti and Khandelwal 2013; Fan et al. 2013; Hallak and Sivadasan 2013).

The main value added of our chapter is twofold. First, we introduce the use of input material into the existing standard heterogeneous firm models à la Bernard et al. (2003), Melitz (2003) and Feenstra and Ma (2008). Our model explains how the use of input materials affects the choices to produce a certain level of product variety and quality in optimising a firm's profit. Second, our chapter provides evidence of how input tariff reductions affect firm's value added through product variety and product quality which are based on firm and product level data. We also detail how importing and exporting firms that use imported inputs respond to tariff reductions, and how this will affect their decision in determining the optimal level of product variety and quality.

Our chapter provides evidence on the role of imported inputs on a firm's value added, product variety and product quality of one growing developing country, Indonesia.[2] The reasons for choosing Indonesia as an exercise object are twofold. First, Indonesia is one of the seven gainers - countries with the increased share of value added to the total world's value added in the manufacturing sector – placing it after China, Korea and India (Baldwin 2013). Second, there has been a growing concern on increasing imports in developing countries, including Indonesia.[3]

Indonesia's manufacturing sector was amongst seven gainers in the share of the world's manufactured products over the past three decades, even though relatively small, in terms of contribution to the world's value added in manufacturing, in which Indonesia's contribution increased from 0.1 per cent in 1970 to 1.8 per cent in 2011.

Based on trade in value added in exports (OECD), Figure 4.2 shows that the ratio of domestic value added embodied in Indonesia's total exports increased from 81 per cent in 2000 to 86 per cent in 2009.[4] This was relatively higher than the average of five Southeast Asian countries (Malaysia, the Philippines, Singapore, Thailand, and Viet Nam), which increased from 57 per cent to 59 per cent over the same period. The main reason for a relatively high domestic value content in total exports of Indonesia, particularly in agriculture and mining, which

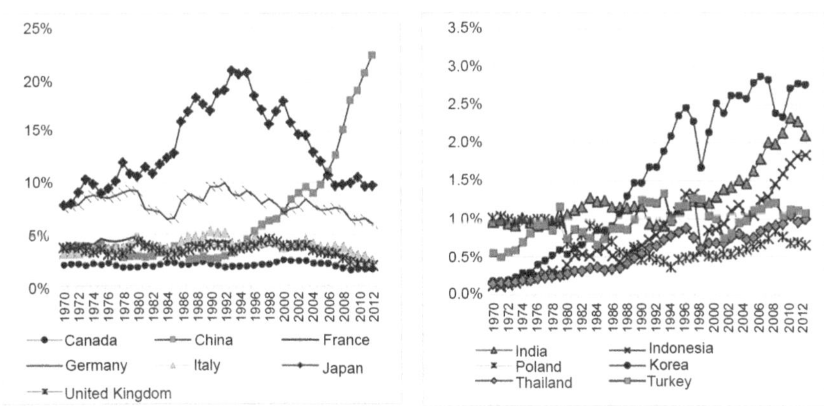

Figure 4.1 Indonesia is one of the top seven gainers in the share of manufacturing value added to the world (1970–2012)

Source: Authors' calculations based on UNSTAT [accessed in March 2013], reconstruction of Baldwin (2013).

had an average of more than 95 per cent from 2000 to 2009, could be relatively natural, in that Indonesia is a resource-abundant country. Nonetheless, domestic value contents were also quite visible in Indonesia's exports of manufactured products. The ratio of that domestic value added to the value of exports of manufactured products was 76 per cent in 2000 and increased to 82 per cent in 2009. It was lower than the ratio of domestic value added to total exports, but was still higher than the average of the other Southeast Asian countries at 50 per cent, which increased to 53 per cent over the same period.

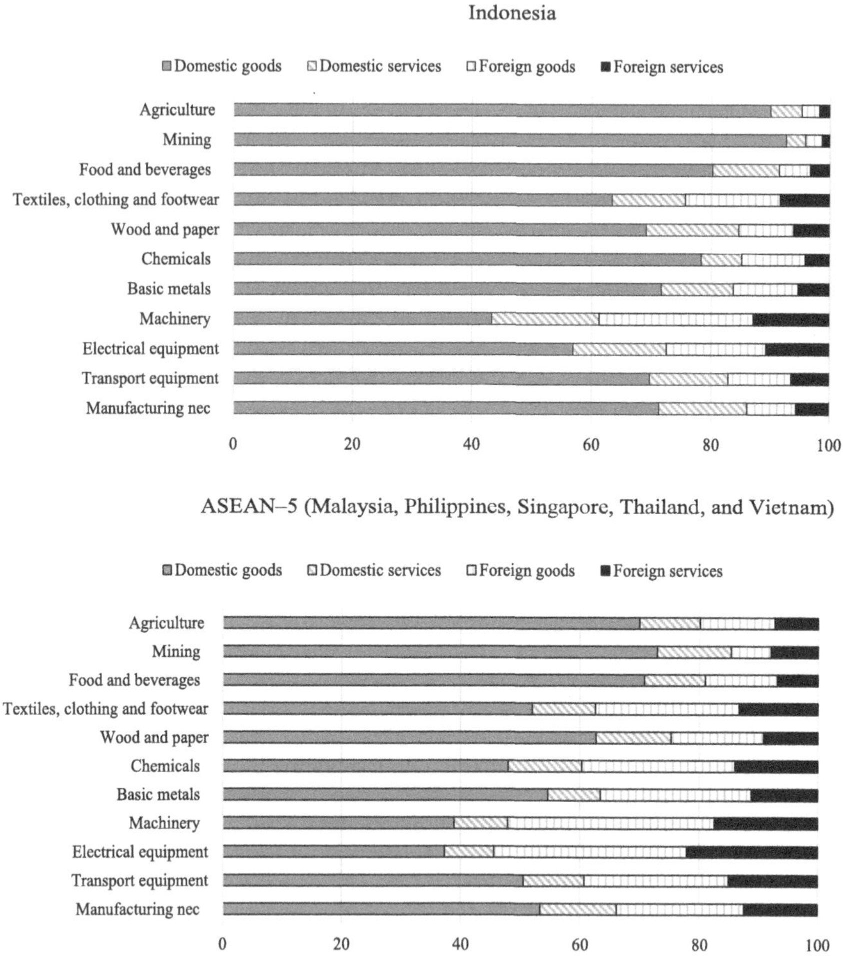

Figure 4.2 The ratio of domestic value added and foreign value added to total exports (2009)

Source: Authors' calculations based on OECD Trade in Value Added, 2013 [accessed in November 2013].

In fact, it is still higher compared with similarly resource-abundant countries like Malaysia, China, and India, whose ratios of domestic value added to exports of manufactured products were 55 per cent, 65 per cent and 71 per cent, respectively, in 2009. It is likewise higher than the developing countries' average of 69 per cent in the same period.

The relatively high domestic contents were also quite visible in most lines of exports of manufactured products. Other than in natural resource-intensive manufacturing industries (for example, leather and footwear, wood and paper and basic metals), domestic contents contributed around 85 per cent, 61 per cent, 73 per cent and 83 per cent to exports of manufactured products in chemical, machinery, electronics and transport, respectively, in 2009.

Apparently, domestic contents were quite visible in most lines of Indonesia's exports of manufactured products; however, if we examine their composition, Indonesia's exports are still relatively reliant on agriculture and natural resource-intensive products. The ratio of the value of exports of manufactured goods to total value of exports decreased from 49 per cent in 2000 to 34 per cent in 2010. Compared with other large emerging countries such as China and Brazil, the ratio of Indonesia's exports in machinery and parts to total exports was lower at about 15 per cent compared with 20 per cent for Brazil and 23 per cent for China in 2011 (Figure 4.3).

This leads us to a question if there is correlation between relatively high domestic contents and composition of exports: What are the roles of imported inputs in value added? What are the roles of imported inputs in product variety and quality?

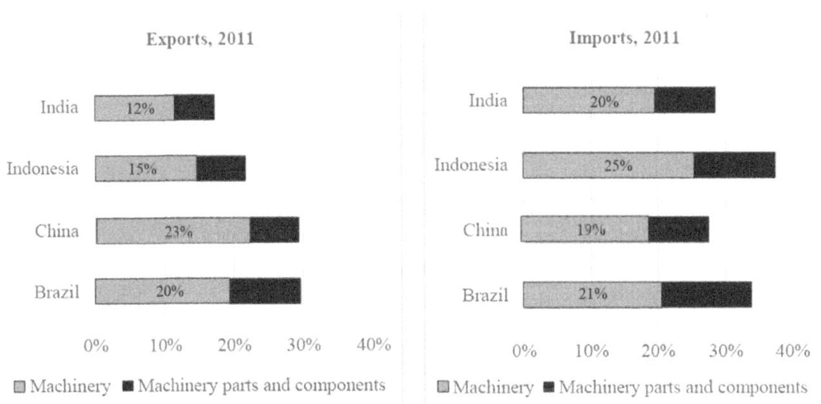

Figure 4.3 Indonesia in production network: Indonesia's exports of machinery and parts were lower than those of its large, emerging peers

Source: Authors' calculations based on UN Comtrade, using SITC Rev 3–5 digit [accessed in April 2014].

Section 2 presents a theoretical framework. Section 3 describes the estimation strategy, data and data sources. Section 4 presents empirical results. Section 5 concludes.

2. Theoretical framework

2.1 Set up

Consumers have non-homothetic preferences of product variety, and firms could enter to produce new varieties of goods if they have an ability to invent a new variety of good and would like to invest a certain amount of fixed costs of innovation that is independent of the number of goods produced and variable cost of a new variety of goods. In producing a new variety of goods, the firms have flexibility to use domestic and/or imported inputs. Our set-up is based on a model of optimal choices of product scope for multiproduct firms developed by Feenstra and Ma (2008). We extend the model by incorporating an intermediate input as a determining factor in the production process. Our model allows us to understand how input tariffs, which will affect the prices of imported intermediate input, will affect a firm's decision on the optimal levels of product variety and product quality.

2.1.1 Consumption

Utility is derived from consuming homogenous and heterogeneous goods. As in Feenstra and Ma (2008), and Fajgelbaoum et al. (2011), individuals have non-homothetic preferences over heterogeneous goods with constant elasticity of substitution.

$$U = y_0 + \rho \ln(Q) \text{ with } \rho < 1 \tag{1}$$

y_0 stands for homogeneous goods, and is treated as numeraire. Q is an index of horizontally heterogeneous goods.

$$Q = \left[\int_0^N q(i)^{\frac{\eta-1}{\eta}} di \right]^{\frac{\eta}{\eta-1}} \text{ with } \eta > 1 \tag{2}$$

q denotes quantity of heterogeneous goods, and $q(i)$ is the quantity consumed of variety i, and $\eta > 1$ is the elasticity of substitution between heterogeneous good varieties, and the variety is continuum $i \in [0, N]$. The quantity of consumed goods of variety i could be rewritten as a function of expenditure:

$$q(i) = \frac{R}{P_Q^{1-\eta}} p_q(i)^{-\eta} \tag{3}$$

R is aggregate expenditure in this sector. The quantity of consumed goods of variety, i could be rewritten as a function of expenditure. Given the quality of goods produced, each firm can choose the continuum of prices, $p(i)$. Suppose that each firm has the same marginal cost for all its varieties, $p_j = p(i)$, the price index can be written as:

$$P_Q = \left(\sum_{j=1}^{J} N_j p_{qj}^{1-\eta} \right)^{\frac{1}{1-\eta}} \tag{4}$$

2.1.2 Production

Suppose there are factors of production, capital, labour and material indexed as K, L and M. Each firm j, $j \in [1, J]$ j uses the three factors in production. In equilibrium, there is full employment, markets clear and firms earn zero profit.

The homogenous goods are freely traded and produced by perfectly competitive firms. The firms in the perfect competition market (PC) have the following costs and profit functions:

$$TC_j^{PC} = FC_j + VC_j = c_{0j} + c_j y_{0j} \tag{5}$$

$$\max_{(p_j, y_{0j}) \geq 0} \Pi_j^{PC} = y_{0j} \left(p_{y_j} - c_j \right) - c_{0j} \tag{6}$$

The optimal choice of price p_{yj} is $p_{yj} = cj$.

The heterogeneous goods are produced in a monopolistic competition (MC) market.

$$q_i = \varphi_i \left(L_i^{\frac{\mu-1}{\mu}} + \int_0^{M_i} x_{m_i}^{\frac{\mu-1}{\mu}} dm_i \right)^{\frac{\mu}{\mu-1}} \quad \text{with } \mu > 1 \tag{7}$$

φ is cumulative productivity of labour and input material. L is the number of unit labour. m is variety of input material, and m_i is the number of varieties of input material used in producing product i. M is the number of input material varieties, $M \in [0, M]$ and M_i is the number of variety of input material used in producing product i. Let x be the quantity of input material, and xm_i the quantity of m-th input material that is used in producing the i-th product. μ is the elasticity of substitution of factors of production (e.g. elasticity of substitution between labour and material or amongst materials). Define an index of all input material used in production of variety i as:

$$X_M = \left(\int_0^M x_m^{\frac{\mu-1}{\mu}} dm \right)^{\frac{\mu}{\mu-1}} \tag{8}$$

We also define a cumulative index of the price of all input materials in production of variety i:

$$P_M = \left(\int_0^M p_m^{1-\mu} dm \right)^{\frac{1}{1-\mu}} \tag{9}$$

p_m stands for the price of material m.

Firms may decide to produce a new product line (in the empirical work, it will be defined as if the firm could produce a new product line at nine-digit product code). If the firms can produce a new variety of goods at a certain level of quality, that requires them to spend on a fixed cost of innovation, $k_{0,}$ and marginal cost of new varieties which is constant across varieties, $k_{1.}$ w stands for the labour wage and pm_i as the price of material used in production of variety i. N_j is the number of varieties produced by firm j and a firm produces at least one variety of good, $N_j > 0$. The cost function:

$$TC_j^{MC} = FC_j + VC_j = c_{0j} + c_j q_j + k_{0j} + k_{1j} q_j N_j \tag{10}$$

The total production and cost functions could be written as follows.

$$N_j q_j = N_j q_i = N_j \varphi_i \left(L_i^{\frac{\mu-1}{\mu}} + \int_0^{M_i} x_{m_i}^{\frac{-1}{}} dm_i \right)^{\frac{\mu}{\mu-1}} \tag{11}$$

$$N_j c_j = k_0 + N_j \left(k_1 + wL_i + \int_0^{M_i} p_{m_i} x_{m_i} dm_i \right) \tag{12}$$

The profit function in the monopolistic competition market is as follows.

$$\max_{(p_{qj}, N_j) \geq 0} \Pi_j^{MC} = N_j q_j \left(p_{qj} - \gamma_j\right) - k_0 - k_1 N_j \tag{13}$$

With γ_j as the marginal cost of labour and material and $_{k1j}$ as the marginal cost of new variety that depends on additional intermediate inputs, both marginal cost of production and that of new variety are constant across variety.

$$\gamma_j = \frac{\left(w^{1-\mu} + P_{M_i}^{1-\mu}\left(_m\right)\right)^{\frac{1}{1-\mu}}}{\varphi_i} \tag{14}$$

Marginal cost of new variety γ_1 depends on index price of all materials, P_M. The prices of input material depend on domestic input prices, $p_{m,}$ and imported input prices which are also affected by import tariffs, $\tau, P_M = \left(p_m, p_m^*\left(p_m^*, \tau\right)\right)$. In an input market, all firms are price takers and thus do not have any power to determine prices on inputs. Domestic and imported inputs are imperfect substitutes. As quality of inputs required in their production is given, firms choose the lowest prices of inputs.

The optimal choice of the number of varieties as a function of output prices and cumulative productivity of labour and input material could be written as follows.

$$N_j = \left[1 - \frac{\gamma_j}{\left(p_{qj} - \gamma_j\right)(\eta - 1)} \right] \alpha^{\eta-1} P_Q^{\frac{1-\eta}{\eta}} \varphi_i^{\frac{\eta-1}{\eta}} \tag{15}$$

$$\text{with } p_{q_j} \geq \gamma_j, p_{qj} \geq \frac{\eta}{(\eta-1)}\gamma_j$$

The optimal choice of price p_{qj}, where α is the share of expenditure to marginal cost of new varieties, can be obtained as follows.

$$p_{qj} = \alpha P_Q^{\frac{\eta-1}{\eta}} \gamma_j^{\frac{1}{\eta}} \text{ where } \alpha \equiv \left[\frac{R}{k_1(\eta-1)}\right]^{\frac{1}{\eta}} \tag{16}$$

The optimal levels of product variety and product quality are directly affected by material prices P_M – and indirectly affected by input tariffs – through the marginal cost of a new variety, γ_j. A reduction in input prices will decrease the price of input material, and thus increase product variety, whereas it will decrease product quality or product prices (see Appendix for derivations). Since import tariffs positively affect input prices, a reduction in input tariff will increase the optimal levels of product variety of average firms and decrease the optimal levels of product prices of importing firms. However, if a firm imports as well as exports, reduced input tariffs may increase product quality because lower input tariffs allow firms to have access to better quality inputs.

3. Linking theory to empirical study

3.1 Estimation strategy

The main contribution of our empirical exercises is twofold. First, we analyse the impacts of output and input tariffs on value added. We also test two mechanisms through which imported inputs could affect value added. Second, we identify the impact of tariff reduction and its interaction with importing firms and exporting firms that use imported inputs. Value added here will be measured by real value added per worker (labour productivity).

$$\ln(VA_{jt}) = \beta_0 + \beta_1 F_{jt} + \beta_2 \tau_{qt} + \beta_{3qt}^{m} + \alpha_j + \alpha_t + \varepsilon_{jt} \tag{17}$$

F_{jt} is time-variant firm characteristics. τ_{qr} is output tariff by ISIC–5 digit. τ_{qt}^{m} is input tariff by ISIC–5 digit. As firms are price takers and do not have any power to determine prices that could affect tariff, $\forall_j$ face the same level of tariffs for the same input. α_j is a fixed effect for time-invariant firm characteristics, and α_t is year-fixed effect.

We will also test two mechanisms through which imported inputs could affect value added. First, imported inputs can act as a catalyst to expand product diversification. N_{jt} is the number of variety of products produced by firm j at time t.

$$\ln(N_{jt}) = \beta_0 + \beta_1 F_{jt} + \beta_2 \tau_{qt} + \beta_{3qt}^{m} + \alpha_j + \alpha_t + \varepsilon_{jt} \tag{18}$$

Second, imported inputs can increase product quality, which is represented by prices of products. Suppose each firm has the same marginal cost for all its varieties, so it will charge the same price for them. p_{jt} is the prices of products produced by firm j at time t. The price is corresponding to unit value, which represents quality of products.

$$\ln(p_{jt}) = \beta_0 + \beta_1 F_{jt} + \beta_2 \tau_{qt} + \beta_{3qt}^{m} + \alpha_j + \alpha_t + \varepsilon_{jt} \qquad (19)$$

3.2 Data and data source

The main data sources are Indonesia's firm-level and product-level data, which are based on manufacturing survey (*Survey Industry*) for large and medium-sized firms with 20 or more employees. We merge the firm level and product level data sets. The data contain very rich information on firm characteristics, such as main firm identities (location, ownership, legal status, starting year of operation), number of workers, share of domestic and foreign ownership, value of total inputs, value of inventories, value of total outputs, value added, list of product varieties at the HS–9 digit level, shares of products sold in domestic and export markets to total sales, value of total sales, share of imported inputs to total inputs and other informative firm characteristic variables. The data are from 2000 to 2010.

The data cover all registered manufacturing firms with around 30,000 firms annually, through a data cleaning process, which results in a total number of observations of 253,911. Of those observations, 21,671 observed firms or 8.5 per cent of the total sample are importing as well as exporting firms at the same time (please note that in our sample we only include manufacturing firms, and exporting and importing firms that are also manufacturers, and exclude distributors).

Value added (*VA*) is defined as the difference between the value of total outputs and the value of total inputs. The real value added is constructed by deflating the value added using a price deflator. The real value added per worker is the real value added divided by the number of total workers.

N stands for product variety. A new variety in the empirical exercises is defined as a new variety at HS–9 digit. A firm is defined to produce a new variety if it could produce a new good with a new product code in the HS–9 digit classification. So the number of varieties of goods is defined as the number of total varieties at HS–9 digit.

p stands for product quality. Product quality is approached by unit value of goods, which is constructed as the value of total outputs divided by the total volume.

F stands for main firm characteristics. We test how ownership, importing firms and exporting firms that use imported inputs affect the decision of selecting the level of product variety and quality in optimising a firm's profit. We use dummy variables for ownership, export and import. Following a number of empirical studies in this field including the one by Amiti and Konings (2007), we define DFDI = 1 if foreign share ≥ 0.1. We also use dummy variables for the share of the value of exports to total sales (DX = 1 if export share > 0) and the share of the value of imported inputs to the value total inputs (DM = 1 if import share > 0). We also estimate equations 17–19 separately by using the share of foreign ownership to total investments, the share of the value of exports to total sales and the share of the value of imported inputs to total inputs.

τ stands for tariffs. τ_{qt} is output tariff. Ideally if we could have input tariff at firm level, τ_{mjt}, we will use this data. Unfortunately, input tariff data at the firm level are not available. Thus, the input tariffs are constructed at the industry level, not at the firm level. Following Amiti and Konings (2007) and Goldberg et al. (2008), the input tariffs are constructed as follows. Input tariffs are defined as $\tau_{qt}^{m} = \sum_{m} s_{mq}\tau_{mt}$ where s_{mq} is the cost share of input *m* in industry *q*. For an illustration, if the *batik* (garment) industry spends 40 per cent of total costs on cloth, 30 per cent on *batik* colouring, 20 per cent on wax and 10 per cent on other materials, we give weights of 40 per cent, 30 per cent, 20 per cent and 10 per cent to *batik* cloth tariff, *batik* colouring tariff, wax tariff and other materials tariff, respectively. Input tariffs are calculated as a weighted average of the output tariffs using Indonesia's input–output data for 2000, 2005 and 2009. The tariff data are from the UN Comtrade database. The concordance of the HS-9 digit to ISIC-5 digit is provided by Statistics Indonesia by request. We also control time-invariant firm characteristics, α_j and fixed effect of year, α_t.

4. Empirical results

Table 4.1 represents summary statistics of variables which consist of output tariff, input tariff, log real value added per worker (labour productivity) which stands for value added, log number of goods which represents product quantity and log output prices which represents product quality. The main firm characteristics are represented by their trading activities which are determined by import and export shares, and by ownership.

Table 4.1 Data summary

Variable	*Observations*	*Mean*	*Standard deviation*
Output tariff	253,911	0.081	0.046
Input tariff	253,911	0.065	0.017
Ln (real value added per worker)	253,911	4.709	1.264
Ln (number of goods)	253,911	0.479	0.650
Ln (output prices)	253,911	5.052	0.325
Import share	253,911	0.093	0.235
DM = 1 if import share > 0	253,911	0.254	0.435
Export share	253,911	0.127	0.295
DX = 1 if export share > 0	253,911	0.127	0.295
Foreign share	253,911	0.070	0.181
DFDI = 1 if foreign share ≥ 0.1	253,911	0.479	0.499

Source: Authors' calculation based on Survey Industry data.

Table 4.2 presents the impacts of tariff reductions on value added. Column 1 shows that output tariff reductions will increase labour productivity by 0.02 per cent. However, once we include input tariffs, as illustrated in Column 2 and further estimations, it can be seen that the impacts of output tariffs on labour productivity are lower than the estimation with output tariff alone, whilst a reduction of 1 per cent in input tariff will raise labour productivity by as much as 0.2 per cent.

Column 3 shows 1 per cent of input tariff reduction will increase the labour productivity by 0.2 per cent, which is a result of the combination of input tariffs and their interaction with importing firms. Columns 4 and 5 also show consistency in findings that a 1 per cent of input tariff reduction will increase labour productivity by 0.2 per cent. They show, on average, that a reduction of 1 per cent in input tariffs alone will increase labour productivity by 0.1 per cent. The magnitude of the effect of tariff reductions on labour productivity is amplified by the interaction between importing firms enjoying the tariff reduction (0.07 per cent) and exporting firms using imported inputs for their exports (0.04 per cent) which means that a reduction of 1 per cent in input tariff will increase labour productivity by 0.2 per cent. One per cent of tariff reductions also provide higher labour productivity gains for importing and exporting firms than for the average of all firms. Importing manufacturing firms have about 1 per cent higher labour productivity, whilst exporting manufacturing firms will have about 0.4 per cent higher labour productivity than the average of all firms.

Our findings are consistent with those of Amiti and Konings (2007), but at a lower magnitude. The reasons are twofold. First, the average input tariffs are lower, with an average of 6.5 per cent, from 2000 to 2010 covered in this study compared with an average of 10.1 per cent in their study over the sample period 1991–2001. Second, the marginal contribution of tariff reductions to labour productivity is decreasing.

Table 4.3 presents the impacts of tariff reductions on product variety. Column 1 shows that output tariff reductions have insignificant impacts on product variety, even when we include input tariffs in the estimations presented in Columns 2–6, whereas input tariff reductions significantly increase product variety. Column 2 shows that 1 per cent of input tariff reduction will increase product variety by 3.15 per cent. When we control for the interaction of input tariff reduction with importing and exporting firms, it is shown that 1 per cent of input tariff reduction will increase product variety by around 2.7–2.8 per cent, as presented in columns 3–5.

Columns 4–5 show 1 per cent input tariff reduction will increase product variety by 3.5 per cent, which is a result of input tariff reductions and their interaction with exporting firms that use imported inputs as their product; their interaction with importing firms per se does not provide any significant contribution to increased product variety. The results are consistent throughout the estimations. Columns 4–5 also show that importing and exporting firms have a tendency to have more product variety than the average firms by 0.18 per cent

Table 4.2 The impacts of tariff reductions on value-added

Dependent variable: value added (in real value added per worker)						
	(1)	*(2)*	*(3)*	*(4)*	*(5)*	*(6)*
Output tariff	−0.024***	−0.017***	−0.017***	0.016***	−0.016***	−0.017***
	(0.006)	(0.002)	(0.002)	(0.003)	(0.003)	(0.003)
Input tariff		−0.129***	−0.108***	−0.105***	−0.098***	−0.116***
		(0.021)	(0.030)	(0.282)	(0.031)	(0.025)
Input tariff x DM			−0.065*	−0.058*	−0.068*	
			(0.038)	(0.034)	(0.038)	
DM = 1 if import share > 0			0.914***	0.915***	0.951***	
			(0.321)	(0.308)	(0.316)	
Input tariff x import share						−0.030*
						(0.018)
Import share						0.899***
						(0.234)
Input tariff x DM x DX				−0.039***	−0.036***	
				(0.006)	(0.010)	
Input tariff x import share x export share						−0.120***
						(0.018)
DX = 1 if export share > 0				0.406***	0.419***	
				(0.023)	(0.073)	
Export share						0.123**
						(0.060)
DFDI = 1 if foreign share ≥ 0.1					0.103***	
					(0.035)	
Foreign share						0.477***
						(0.059)
Island and year effect	yes	yes	yes	yes	yes	yes
Firm fixed effect	yes	yes	yes	yes	yes	yes
Observations	253,911	253,911	253,911	253,911	253,911	253,911
F-test	89	83.0	73.7	417	384.7	88.6
R-squared	0.60	0.64	0.70	0.71	0.72	0.65

Note: Robust standard errors in the parenthesis. ***Significant at, or below, 1%. **Significant at, or below, 5%. *Significant at, or below, 10%.

Source: Authors' estimation.

and 0.05–0.06 per cent, respectively, but multinational firms do not necessarily have more product variety.

South Korea's manufacturing firms in 1990–1998 showed a likelihood to increase more product varieties after entering an export market which was due to the effects of exporting activities on product development being larger than the effects of them on cutting the number of product (Hahn 2012). In the case of firms in Mexico, it is claimed that a new exporting firm usually starts 'small' in terms of both values and number of exported products. This could be a reason that there is a substantial degree of product turnover at the firm product level in

Table 4.3 The impacts of tariff reductions on product variety

Dependent variable: product variety (in number of goods)						
	(1)	*(2)*	*(3)*	*(4)*	*(5)*	*(6)*
Output tariff	−0.133	0.051	0.057	0.063	0.059	0.0637
	(0.140)	(0.115)	(0.120)	(0.121)	(0.120)	(0.122)
Input tariff		−3.154***	−2.764***	−2.711***	−2.773***	−2.816***
		(0.678)	(0.859)	(0.822)	(0.836)	(0.782)
Input tariff × DM			−1.050	−0.858	−0.859	
			(0.818)	(0.854)	(0.852)	
DM = 1 if import share > 0			0.185***	0.182***	0.185***	
			(0.062)	(0.062)	(0.062)	
Input tariff × import share						−1.388
						(1.218)
Import share						
Input tariff × DM × DX				−0.782**	−0.812**	
				(0.396)	(0.413)	
Input tariff × import share × export share						−3.584***
						(0.934)
DX = 1 if export share > 0				0.051**	0.060**	
				(0.026)	(0.029)	
Export share						0.030
						(0.059)
DFDI = 1 if foreign share ≥ 0.1					−0.029	
					(0.020)	
Foreign share						−0.007
						(0.015)
Island and year effect	yes	yes	yes	yes	yes	yes
Firm fixed effect	yes	yes	yes	yes	yes	yes
Observations	253,911	253,911	253,911	253,911	253,911	253,911
F-test	89.9	43.9	49.1	81.2	77.2	85.2
R-squared	0.20	0.20	0.20	0.27	0.28	0.27

Note: Robust standard errors in the parenthesis. ***Significant at, or below, 1%. **Significant at, or below, 5%. *Significant at, or below, 10%.

Source: Authors' calculation.

response to declining trade costs (Iacovone and Javorcik 2010). In a dynamic setting, Costantini and Melitz (2008) predict that a new exporter is likely to increase product variety one year prior to entering an export market and after two years in the export market, and the firm subsequently tends to specialise in a certain number of products that give it optimal profit.

Table 4.4 presents the impacts of tariff reductions on product quality. Column 1 shows 1 per cent output tariff reduction will increase product quality by 0.15 per cent. But when we include input tariffs and their interaction with

Table 4.4 The impacts of tariff reductions on product quality

Dependent variable: product quality (in unit value of goods)						
	(1)	*(2)*	*(3)*	*(4)*	*(5)*	*(6)*
Output tariff	−0.153***	−0.057	−0.059	−0.056	−0.066	−0.047
	(0.061)	(0.046)	(0.047)	(0.047)	(0.049)	(0.044)
Input tariff		−1.450***	−1.539***	−1.504***	−1.670***	−1.459
		(0.193)	(0.177)	(0.180)	(0.149)	(0.179)
Input tariff × DM			0.324***	0.325***	0.287***	
			(0.120)	(0.121)	(0.094)	
DM = 1 if import share > 0			−0.033***	−0.033***	−0.024***	
			(0.009)	(0.009)	(0.007)	
Input tariff × import share						0.468**
						(0.235)
Import share						−0.060**
						(0.016)
Input tariff × DM × DX				−0.066**	−0.115**	
				(0.033)	(0.058)	
Input tariff × import share × export share						−0.131
						(0.066)
DX = 1 if export share > 0				0.037***	0.059***	
				(0.021)	(0.022)	
Export share						0.006**
						(0.003)
DFDI = 1 if foreign share ≥ 0.1					−0.071	
					(0.044)	
Foreign share						0.142
						(0.011)
Island and year effect	yes	yes	yes	yes	yes	yes
Firm fixed effect	yes	yes	yes	yes	yes	yes
Observations	253,911	253,911	253,911	253,911	253,911	253,911
F-test	49.3	84.0	105.4	110.6	136.8	110
R-squared	0.38	0.38	0.38	0.39	0.39	0.31

Note: Robust standard errors in the parenthesis. ***Significant at, or below, 1%. **Significant at, or below, 5%. *Significant at, or below, 10%.

Source: Authors' calculation.

firm characteristics that determine product quality, our tariff reduction does not significantly affect product quality. Further, comparison of Columns 1 and 2–6 suggests there is an omitted variable bias in column 1. Column 2 shows that 1 per cent of input tariff reduction will increase product quality by 1.45 per cent. Columns 2–6 show that output tariff alone does not significantly affect product quality.

Column 3 shows that 1 per cent of input tariff reduction will increase product quality by 1.5 per cent, which is consistent throughout estimation results presented in Columns 4 and 5.

Column 5 shows that 1 per cent of input tariff reduction alone will increase product quality by 1.67 per cent and the interaction of input tariff reductions and exporting firms that use imported products as their inputs will increase product quality by 0.12 per cent, whilst the interaction of input tariff reduction and importing firms will reduce product quality by 0.29 per cent. Tariff reduction and its interaction with importing and exporting firms that use imported inputs all bring together the effect of 1 per cent of input tariff reductions will increase product quality by 1.5 per cent.

It is interesting to see that the interaction of input tariff reduction and importing firms will decrease product quality, but the interaction of input tariff reduction and exporting firms that use imported products as their inputs will increase product quality.

One explanation could be that importing firms tend to import goods due to lower prices and sell their products in the domestic market. The importing firms that use imported products as their inputs, however, the importing firms that are also exporting firms, tend to import products not only due to lower prices but also taking into consideration the quality of imported products to be able to produce output products of a certain quality level, and thus the interaction of input tariff reductions and exporting firms that use imported products as their inputs will increase product quality.

Our findings are consistent with studies conducted by Fan et al. (2013) and Bas and Strauss-Kahn (2015). They explained that reductions in import tariffs will encourage firms to choose higher-quality imported inputs. The higher-quality inputs will induce firms to produce higher-quality products (i.e. higher priced-products).

5. Conclusions

The rich information on firm- and product-level data of Indonesia's manufacturing surveys allow us to test the hypotheses on how input tariff reductions affect value added (labour productivity), via product variety and quality.

A reduction of 1 per cent in input tariff will increase labour productivity by 0.02 per cent. The impacts of input tariff reductions on product variety and quality are not only driven by their own channels, but also magnified by their interaction with exporting firms that use imported products as their inputs. A reduction of 1 per cent in input tariff will increase product variety by 3.5 per cent and product quality by 1.5 per cent.

It is interesting to see that tariff reductions will decrease product quality (product prices) of importing firms, but will increase product quality of exporting firms that use imported inputs. The reason could be that importing firms tend to import goods due to lower prices; however, exporting firms that use imported inputs tend to import products not only because of lower prices, but also because of the better quality of imported inputs that could ensure producing good quality outputs.

We hypothesised that a reduction in input tariff could affect final product variety and quality due to greater variety of inputs or better quality of inputs. Differentiating complementarity amongst domestic and foreign inputs and quality of inputs would be an area for future research.

Appendix

a. Imported inputs, input tariffs, product variety and product quality

Correlation between imported inputs and product variety

$$N_j = \left[1 - \frac{\gamma_j}{(p_{qj} - \gamma_j)(\eta - 1)}\right] \alpha^{\eta-1} P_Q^{\frac{1-\eta}{\eta}} \varphi_i^{\frac{\eta-1}{\eta}} \tag{20}$$

with $p_{q_j} \geq \gamma_j, p_{qj} \geq \frac{\eta}{(\eta-1)}\gamma_j$

$$\frac{\partial N_j}{\partial P_{M_i}} = \left[- \frac{\gamma_j p_{qj}}{(p_{qj} - \gamma_j)^2 \eta P_{M_i}^{\mu}(_m)(w^{1-\mu} + P_{M_i}^{1-\mu}(_m))}\right] \alpha^{\eta-1} P_Q^{\frac{1-\eta}{\eta}} \varphi_i^{\frac{\eta-1}{\eta}} < 0 \tag{21}$$

Correlation between imported inputs and product quality

$$p_{qj} = \alpha P_Q^{\frac{\eta-1}{\eta}} \gamma_j^{\frac{1}{\eta}} \text{ where } \alpha \equiv \left[\frac{R}{k_1(\eta-1)}\right]^{\frac{1}{\eta}} \tag{22}$$

$$\frac{\partial p_{qj}}{\partial P_{M_i}} = \frac{\alpha P_Q^{\frac{\eta-1}{\eta}} \gamma_j^{\frac{1}{\eta}} P_{M_i}^{-\mu}(_m)}{\eta(w^{1-\mu} + P_{M_i}^{1-\mu}(_m))} > 0 \tag{22}$$

$$\frac{\partial p_{qj}}{\partial P_Q} = \frac{\eta-1}{\eta} \alpha P_Q^{\frac{\eta-2}{\eta}} \gamma_j^{\frac{1}{\eta}} > 0 \tag{23}$$

Notes

1 The authors thank Statistics Indonesia for providing Indonesia's manufacturing data at the firm and product levels. This chapter has benefited from the comments provided by Miaojie Yu, Xiao Jiang, Fukunari Kimura, Nina Pavcnik, Hal Hill, Kazunobu Hayakawa, Toshiyuki Matsuura, Sadayuki Takii and Ari Kuncoro.

2 The discussion on variety and quality of a country's exports is in Hummels and Klenow (2005) and Helpman et al. (2008).

3 Moreover, the ratio of the value of exports of manufactured goods to the total value of exports decreased from 49 per cent in 2000 to 34 per cent in 2010 (Ing et al. 2015).

4 The ratio of domestic value added to exports is measured by the ratio of the value contributed by domestic factors of production to the value of total exports.

References

Amiti, M and Khandelwal, A 2013, 'Import competition and quality upgrading', *Review of Economics and Statistics*, vol. 92, no. 2, pp. 476–490.

Amiti, M and Konings, J 2007, 'Trade liberalization, intermediate inputs, and productivity: evidence from Indonesia', *American Economic Review*, vol. 97, no. 5, pp. 1611–1638.

Baldwin, R 2013, 'WTO 2.0: Thinking Ahead on Global Trade', in *The Future of the World Trading System: Asian Perspectives*, eds. R Balwind, M Kawai and G Wignaraja, Centre for Economic Policy Research, London, pp. 23–28.

Bas, M and Strauss-Kahn, V 2015, 'Input-trade liberalization, export prices and quality upgrading', *Journal of International Economics*, vol. 95, no. 2, pp. 250–262.

Bernard, A, Eaton, J, Jensen, J and Kortum, S 2003, 'Plants and productivity in international trade', *American Economic Review*, vol. 93, no. 4, pp. 1268–1290.

Bernard, A, Redding, B and Schott, P 2006, 'Multi-product firms and trade liberalization', *Working Paper* 12782, NBER, Cambridge, MA.

Costantini, J and Melitz, M 2008, 'The Dynamics of Firm-Level Adjustment to Trade Liberalization', in *The Organization of Firms in a Global Economy*, eds. E Helpman, D Marin and T Verdier, Harvard University Press, Cambridge, MA, pp. 107–141.

Fajgelbaum, P, Grossman, G and Helpman, E 2011, 'Income distribution, product quality, and international trade', *Journal of Political Economy*, vol. 119, no. 4, pp. 721–765.

Fan, H, Li, Y and Yeaple, S 2013, 'Trade liberalization, quality, and export prices', *Paper 51370*, MPRA, Munich.

Feenstra, R and Ma, H 2008, 'Optimal Choice of Product Scope for Multiproduct Firms under Monopolistic Competition', in *The Organization of Firms in a Global Economy*, eds. E Helpman, D Marin and T Verdier, Harvard University Press, Cambridge, MA, pp. 173–199.

Fernandes, A 2007, 'Trade policy, trade volumes and plant-level productivity in Columbian manufacturing industries', *Journal of International Economics*, vol. 71, no. 1, pp. 52–71.

Goldberg, P, Khandelwel, A, Pavnick, N and Topalova, P 2008, 'Imported intermediate inputs and domestic product growth: evidence from India', *Working Paper* 14416, NBER, Cambridge, MA.

Hahn, C 2012, 'Learning-by-exporting, introduction of new products, and product rationalization: evidence from Korean manufacturing', *Journal of Economic Analysis and Policy*, vol. 12, no. 1, Article 24.

Hallak, J and Sivadasan, J 2013, 'Firm's exporting behaviour under quality constraints', *Journal of International Economics*, vol. 91, no. 1, pp. 53–67.

Helpman, E, Melitz, M and Rubinstein, Y 2008, 'Estimating trade flows: trading partners and trading volumes', *Quarterly Journal of Economics*, vol. 123, no. 2, pp. 441–487.

Hummels, D and Klenow, P 2005, 'The variety and quality of a nation's exports', *American Economic Review*, vol. 95, no. 3, pp. 704–723.

Iacovone, L and Javorcik, B 2010, 'Multi-product exporters: product churning, uncertainty and product discoveries', *Economic Journal*, vol. 120, no. 544, pp. 481–499.

Ing, L, Pangestu, M and Rahardja, S 2015, 'Managing Indonesia's Trade Policy: How to Remove the Agenda?', in *Macroeconomic Policies in Indonesia*, ed. A Nasution, Routledge, London and New York, pp. 224–244.

Kee, H and Feenstra, R 2011, 'Trade liberalization and export variety: a comparison of Mexico and China', *Policy Research Working Paper* 66617, The World Bank, Washington, DC.

Kugler, M and Verhoogen, E 2012, 'Prices, plant size and product quality', *Review of Economic Studies*, vol. 79, no. 1, pp. 307–339.

Melitz, M 2003, 'The impact of trade on intra-industry reallocations and aggregate industry productivity', *Econometrica*, vol. 71, no. 6, pp. 1695–1725.

UN Comtrade 2013, UN Comtrade Database. Available from: http://comtrade.un.org/db.

5 Trade in value added

The case of Malaysia

Siew Yean Tham and Andrew Jia Yi Kam

1. Introduction

Malaysia is a small open economy that is part of the regional production networks that have emerged in Southeast Asia with the influx of foreign direct investment (FDI), especially from Japan, since the mid-1980s. Although Malaysia had early-mover advantage in using FDI for developing its manufacturing sector, it has increasingly been losing its advantage in low-cost manufacturing to other competitors that have also tapped into the same FDI-driven growth strategy. This includes competitors from other Association of Southeast Asian Nation (ASEAN) member states such as Thailand and Viet Nam, as well as non-ASEAN members such as China. At the same time, Malaysia aspires to move beyond its current middle-income status to become a high-income nation by 2020. To achieve this, it is important for the country to upgrade its industrial structure and to shift from a production-based to a knowledge-based, innovation-led economy.

Given these aspirations, it is timely to examine the nature of Malaysia's trade in value-added exports to discern changes in its trends and patterns in the face of changing domestic and external circumstances and policies. Internally, Malaysia has undergone three economic crises – in 1985, 1997, and 2008. Although the country has recovered from each of these crises, restructuring the economy from low to high value-added manufacturing production is a continuing challenge. Externally, the lingering after-effects of the global financial crisis continue to pose challenges due to Malaysia's open economy and extensive trade and investment linkages with the rest of the world.

This chapter examines the imported content of Malaysia's manufacturing sector by using Hummel's vertical specialization method (Hummels et al. 2001), based on the input–output (I/O) tables of Malaysia from 2000 to 2010. A decrease in direct and indirect imported content based on Hummel's vertical specialization index is interpreted as an increase in the local content of Malaysia's exports. The results will be compared with the Organisation for Economic Co-operation and Development (OECD) trade in value-added data for Malaysia for 1995, 2000, 2005, 2008, and 2009. We use the information and communications technology (ICT) manufacturing sector as a case study.

2. Literature review

The emergence of global value chains with production fragmentation has progressively been shifting trade research from trade based on gross values to trade in value added. Various approaches are used to measure value added in trade, including firm-level studies and surveys on outsourcing and offshoring, measuring trade in parts and components, and the share of foreign imports in exports as in the vertical specialization (VS) ratio, as well as other methods based on inter-country I/O tables.

Firm-level studies provide rich micro information, but they cannot reveal the extent to which domestic value added is created at a macro level (Inomata 2013). Trade in intermediate inputs or components trade, especially in East Asia, is therefore used as a proxy for measuring VS since cross-border VS cannot be measured from trade statistics (Ng and Yeats 2001; Athukorala 2003). However, trade in intermediate goods is an imperfect proxy since these products can be used as consumption as well as intermediate inputs. A better way of measuring VS will require the use of either firm-level data or I/O tables that can distinguish between imported and locally sourced inputs used in the production of a good.

The vertical specialization (VS) index proposed by Hummels et al. (2001) to capture trade in value added defines the use of imported intermediate goods in products that are subsequently exported as international VS. Their measurements of VS for 13 OECD countries, including Taiwan, indicate the share of VS in exports increased from 15 per cent in 1970 to 20 per cent in 1990. Small countries like Singapore and Malta are found to have larger shares of VS in their exports compared with large countries, as more links in the global supply chain can be located within the country in large countries compared with small ones (Nordås 2005). A study by the World Trade Organization (WTO) and the Institute of Developing Economies (IDE–JETRO) (2011) shows VS activity in the Asian region is still very prevalent in manufacturing activities, even though VS activity in services has been increasing. Using the IDE–JETRO Asian I/O tables, the findings of the study further indicate that the degree of VS increased for all the selected Asian economies as well as the United States (US) from 1985 to 2008. Consistent with findings in Hummels et al. (2001), Singapore is found to have the highest VS share of exports amongst the countries investigated throughout the entire period studied. A high VS share is also observed for Taiwan, the Republic of Korea (henceforth, Korea), and Thailand, whereas the VS share for Japan, the United States, and Indonesia are far below Asia's average.

Alternative methods are based on inter-country I/O tables, as these tables show the international sources of inputs in goods produced in a region or throughout the world (Powers 2012). They are, therefore, able to reveal a country's position and role in global supply chains by distinguishing between national and global sources of inputs.

3. Overview of industrialization and industrial policies in Malaysia

Malaysia was mainly a rubber- and tin-producing economy at the time of independence in 1957. In 1960, agricultural production contributed to 40 per cent of Malaysia's gross domestic product (GDP) (Leete 2007), but it had fallen to just 7.0 per cent by 2014. The share of employment in agricultural production also declined continuously, whereas the share of manufacturing grew steadily from 9 per cent of the country's GDP in 1960 to a peak of 32 per cent in 2000, before falling to 25 per cent in 2014. Similarly, the share of employment in manufacturing grew from less than 10 per cent in the 1960s to a peak of 30 per cent in 2006 and 2007, before falling to 16 per cent in 2014. The services sector continued to be the largest contributor to Malaysia's GDP, growing to 55 per cent in 2014. It also had the largest share in terms of employment, accounting for 59 per cent of total employment in 2014.

The shift in economic structure from agriculture to manufacturing production can be at least partially attributed to deliberate government policies, as Malaysia's underlying development philosophy encompasses the use of active government support and direction together with free enterprise (Kaziah 2005). An FDI-led strategy was utilized for developing its manufacturing sector to reduce the country's dependence on primary commodities. FDI was initially courted for import substitution and later for export promotion, when the country abandoned import substitution as the domestic market was too small. The reintroduction of import substitution in heavy industries in 1980 did not mean that FDI and export promotion were abandoned. Rather, the promotion of heavy industrial development was prioritized at that point, with the state engaging with foreign partners for the development of heavy industries by state-owned enterprises, rather than the

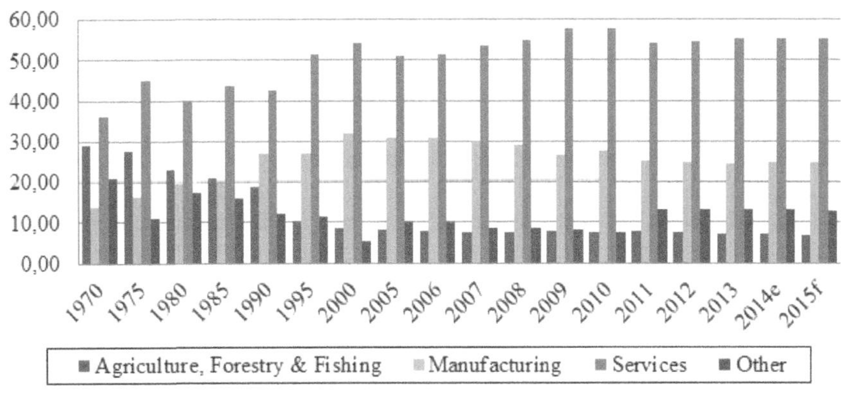

Figure 5.1 Share of sectors in Malaysia's GDP (%), 1970–2015

Source: Malaysia, Economic Report, various years.

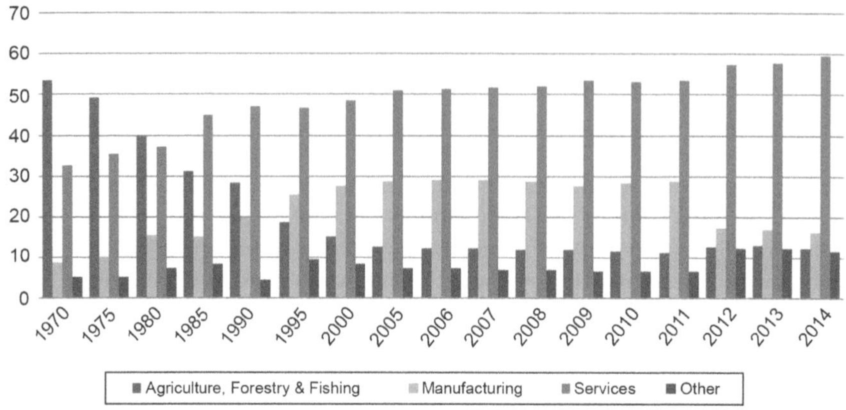

Figure 5.2 Share of sectors in Malaysia's total employment (%), 1970–2015

Source: Malaysia, Economic Report, various years.

private sector-led development strategy in the first round of export promotion (1968–1980). Similarly, when export promotion was reintroduced in 1986, due to the twin deficits (fiscal and external deficit as a result of increased government expenditure and imports of intermediate and capital goods needed to support the development of heavy industries), the protection of the heavy industries was not abandoned but continued.

Import substitution used tariff protection and import licenses for the protection of import substitution and infant industries, and fiscal incentives were used to attract FDI. But the focus of the incentives depended on the industrialization objectives in each period. Special zones were developed, for example, free trade zones (FTZs) where export is promoted through the use of duty-free imports. Specialized infrastructure in the form of technology parks was also provided (OECD 2011). The government also developed both hard and ICT infrastructure. In the case of the former, the development of airports and roads facilitated the movement and export of electronics goods and lowered the trade costs of the country (Tham et al. 2009). As for the latter, government investments improved Malaysia's broadband infrastructure and raised the household penetration rate from 1 per cent in 2004 to 32 per cent in 2009 (OECD 2011). But the capacity and speed of broadband in Malaysia remains behind its neighbour Singapore, and the cost is higher (Yusuf and Nabeshima 2009).

There has also been considerable investment in education to provide human resources for the country. This provided human resources in terms of the *quantity* needed, especially in the early days of industrial development, for joining global value chains through hosting low-value-added, assembly-type operations. However, there are signs that Malaysia's educational and training resources are inadequate for moving up the value-added chain as the *quality* of human resources grows increasingly important.

Moreover, there have been few new policies aimed at moving Malaysia up the value chain. In part, this is due to a lack of detailed assessments of the effectiveness of industrial development policies. Although the cluster approach to industrial development proposed at the time of the Second Industrial Master Plan (IMP2: 1996–2005) targeted eight clusters (Ohno 2006), its implementation was deemed poorly executed (Danaraj 2011), and Malaysia did not succeed in developing the clusters or a critical mass of firms needed to move up the global value chain. Moreover, as industries move to more complex production structures, greater policy coordination is needed. Unfortunately, the policy framework needed to coordinate the necessary relationship between macro-institutions, meso-organizations, and micro-agents for stimulating innovation in these countries are either weak or lacking (Rasiah 2014).

4. Development of the ICT sector in Malaysia

The Malaysian government targeted the ICT sector from the Seventh Malaysia Plan (1996–2000) onwards. In that plan, ICT is seen as a production sector and an enabler of the country's growth strategy. The plan adopted McKinsey's suggestion to build an IT hub, modelled after Silicon Valley (Vicziany and Puteh 2004), and the Multimedia Super Corridor (MSC) was established in 1996 to enable the country to 'leapfrog' into a knowledge economy. The ICT sector continued to be emphasized in all of Malaysia's subsequent five-year development plans. For example, ICT was identified in the Eighth Malaysia Plan (2001–2005) as a key driver to support and contribute to economic growth as well as to enhance quality of life in the country. It also aims to turn Malaysia into a major IT and multimedia hub. The ICT sector is viewed as a key determinant to facilitate the economy's shift up the value chain in the Ninth Malaysia Plan (2006–2010), and it is one of the 12 sectors targeted under the Tenth Malaysia Plan (2011–2015). The importance of this sector is shown by its contribution to GDP, which was 9.8 per cent in 2009, and it was targeted to contribute 10.2 per cent by 2015.

Several initiatives were undertaken to develop the MSC, which is a long-term strategic initiative implemented from 1996 to 2020. Fiscal incentives, subsidized electricity, water, and land facilities, as well as the ability to hire foreign knowledge workers to meet the shortage of skilled workers, are used to attract local and global investors to the MSC. Other government support includes the provision of ICT infrastructure, including expanding broadband services. In addition, a suite of new cyber laws was enacted to support communication, commerce, and trade in the digital medium such as the Digital Signature Act (1997), the Computer Crime Act (1997), and the Communications and Multimedia Act (1998) to cater for the convergence of the telecommunications, broadcasting, and computing industries in Malaysia, as well as the Data Protection Act (2002) (Kuppusamy et al. 2009). Demand for ICT products and services is generated through various flagship projects such as electronic government, development of multipurpose cards, telemedicine and health, smart schools, e-business, techno-preneur development, and the development of an R&D cluster.

The ICT sector in Malaysia is classified into five main subsectors – ICT manufacturing, trade (which is essentially wholesale and retail trade), services, content and media, and other industries. There were some changes in the ICT sector in Malaysia from 2005 to 2013 based on the ICT satellite data from the Department of Statistics (DOS). There was a shift in the relative importance of ICT manufacturing to ICT trade (wholesale and retail) and services. In 2005, the shares of ICT manufacturing, trade, and services in the ICT sector were 52 per cent, 10 per cent, and 28 per cent, respectively, but this changed to 32 per cent, 15 per cent, and 41 per cent in 2013 (DOS 2014). This was due to the decline in the electronics sector, whereas the ICT services sector continued to grow with increasing demand for these services. Nevertheless, the bulk of ICT exports is still in manufacturing, comprising 97 per cent of total ICT exports in 2005, although this share had fallen to 93 per cent by 2013. Conversely, the share of ICT services exports rose from 1 per cent to 3 per cent over the same period. Similarly, the export of content and media products increased from 2 per cent to 5 per cent over the same period. But, trade in ICT manufacturing generated a surplus for each year from 2005 to 2013; the converse holds for trade in ICT services for the same period.

5. Methodology

In view of the objectives of this chapter, we employ the concept of vertical specialization (VS) as set out in Hummels et al. (2001) to identify the share of imported input content for exports. The VS index Hummels et al. introduced was originally used to measure the VS phenomenon as well as the extent of international fragmentation of production chains in the world. It can also be used to profile the pattern of VS in Malaysia.

Based on Hummels et al. (2001), VS is defined based on the three following conditions: first, a good is produced in two or more sequential stages. Second, two or more countries generated value added during the production of the good; and finally, imported inputs are used in its production stages with some of the resulting output exported.

Based on these three criteria, VS of country *k* in good/sector i, VS_{ki} is defined as follows:

$$VS_{ki} = \left(\frac{imported\ intermediates}{gross\ output} \right) .exports \qquad (1)$$

The overall VS for country *k* is the sum of VS across all *i*, $VS_k = \sum_i VS_{ki}$. Dividing this by total exports of country, X, gives the VS share of total exports, that is:

$$\frac{VS_k}{X_k} = \frac{\sum_i VS_{ki}}{\sum_i X_{ki}} \qquad (2)$$

In matrix notation, $VS_k/X_k = uA^M X/X_k$, where u is a 1 × n vector of 1's, A^M is the n × n imported coefficient matrix, X is an n × 1 vector of exports, n is the number of sectors/subsectors, and X_k is the sum of exports across the n sectors/subsectors.

This formula measures the value of imported inputs used *directly* in the production of an exported good. As pointed out in Hummels et al. (2001), using the I/O tables allows us to calculate the value of imported inputs used *indirectly* as well. This can be done by adding the Inverse Matrix of Domestic Production $[I - A^D]^{-1}$ into our formula. The VS share of total exports that includes both direct and indirect value of imported inputs used becomes:

$$VS_k/X_k = uA^M [I - A^D]^{-1} X/X_k \tag{3}$$

where I is the identity matrix, and A^D is the n × n domestic coefficient matrix.

Hence, VS_k/X_k is the imported VS share of output used directly and indirectly in the production of an exported good. This can also be interpreted as the share of imported content (or foreign content) of the output used for exports. At the aggregated country level, Equation (3) is defined as the *Hummels VS* in this study. However, at disaggregated levels for all the manufacturing subsectors, VS share of exports is equivalent to the input share of gross output (Hummels et al. 2001), which will be defined only as *VS* in this study.

The value of VS ranges from 0–1. If a sector does not source any imported inputs or has no exports at all, VS = 0. The higher the VS value of a sector, the more the country sources imported inputs for its production. This also implies that a country is more integrated with the global production chain in that sector.

We use both published and unpublished data from Malaysia's I/O tables 2000, 2005, and 2010 (Department of Statistics, Malaysia).[1] Obtaining comparable time-series data from the Malaysian I/O table is challenging as the number of industries included in the I/O tables has progressively increased over time. For example, the 2000 I/O table had 94 industries, and this increased to 120 in 2005 and expanded to 124 in 2010. It is further complicated by the changing of industrial classifications over time. Therefore, to facilitate the analysis, the study also uses an unpublished set of compressed I/O tables constructed by the Department of Statistics (DOS) into identical industries/commodities at the request of the authors. There are 95 commodities for both 2005 and 2010 I/O tables[2] with the ICT sectors specially created by DOS into the compressed I/O tables.[3] At the disaggregated level, the study will analyze the uncompressed (original I/O) table, followed by the compressed table with the caveat that the definition of ICT DOS gives includes ICT manufacturing, trade, and services industries.[4]

6. Results and analysis

Malaysia has a high VS compared with other countries, as shown in Table 5.1, with the exception of Singapore. In contrast, Indonesia has a low VS due to its export structure, which is dominated by primary goods such as agricultural goods, fuels, and mining. Resource-based exports are generally less dependent on imported inputs, especially when there is less processing before it is exported. At the country level, the overall Hummels VS has been declining for Malaysia. Using the compressed or uncompressed table, Table 5.2 indicates Malaysia depended less on foreign inputs from 2000 to 2010.

Table 5.3 reports the foreign value-added (FVA) share in gross exports for selected ASEAN member states and China from the OECD TiVA database. Singapore has the highest FVA for the period shown, followed by Malaysia, the Philippines, Thailand, Viet Nam, and China. Malaysia's FVA in its gross exports increased from 1995 to 2000 and subsequently fell marginally from 2000 to 2005, before continuing to fall further in 2008 and 2009.

Table 5.1 Hummels VS – selected countries

Year	*Malaysia*	*China*	*Singapore*	*Indonesia*
1997		0.29		
2000	0.52			
2002		0.36		
2005	0.45			
2008			0.58	0.13

Source: Loke and Tham (2014).

Table 5.2 Hummels VS for Malaysia, 2005–2010

Year	*Uncompressed table*	*Compressed table*
2000	0.52	n.a.
2005	0.45	0.43
2010	0.40	0.31

Source: Calculated based on I/O tables 2000, 2005, and 2010.

Table 5.3 Total foreign value added in gross exports (%), 1995–2009

Country	*1995*	*2000*	*2005*	*2008*	*2009*
China	12	19	36	33	33
Indonesia	15	19	18	17	15
Malaysia	40	43	42	38	38
Philippines	31	46	46	42	38
Singapore	47	51	52	53	50
Thailand	30	35	38	38	35
Viet Nam	24	30	35	40	37

Source: OECD–WTO TIVA, downloaded 12 April 2014.

Disaggregating based on the sectoral classifications given in Malaysia's I/O tables indicates that the average VS share is highest for manufacturing and construction in 2005 and 2010, whilst agriculture, fishery, and forestry; mining and quarrying; as well as services have a lower average VS share (Table 5.4). The average VS share for manufacturing increased slightly from 2005 to 2010, whereas that of construction and other sectors fell. The improvement in Malaysia's overall VS can be attributed, therefore, to the improvement in the VS share of construction and sectors other than manufacturing.

The VS share of manufacturing by subsectors is shown in Table 5.5. In general, manufacturing ICT products have relatively high imported components for all three years shown. Whilst noting that the computed VS shares are not comparable on a strictly one-to-one basis over time due to the differences in commodity/industry classifications and disaggregation in the I/O tables, the disaggregated ICT products over the three years generally occupy the top ten positions in the computed VS. This reflects the nature of production of these products, which is linked with the regional production networks where imports are used to facilitate exports. Loke and Tham (2014) also found electrical and electronics-related

Table 5.4 Uncompressed industry average VS 2005 and 2010

Industry	*Code*	*VS 2005*	*Industry*	*Code*	*VS 2010*	*VS difference*
1	Agriculture, fishery, and forestry	0.189	1	Agriculture, fishery, and forestry	0.181	0.009
2	Mining and quarrying	0.205	2	Mining and quarrying	0.140	0.065
3	Manufacturing	0.424	3	Manufacturing	0.447	–0.023
4	Electricity, gas, and water	0.233	4	Electricity, gas, and water	0.267	–0.034
5	Construction	0.434	5	Construction	0.353	0.081
6	Wholesale and retail trade	0.350	6	Wholesale and retail trade	0.210	0.140
7	Hotel and restaurants	0.154	7	Hotel and restaurants	0.196	–0.042
8	Transport and communication	0.216	8	Transport and communication	0.264	–0.048
9	Finance and insurance	0.126	9	Finance and insurance	0.100	0.025
10	Real estate and ownership of dwellings	0.151	10	Real estate and ownership of dwellings	0.080	0.071
11	Business and private services	0.220	11	Business and private services	0.165	0.055
12	Government services	0.241	12	Government services	0.198	0.043

Source: Calculated based on I/O tables 2005 and 2010.

Table 5.5 Ranked VS of selected commodities in the input–output table 2000, 2005, and 2010 (uncompressed)

Code	*Commodities*	*VS 2000*	*Code*	*Commodities*	*VS 2005*	*Code*	*Commodities*	*VS 2010*
56*	Household machinery	0.750	75*	TV, radio receivers and transmitters, and associated goods	0.776	72	Insulated wires and cables	0.757
22	Animal feeds	0.703	77	Measuring, checking, and industrial process equipment	0.664	78*	Optical instruments and photographic equipment	0.740
57*	Radio, TV, and com. equipment	0.676	72	Insulated wires and cables	0.658	61	Basic precious and non-ferrous metals	0.685
50	Iron and steel	0.660	74*	Semiconductor devices, tubes, and circuit boards	0.652	69*	Office, accounting, and computing machinery	0.641
51	Non-ferrous metal	0.645	65	Industrial machinery	0.623	80	Motor vehicles	0.639
59	Other electrical machinery	0.590	69*	Office, accounting, and computing machinery	0.616	74*	Semiconductor devices, tubes, and circuit boards	0.639
64	Instruments and clocks	0.568	68	Domestic appliances	0.608	70	Electrical machinery and apparatus	0.632
52	Other fabricated metal and fixtures	0.549	91	Special trade works	0.586	77	Measuring, checking, and industrial process equipment	0.632

61	Motor vehicles	0.542	64	Other fabricated metal products	0.576	62	Casting of metals	0.606
58	Elect. appliances and houseware	0.532	66	General purpose machinery	0.572	79	Watches and clocks	0.582
53	Structural metal products	0.525	80	Motor vehicles	0.572	42	Publishing	0.578
29	Wearing apparel	0.518	26	Animal feeds	0.541	71	Other electrical machinery	0.566
27	Knitted fabrics	0.517	70	Electrical machinery and apparatus	0.540	60	Iron and steel products	0.555
25	Tobacco	0.503	24	Confectionery	0.530	81	Motorcycles	0.547
24	Soft drinks	0.502	62	Casting of metals	0.525	55	Plastics products	0.543
62	Cycles and motorcycles	0.496	61	Basic precious and non-ferrous metals	0.515	51	Tyres	0.542
26	Yarns and cloth	0.495	79	Watches and clocks	0.510	68	Domestic appliances	0.541
28	Other textiles	0.493	30	Yarn and cloth	0.509	65	Industrial machinery	0.535
60	Ships and boats	0.492	59	Concrete and other non-metallic mineral products	0.502	52	Rubber processing	0.527
65	Other products	0.492	42	Publishing	0.498	75*	TV, radio receivers and transmitters, and assoc. goods	0.521

(*Continued*)

Table 5.5 (Continued)

Code	*Commodities*	*VS 2000*	*Code*	*Commodities*	*VS 2005*	*Code*	*Commodities*	*VS 2010*
54	Other metal products	0.476	78	Optical instruments and photographic equipment	0.484	53	Rubber gloves	0.520
19	Confectionery	0.471	63	Structural metal products	0.482	54	Rubber products	0.519
45	Plastic products	0.471	55	Plastic products	0.479	64*	Other fabricated metal products	0.502
14	Preserved fruits and vegetables	0.471	73	Electrical lamps and lighting equipment	0.478	66	General purpose machinery	0.496
31	Footwear	0.471	60	Iron and steel products	0.476	82	Building and repairing of ships and boats, manufacture of bicycles	0.490

Note: *Represents ICT manufacturing commodities. For years 2000 and 2005, the MSIC 2000 classifications for manufactured ICT product codes are 32101, 32102, 32109, 30002, 32200, 32300; for year 2010, the MSIC 2008 is used and the codes are 20101–20105, 20109, 26201, 26202, 26300, 26400, and 26800.

Source: Calculated from the 2000, 2005, 2010 I/O table from DOS.

subsectors to have relatively high VS, especially when compared with resource-based subsectors.

When the industries are disaggregated further in 2005 and 2010, it appears that some ICT products – TVs, radio receivers and transmitters, and associated goods – were made with less imported input when we compare their VS shares in these two years (Table 5.5). Semiconductor devices, tubes, and circuit boards, as well as other fabricated metal products, also indicated a decrease in their respective imported content when we compared 2005 with 2010. But other ICT products indicated the converse, such as insulated wires and cables and office, accounting, and computing machinery. There appears to be industrial upgrading in some electronics products. Firm-level interviews of the electronics cluster in Penang indicated that progressive industrial upgrading occurred after the Asian Financial Crisis (AFC), from 2000 to 2006, with multinational companies (MNCs) increasingly embedding their operations in situ, through using local skilled labour and supply firms (Kharas et al. 2010; Edgington and Hayter 2013). Kharas et al. (2010), for example, showed that the number of products exported from Malaysia stagnated at 740 from 1995 to 2000, before increasing to 746 in 2006–2007. The number of products with revealed comparative advantage (RCA) greater than one declined from 103 to 101 from 1995 to 2000, but increased to 122 in 2006–2007.

Motor vehicles showed increasing import content, with the VS share increasing from 0.542 in 2000 to 0.572 in 2005 and to 0.639 in 2010. One possible explanation for the increase in FVA in Malaysia's exports is the implementation of the ASEAN Free Trade Area (AFTA) in 1993. For the original ASEAN–6 members, initially, tariff rates on the products in the Inclusion List (IL) were scheduled to be reduced to between 0 per cent and 5 per cent by 2008. The tariff reduction schedule was revised in 1994 and 1998, and the due date of tariff reduction to the 0–5 per cent range for the products in IL was moved forward to 2002. By 2000, the average Common Effective Preferential Tariff Rate (CEPT) had fallen by more than half, from 10.0 per cent in 1995 to 3.64 per cent in 2000 (Austria 2004). This increased the incentive to import from other ASEAN member states in Malaysia's production and exports, especially when processing trade is involved. Tariffs on 99.55 per cent (44,160 tariff lines out of total 44,361 tariff lines) of products in the 2003 IL of the ASEAN–6 were reduced to the 0–5 per cent tariff range, so that AFTA had almost been achieved by then. Market liberalization in the automotive industry has allowed the industry to procure more competitive inputs from countries in the region, such as Thailand.

Similarly, the OECD TiVA data by sectors also show that the highest FVA is found in non-resource manufacturing sectors such as basic metals, machinery and equipment, and electrical and optical equipment as compared with resource-based sectors such as food products and wood and paper products (Table 5.6). Transport equipment has lower foreign value added, due in part to its inward-looking policies and local content requirements before it was removed under the Trade-Related Investment Measures (TRIMs) under WTO commitments. Liberalization under AFTA has also contributed to the increase in FVA of this sector

Table 5.6 Foreign value added of gross exports of Malaysia (%), 1995–2009

Product	*1995*	*2000*	*2005*	*2008*	*2009*
Manufacturing					
Food products	27.75	33.34	31.89	26.00	25.22
Textiles	41.42	43.22	42.9	39.92	37.91
Wood, paper products	23.38	28.88	27.94	22.39	21.88
Chemicals	39.94	42.21	39.75	34.26	33.86
Basic Metals	49.77	50.88	51.08	50.39	49.07
Machinery and equipment	54.59	56.44	56.68	56.23	55.66
Electrical and optical equipment	56.85	58.66	58.84	56.75	56.14
Transport equipment	47.5	50.89	50.83	48.18	47.36
Manufacturing nec; recycling	45.46	48.71	48.98	46.37	44.77

Note: nec = not elsewhere classified.

Source: OECD–WTO TIVA, downloaded 12 April 2014.

over time. Electrical and optical equipment had the highest FVA in its exports, increasing from 57 per cent in 1995 to 59 per cent in 2005, before dropping marginally to 56 per cent in 2009. Unfortunately, due to differences in the classification of the industries used, we are unable to compare the results with the disaggregated VS results shown in Table 5.5.

Although the uncompressed I/O table points to a certain extent to an upgrade in local content of certain ICT products, we are unable to compare across years for each subsector due to the change in sectoral coverage over time. To overcome this limitation, we use the compressed I/O table constructed by DOS. Based on the findings in the compressed I/O table (Table 5.7), the VS share for ICT goods decreased between 2005 and 2010. It should, however, be noted again that the definition of ICT used in the compressed I/O tables comprises three components – manufacturing, trade, and services. Therefore, the decrease in the foreign input share of output may be due to changes in ICT manufacturing, trade, or the increasing share of services in the ICT sector, as services in general has higher local content.

7. Case study: ICT manufacturing

Since the I/O data are dated, we use the most recent industry and trade data – from 2000 to 2013 – to examine recent changes in ICT manufacturing in this section. Moreover, most of the studies on this sector are based on information before the global financial crisis (GFC), as post-GFC analysis is still lacking.

Table 5.7 Ranked by VS (> 0.50) in the input–output table 2005 and 2010 (compressed)

Rank		*VS 2005*		*VS 2010*
1	Other metal products	0.61	Non-ferrous metal	0.691
2	Soft drinks	0.61	Other electrical machinery	0.665
3	Elect. appliances & houseware	0.59	Motor vehicles	0.639
4	ICT	0.586	Instruments and clocks	0.633
5	Instruments and clocks	0.58	Iron and steel	0.57
6	Motor vehicles	0.567	Knitted fabrics	0.567
7	Household machinery	0.549	Cycles and motorcycles	0.557
8	Animal feeds	0.545	Plastic products	0.551
9	Confectionery	0.536	Processed rubber	0.539
10	Paper and board	0.515	Elect. appliances and houseware	0.538
11	Non-ferrous metal	0.512	Rubber products	0.536
12	Industrial machinery	0.502	Other metal products	0.511
13	Other non-metal products	0.501	ICT	0.505
			Other fabricated metal and fixtures	0.504

Source: Calculated from the compressed I/O table.

Historically, the development of this sector can be traced back to the Free Trade Zone (FTZ) Act (1971) that was established to encourage FDI into the country. This helped to attract the relocation of foreign electronics firms to Penang with the opening of the Bayan Lepas FTZ in 1972 (Rasiah 2012). The sector grew rapidly to become the leading manufacturing sector in the country, with the help of MNCs. By 2000, Malaysia was the fourth leading developing country for the production of radio, television, and communications equipment, contributing 6.6 per cent of the total share of developing countries' value added in this sector (UNIDO 2009). However, by 2006, although Malaysia managed to maintain its position as the fourth largest developing country contributor, its share in the total share of developing countries' value added had fallen to 4.3 per cent (Kharas et al. 2010).

The ICT and electrical and electronics (e&e) sectors are dominated by foreign companies, as seen by their large share of total foreign manufacturing establishments (21 per cent in 2000), although this also fell from 2000 to 2008 (Tham and Kam 2013). The two sectors accounted for 33 per cent and 30 per cent, respectively, of total manufacturing value added and employment in 2000. These shares declined steadily, to 24 per cent and 17 per cent, respectively, by 2007, before the onset of the GFC. Both the ICT and e&e subsectors continued to contract after the GFC, to 19 per cent and 16 per cent, respectively, in 2009. By 2012, there had been a small improvement as the respective shares of ICT and e&e increased slightly, to 21 per cent and 18 per cent.

The contraction in the ICT sector, including e&e, can be attributed to labour shortages in Malaysia and a loss in competitiveness of labour-intensive operations in relation to other ASEAN member countries and China that had lower labour costs. Anecdotal evidence suggests the relocation of labour-intensive operations out of Penang, which is the strongest electronics cluster in Malaysia. The increasing use of foreign unskilled workers slowed down this exit, but could not prevent it, so that the sectors continued contracting, as shown in Table 5.6.[5]

Both sectors use relatively skilled workers.[6] Despite the decline in the respective VA shares in total manufacturing, the shares of skilled workers for both sectors have increased for the time period shown. For the ICT sector, the share of skilled workers increased from 20 per cent in 2000 to 28 per cent in 2008, before declining marginally, to 27 per cent, in 2012. As for the e&e sector, the share of skilled workers increased steadily from 19 per cent in 2000 to 28 per cent in 2012. The shrinking of the sector and the increase in the demand for skilled workers indicate that the sector is moving into niche industries that require highly skilled workers as the labour-intensive segments shift out of the country. This assertion is further supported by the increase in labour productivity in both sectors. In addition, wages per worker in the ICT and e&e sectors are higher compared with the general manufacturing average.

In terms of trade, using the OECD's (2011) classifications of the ICT sector,[7] we can see ICT exports increasing steadily from 2003 to 2006 (Figure 5.3) (see also Bourassa 2011). After the GFC, there was a dip in exports before they increased again from 2012 to 2013. Imports followed approximately the same trend, and this sector has a positive trade balance for the period of this study. However, the major components of ICT exports changed after the GFC (Figure 5.4). Prior to 2006, the largest component was computers and peripheral equipment, but exports of this group of products have dropped steadily since 2007. After the GFC, electronic components have dominated ICT exports.

The RCA[8] of electronics components is the highest for most of the period of this study (except for the period of the GFC), followed by computer and peripheral equipment (Figure 5.5). The latter's RCA, however, dropped after the GFC, although it remained above one. Communication equipment fell from an RCA greater than one before the GFC to less than one after the GFC. Consumer electronic equipment saw some improvement after the GFC, but this subsequently deteriorated from 2011 to 2013.

We further investigate the electronics sector using an extended definition by Ando and Kimura (2015), as cited in Aldaba (2015), to ascertain the changes within electronics for a broader spectrum of goods (Appendix 4). Two main shifts can be observed in electronics exports from 2000 to 2013. First, we observe a shift towards finished goods in terms of number of products, but not in terms of

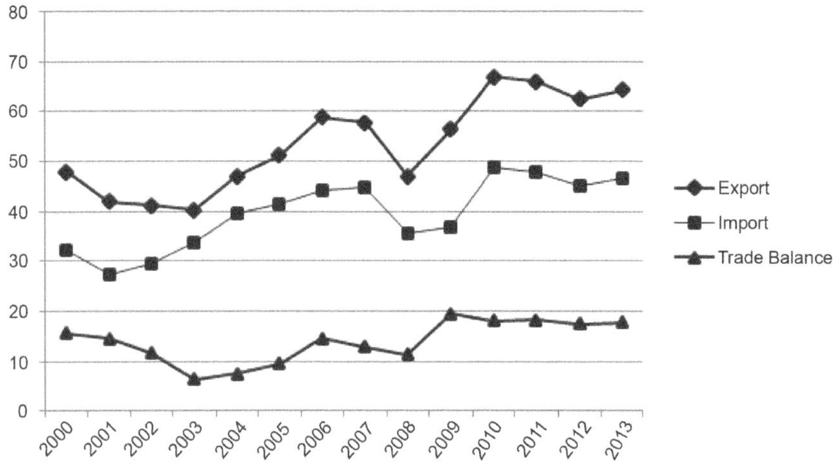

Figure 5.3 Malaysia's exports, imports, and trade balance of ICT goods (USD billion)

Source: Calculated using OECD (2011) and Comtrade data.

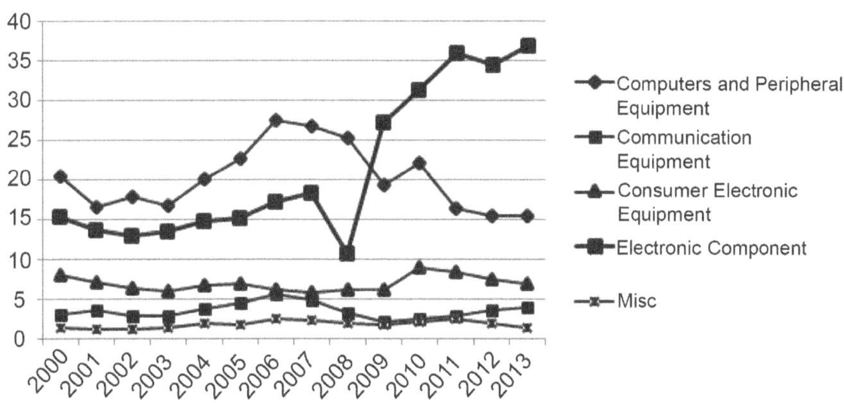

Figure 5.4 Major components of ICT exports, 2000–2013 (USD billion)

Source: Calculated using OECD (2011) and Comtrade data.

export values. Second, there is also a shift towards semiconductor manufacturing service (SMS) away from electronics manufacturing service (EMS).[9] Within the electronics sector, components/devices semiconductors is the largest export sub-sector. It is exported mostly as parts and components (PNC), as compared with auto electronic and consumer electronics, which are mostly exported as finished goods.

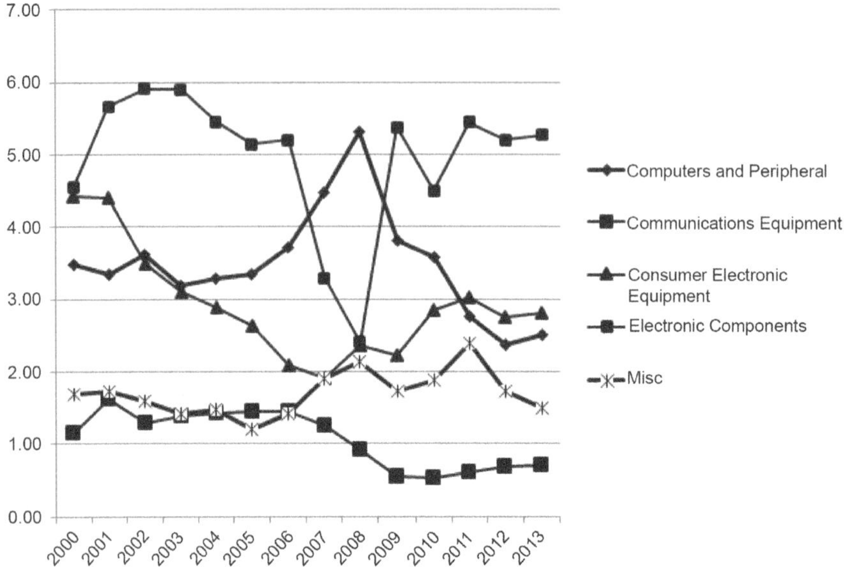

Figure 5.5 RCA of major components in ICT

Source: Calculated using OECD codes and Comtrade data.

In terms of the RCA greater than one for the whole period, the product groups with the highest RCA were auto electronics finished goods, followed by EDP and components/devices semiconductor parts (Figure 5.6). Not many electronics goods had increasing RCA. Only control instrumentation parts finished goods showed consistent improvements in their RCA from 2000 to 2013. Components/devices semiconductors did not show significant improvements in its RCA after 2003. The second largest exported electronic goods category, electronic data processing, showed significant decreases in its RCA after the GFC from 2008 to 2013, be it for parts or finished goods.

Table 5.8 shows that there was a shift from EMS into SMS activities as it can be observed that export values of SMS increased steadily over time, barring the period of the GFC. Export values of EMS fluctuated with a downward trend observed after 2008. The share of SMS in total Malaysian manufactured goods exports increased, whereas the converse can be observed for EMS. Analysis of Malaysia's position in the electronics global value chain appears to support this result, as Malaysia's firms in electronics focus more on fabrication and fabless semiconductor production, including components, whereas EMS contract manufacturing and original design manufacturing is lacking compared with Singapore and Thailand (AIEN 2014). The RCAs of both were over unity for the period of the study, the RCA of SMS was larger than the RCA for EMS, and the RCA of EMS showed a downward trend after 2008.

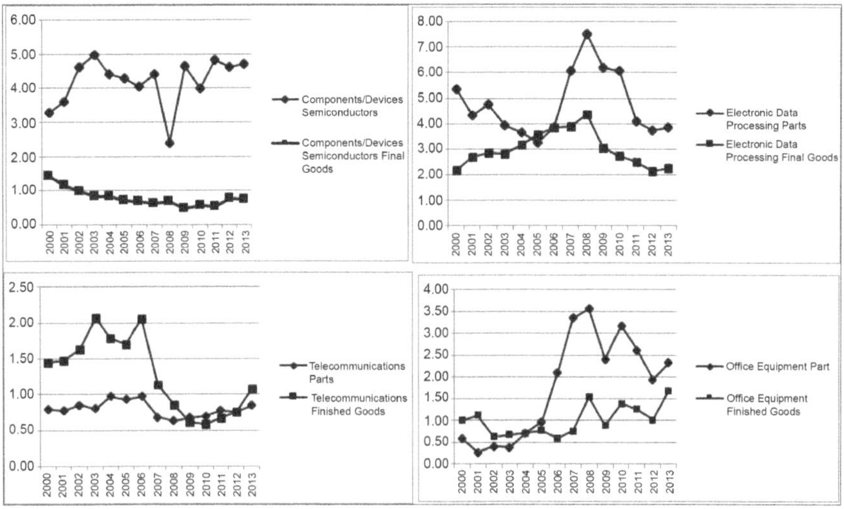

Figure 5.6 RCA for electronic parts and finished goods

Notes: Y-axis are values of RCA.

Sources: Calculated using HS 1996 codes from UN Comtrade. Classifications by Aldaba (2015).

Table 5.8 Exports and RCA of SMS and EMS, 2000–2013

	Export values (USD billions)		*Share of malaysian Manufacturing export (%)*		*RCA*	
	SMS	*EMS*	*SMS*	*EMS*	*SMS*	*EMS*
2000	18.0	36.4	22.8	46.1	3.12	2.61
2001	14.4	31.2	20.4	44.3	3.37	2.62
2002	21.3	30.9	28.4	41.4	4.32	2.56
2003	24.4	29.8	30.4	37.1	4.64	2.35
2004	25.6	36.1	26.7	37.8	4.13	2.38
2005	26.2	40.4	24.8	38.2	3.97	2.36
2006	27.8	47.0	23.6	39.8	3.74	2.45
2007	30.6	44.9	24.6	36.0	4.01	2.60
2008	13.3	41.2	12.4	38.3	2.19	2.88
2009	28.2	33.4	25.8	30.5	4.15	2.19
2010	33.0	41.3	24.7	31.0	3.61	2.24
2011	37.8	35.8	26.9	25.4	4.32	1.96
2012	36.6	34.7	26.2	24.8	4.18	1.82
2013	39.3	33.1	28.4	23.9	4.28	1.77

Source: Calculated using Comtrade data, based on classification in Aldaba (2015).

8. Conclusion and policy implications

The key finding of this chapter is that Malaysia's domestic value-added content increased from 2000 to 2010. A decreasing vertical specialization (VS) share in construction made a contribution, whereas the VS share of manufacturing increased. When we disaggregated within manufacturing, some ICT subsectors showed a decrease in their imported content, whereas others showed an increase. The compressed I/O table showed that the ICT sector's overall use of imported inputs decreased. But as the definition of ICT includes manufacturing, trade, and services, it is unclear whether manufacturing decreased its use of imported inputs, as its share in overall ICT shrank. In contrast, the share of ICT services increased and services in general used less imported inputs.

More recent industry-level data show that the ICT and electronics sectors contracted from 2000 to 2012 in terms of their respective shares in total manufacturing value added and employment. Both sectors also showed a shift towards the use of skilled workers, as labour-intensive production continued to relocate to other countries. The contraction of these two sectors is inevitable given the loss in competitiveness in labour-intensive manufacturing in Malaysia. The increase in usage of skilled labour can be further encouraged by improving education and labour policies in the country to produce more of the skilled labour needed. It has already been identified that the key talent shortage is in R&D and engineering support functions for addressing the gap between the demand and supply of skills in these two sectors (IPSOS Business Consulting 2012). Shifting to ICT services will also require addressing the skills gap issue.

Using the OECD classification of ICT, ICT exports increased from 2003 to 2006. After the GFC, exports dipped before increasing again from 2012 to 2013. Imports followed approximately the same trend, and this sector had a positive trade balance for the period of this study. However, the major components of ICT exports changed after the GFC. Prior to 2006, the largest component was computers and peripheral equipment, but exports of this group of products dropped steadily from 2007 to 2013.

After the GFC, electronic components dominated ICT exports. Using a more detailed classification of the electronics sector, two main shifts can be observed in electronics exports in 2000–2013. First, there was a shift towards finished goods in terms of the number of products, but not in terms of export values. Second, there was a shift towards semiconductor manufacturing service (SMS) away from electronics manufacturing service (EMS). It would appear that the contraction in the ICT and electrical and electronics sectors left this sector focusing more on semiconductor manufacturing. The shift towards higher value added in this sector will require a greater supply of the skilled workers needed for this sector to address Malaysia's skills gap, as mentioned earlier. Although the tools of industrial policy have facilitated Malaysia's entry into global value chains, they have not been able to generate a shift up the chain to the aspired higher value-added activities such as R&D and design. In view of this, the government has to reconsider its industrial policies. Instead of using the same policy tools, it needs to come up with new policies to facilitate the shift towards greater emphasis on quality – quality of human capital as well as quality of institutions – to support a shift to higher value-added activities.

Notes

1 The 2010 I/O table is the latest publication at the time of writing (Department of Statistics 2015a).
2 However, for the 2000 I/O table, the study is unable to obtain the compressed ICT table due to difficulties in matching the changes in codes. Hence, the analysis for the 'compressed' table will only discuss the years 2005 and 2010.
3 Details of the I/O tables can be found in Tham and Kam (2013).
4 Hence, there may be disparity in the magnitude of VS figures in the compressed and uncompressed tables.
5 The share of foreign workers in manufacturing is reported to have doubled from 2000 to 2010. They made up 30 per cent of the manufacturing workforce in 2010 (World Bank 2013).
6 This is defined in terms of occupational groups. Skilled workers comprise two groups: (1) managers, professionals, and executives; and (2) technicians and associate professionals.
7 This is because it has the same codes as adopted by the DOS for the ICT satellite data, as well as for the I/O tables used in this study (OECD 2011; DOS 2014).
8 RCA = $(E_{mal,j}/E_{mal,t})/(E_{world,j}/E_{world,t})$ with $E_{mal,j}$ = Malaysia exports of j, $E_{mal,t}$ = Malaysia total manufacturing exports, $E_{world,j}$ = world exports of j, $E_{world,t}$ = world total manufacturing exports.
9 SMS consists of semiconductor companies that manufacture integrated circuits (ICs), transistors, diodes, resistors, capacitors, coils, transformers, printed circuit boards (PCBs), and other components. EMS activities include subsectors such as electronic data processing, office equipment, telecommunications equipment, communications and radar, control and instrumentation, medical and industrial, automotive electronics, and consumer electronics (Aldaba 2015).

References

Aldaba, RM 2015, 'The Philippines in the electronics global value chain', paper presented to the Second EIA Workshop on 'Levelling up Southeast Asia's Value Added', Tokyo, 24–26 April.

Asian International Economists Network (AIEN) 2014, 'Enabling GVC Participation: Industry-specific Policy – The Electronics Industry'. Available from: www.aienetwork.org/training/2014/role-of-trade-and-investment-policy/presentations/electronics-industry.pdf [2 June 2015].

Athukorala, P 2003, 'Product fragmentation and trade patterns in East Asia', *Working Paper* No. 2003/21, Australian National University, Canberra.

Austria, M 2004, 'ASEAN free trade area: lessons learned and the challenges ahead'. Center for Business and Economics Research and Development (CBERD) *Working Paper Series* 2004–1. Available from: www.dlsu.edu.ph/research/centers/cberd/pdf/papers/Working%20Paper%202004–01.pdf [26 August 2014].

Bourassa, F 2011, 'Measuring Trends in ICT Trade: From HS2002 to HS2007 / ICT Product Definition', *Committee for Information, Computer and Communications*

Policy, United Nations Department of Economic and Social Affairs Statistics Division, New York, NY.
Danaraj, N 2011, 'Technology Policy: A Critical Appraisal', chapter 13 in *Malaysia: Polices and Issues in Economic Development*, Institute of Strategic and International Studies (ISIS), Kuala Lumpur. Available from: www.isis.org.my/index.php/research-a-publications/publications/books/1658-malaysia-policies-a-issues-in-economic-development.
Department of Statistics (DOS) Malaysia 2014, *Information, Communication and Technology (ICT) Satellite Account (ICTSA)*, DOS, Putrajaya, Malaysia.
Department of Statistics (DOS) Malaysia 2015a, Input–Output Table with ICT sector, unpublished data.
Department of Statistics (DOS) Malaysia 2015b, Annual Manufacturing Survey (Multiple years), unpublished data.
Edgington, D and Hayter, R 2013, 'The in situ upgrading of Japanese electronics firms in Malaysian industrial cluster', *Economic Geography*, vol. 89, no. 3, pp. 227–259.
Hummels, D, Ishii, J and Yi, KM 2001, 'The nature and growth of vertical specialization in world trade', *Journal of International Economics*, vol. 54, pp. 75–96.
Inomata, S 2013, 'Trade in value added: an East Asian perspective', *ADBI Working Paper* No. 451, ADBI, Tokyo.
IPSOS Business Consulting 2012, 'E&E Sector Study on the Supply and Demand of Talent in Malaysia', final report. Available from: www.talentcorp.com.my/ . . . /E&E%20Sector%20Study%20-%20A [20 April 2015].
Kaziah, AK 2005, 'Investment Incentives', PowerPoint presentation at Workshop on 'Improving Investment Climate', Indonesia, 16–17 November.
Kharas, H, Zeufack, A and Majeed, H 2010, *Cities, People and the Economy: A Study on Positioning Penang*. The World Bank, Washington, DC. Available from: http://documents.worldbank.org/curated/en/2010/09/13062998/cities-people-economy-study-positioning-penang [22 September 2014].
Kuppusamy, M, Raman, M, and Lee, G 2009, 'Whose ICT investment matters to economic growth: private or public? The Malaysian perspective', *The Electronic Journal on Information Systems in Developing Countries*, vol. 37 no. 7, pp. 1–19.
Leete, R 2007, *Malaysia: From Kampung to Twin Towers*. Oxford Fajar Sdn. Bhd., Shah Alam, Malaysia.
Loke, WH and Tham, SY 2014, 'Vertical specialization and backward linkages: reconsidering Malaysian manufacturing development', *Malaysian Journal of Economic Studies*, vol. 51 (Special Issue), pp. 57–76.
Ng, F and Yeats, A 2001, 'Production Sharing in East Asia: Who Does What for Whom and Why?' in *Global Production and Trade in East Asia*, eds. L Cheng and H Kierzkowski, Kluwer Academic Publishers, Boston, MA, pp. 63–109.
Nordås, HK 2005, 'International production sharing: a case for a coherent policy framework', *WTO Discussion Paper* No. 11, World Trade Organization, Geneva, Switzerland.
OECD n.d., OECD–TIVA Data. Available from: www.google.com/?gws_rd=ssl#q=oecd+tiva+database [20 September 2014].
OECD 2011, *OECD Guide to Measuring the Information Society 2011*, OECD, Paris. Available from: www.oecd.org/sti/ieconomy/oecdguidetomeasuringtheinformationsociety2011.htm [29 May 2015].
Ohno, K 2006, *Industrial Policy Formulation in Thailand, Malaysia and Japan: Lessons for Vietnamese Policy Makers*, Vietnam Development Partnership Forum, Hanoi.

Powers, W 2012, 'The value of value added: measuring global engagement with gross and value added trade', *Office of Economics Working Paper* No. 2012–11A, International Trade Commission, Washington, DC.

Rasiah, R 2012, 'Driving technological catch up in IC industry in Malaysia', *MyForesight* (January), pp. 233–236.

Rasiah, R 2014, 'Stimulating innovation in ASEAN: institutional support, R&D activity and intellectual property rights'. *Economic Research Institute for ASEAN and East Asia (ERIA) Policy Brief* No. 2014–06. Available from: www.eria.org/ERIA-PB-2014–06.pdf [20 May 2015].

Tham, SY, Devadason, E, and Heng, LW 2009, 'Infrastructure and Trade Costs in Malaysia: The Importance of FDI and Exports', in *Infrastructure's Role in Lowering Asia's Trade Costs: Building for Trade*, eds. DH Brooks and D Hummels, Asian Development Bank Institute and Edward Elgar, Cheltenham, UK, pp. 148–181.

Tham, SY and Kam, AJY 2013, 'Contribution of Foreign Direct Investment to Employment Generation and Skills Development in Electronic Component and ICT Goods Industries in Malaysia', final unpublished research report submitted to UNESCAP, Bangkok.

UNIDO 2009, *Industrial Development Report 2009: Breaking In and Moving Up: New Industrial Challenges for the Bottom Billion and the Middle-Income Countries*, United Nations Industrial Development Organization, Vienna.

Vicziany, M and Puteh, M 2004, 'Vision 2020, the Multimedia Supercorridor, and Malaysian Universities', Conference Proceedings of the 15th Biennial Conference of the Asian Studies Association of Australia, Canberra, 29 June–2 July.

World Bank 2013, 'Immigration in Malaysia: Assessment of Its Economic Effects, and a Review of the Policy and System', a report completed with ILMIA–Ministry of Human Resources, Malaysia. Available from: http://psu.um.edu.my/images/psu/doc/Recommended%20Reading/Immigration%20in%20Malaysia.pdf [20 April 2015].

WTO and IDE–JETRO 2011, *Trade Patterns and Global Value Chains in East Asia: From Trade in Goods to Trade in Tasks*, WTO, Geneva, Switzerland.

Yusuf, S and Nabeshima, K 2009, *Tigers under Threat: The Search for a New Growth Strategy by Malaysia and its Southeast Asian Neighbors*, The World Bank. Washington, DC.

6 The Philippines in the electronics global value chain

Upgrading opportunities and challenges

Rafaelita Aldaba

1. Introduction

With increasing globalization and economic integration, global value chains (GVCs) have rapidly emerged and, as described by the Organisation for Economic Co-operation and Development (OECD) (2013), they have deepened the process of globalization not only geographically (by including more countries), but also sectorally (by affecting both manufacturing and services sectors) and functionally (by including production, distribution, research and development [R&D], and innovation). The participation of developing countries in GVCs and global production networks can provide benefits such as access to export markets and foreign direct investment (FDI), newer technology, and greater attention to quality control, cost control, timely delivery, and human resource development. Successful participation in GVCs can also result in stability and growth opportunities for small and medium enterprises (SMEs). SMEs involved in GVCs have greater exposure and access to information, business practices, and technologies that can lead to an upgrading of technological and human capital. As Pietrobelli and Rabellotti (2011) pointed out, for firms in developing countries, inclusion in GVCs not only provides new markets for their products, but also plays an important role in access to knowledge and enhanced learning and innovation. In East Asia, the formation of regional production/distribution networks has enhanced regional manufacturing competitiveness and contributed to the rapid economic growth of countries in the region (Ando and Kimura 2013a).

This chapter examines the participation of the Philippines in the electronics GVC, focusing on the extent of its GVC participation and the policies it has pursued in response to the changing international trade landscape. The electronics industry is the country's most important export sector, accounting for an average share of around 60 per cent of total exports. Within the electronics industry, exports are highly concentrated in semiconductors, representing an average share of 73 per cent of total electronics exports.

This chapter is divided into four main parts. After the introduction, Section 2 reviews the trade, investment, and industrial policies implemented by the government to develop and upgrade the industry. Section 3 analyzes the current state and economic performance of the industry. Section 4 discusses the GVC framework and examines the position of the industry in the GVC along with the challenges that it confronts using revealed comparative advantage as a measure of export competitiveness. To assess the extent and depth of Philippine participation in the GVC, extensive margin and trade-in-value-added (TiVA) indicators are examined along with a value chain analysis to identify the gaps. Section 5 provides recommendations on how to upgrade the industry and move up the value chain, underscoring the need for well-functioning and competitive markets to attract more investment and build an agglomeration of activities in the electronics industry.

2. Government policies on the development of the electronics industry

2.1 Import substitution, liberalization, export promotion, and new industrial policy

From the 1950s up to the 1970s, the Philippines pursued an industrialization strategy characterized by protectionism and import substitution. Manufacturing was one of the industries most favoured by policy makers. Through its regulatory policies, the government effectively controlled prices, domestic supply, and market entry to promote growth and development of sectors such as electrical appliances, motor vehicles, steel, cement, sugar, flour, textile, synthetic fibre, and paper.

However, the complex array of protective policies, investment incentive measures, and administrative and regulatory controls failed to promote an efficient mechanism for allocating domestic resources in the economy. More than three decades of import substitution have left a legacy of not only high levels of industrial concentration and concentration of economic wealth amongst a small number of families and groups, but also the lack of a culture of competition in the country.

Beginning in the 1980s, the government carried out trade liberalization by removing tariff and non-tariff barriers. This was accompanied by privatization and deregulation policy reforms that changed the set of rules governing economic activities in the country. Reforms were initiated not only in the financial sector, but also in utilities covering telecommunications, power, water, air transport, and shipping.

As the country's market-oriented reforms deepened during the 1990s, the attitude and policy direction of the Philippines towards FDI also changed considerably. Recognizing the need to expand exports and the potential economic contribution of FDI through the transfer of knowledge and experience, a policy shift transpired as the government adopted more open and flexible policies towards FDI. Investment liberalization was implemented by allowing foreign investment in sectors that were not specified in the Negative List. A new Omnibus Investment Code was legislated to simplify the investment incentive system.

At the same time, the government instituted changes in its economic zone (ecozone) policies away from government-owned, export-processing zones towards private industrial zones. Given the trade and investment policy changes instituted since the mid-1990s, ecozones have served as the centrepiece of the government's export and FDI strategy. Ecozones have provided a simulated free trade environment in which domestic and foreign locators – companies located in ecozones – can import raw material inputs at world prices and avoid the problems associated with customs facilitation, weak infrastructural facilities, and the bureaucracy that have characterized the rest of the domestic economy.

The Philippines has continued to maintain an open trade and investment policy. The government sees free trade agreements (FTAs) not only as a tool for market access, but also as a means to access opportunities to enhance investment, services, trade facilitation, institution building, and technology upgrading. Currently, the country has a total of seven concluded FTAs – Japan–Philippines, ASEAN Free Trade Area (AFTA), ASEAN–Republic of Korea (henceforth, Korea), ASEAN–China, ASEAN–Japan, ASEAN–India, and ASEAN–Australia and New Zealand. The government continues to engage in further discussions on possible bilateral partnership with the European Union and Taiwan along with regional cooperation such as the Regional Comprehensive Economic Partnership (RCEP) and the Trans-Pacific Partnership (TPP).

The government has pursued these market-oriented reforms to remove barriers to competition and promote factor mobility and firm growth, and to secure high, sustained, and inclusive economic growth and rapid poverty alleviation. With the challenges and opportunities arising from increasing globalization and regional economic integration in East and Southeast Asia, the government has adopted a new industrial policy that focuses on transforming and upgrading industries, deepening the participation of industries in higher segments of the global/regional value chain, and creating an environment conducive to the generation and expansion of investment spillovers. Under the new policy, FDI is an important growth pillar, especially in terms of bringing in new technology. The strategy is to help markets work better by creating a business environment conducive to private-sector development and investment, addressing gaps in value and supply chains, implementing reforms for effective regulation, and promoting human resources development and industrial clusters/agglomeration.

It is within this context that policies and programmes to support the development of the electronics industry have evolved. Currently, the regional economic integration process through the Association of Southeast Asian Nations (ASEAN) Economic Community (AEC) and the Regional Comprehensive Economic Partnership (RCEP) is seen as important in facilitating the establishment and development of regional production networks. This is illustrated by the cases of the electronics and automotive industries whose operations have been governed by global/regional production networks.

2.2 Review of government policies to develop the Philippine electronics industry import substitution

In the earlier years, government efforts to develop the electronics parts and components sector were carried out by enforcing localization through legislation that specified differential sales taxes based on local content. The government's import substitution policy on the electronics industry was focused mainly on the manufacture of finished products and failed to develop specific industry support programmes to directly cater to the needs of domestic parts makers and components suppliers. At the heart of the localization process is a network of efficient suppliers of parts and components capable of delivering these at the cost, quality, and time specified by manufacturers. To develop parts and components manufacturing in the country, the government adopted local content and progressive export programmes supported by high tariffs, import restrictions, and differentiated sales taxes. In 1995, the government implemented a Backward Linkage Program for electronics supplier firms to match suppliers and assemblers and to raise local supplier capabilities in the metal press and plastic injection sectors. However, these efforts were not sustained due to lack of resources, particularly technical manpower to effectively institutionalize the programme.

In the case of Malaysia, the transformation plan for its electrical and electronics industry started as early as 1995, with the industry plan being an integral part of Malaysia's Second Industrial Master Plan (IMP2) 1996–2005. IMP2 covered eight manufacturing industries, including electrical and electronics, with the goal of transforming the manufacturing industry into a resilient, broad-based, and internationally competitive sector focusing on cluster-based industrial development, and lifting and broadening activities along the value chain (manufacturing plus strategy).

Similarly, Thailand formulated its own Master Plan for the Electronics Industry in the late 1990s to create world-class quality systems of support and production, and increase the value added processes of the Thai portion of assembly and production.

2.2.1 Trade liberalization

With the government's trade liberalization programme that started in the 1980s, tariff rates on both electronics finished goods and material inputs continuously declined from 1981 to 2005. In 1988, the discriminatory taxes on imports were removed and replaced by the value-added tax system. Import restrictions on consumer durables, both final goods and inputs, were abolished in 1991 and 1992. Consumer electronics were the first to be liberalized in 1991, followed by TV sets and radios along with the majority of parts and components for consumer electronics that were restricted under the Progressive Export Program of the Consumer Electronics Program. In 1990, tariff rates for most electronics products were in the 20 per cent to 50 per cent range. In 2000, these were eliminated for all products of electronic data processing, office equipment (except for one product), and semiconductors (except for 27 per cent of the total number of product lines). In 2004, most products under automotive electronics, consumer electronics, telecommunications, and control and instrumentation had 3 per cent, 10 per cent, and 15 per cent tariff rates, respectively.

Through the Information Technology Agreement (ITA),[1] the Philippines committed to bind its tariffs to zero on 188 IT product lines by 2000 and 47 product lines by 2005. The ITA is an important agreement that has significantly affected the IT industry, particularly the increase in trade in IT products.

2.2.2 *Investment incentives*

The Philippines has various investment incentive schemes administered by the BOI, the Philippine Economic Zone Authority (PEZA), the Subic Bay Metropolitan Authority (SBMA), the Clark Development Corporation (CDC), and other investment promotion bodies mandated by various laws to establish, maintain, and manage special economic or free port zones. BOI-registered enterprises are allowed income tax holidays of up to eight years, tax and duty-free importation of spare parts, and tax credit on raw materials (Table 6.1). PEZA grants the most generous incentives, including income tax holidays, basic income tax rate of 5 per cent of gross income, and tax- and duty-free importation of capital equipment, spare parts, and raw material inputs. Except for the income tax holidays, Subic and Clark enterprises enjoy the same incentives available to PEZA enterprises.

Under the new Omnibus Investments Code, foreign and domestic investors may avail themselves of fiscal and non-fiscal incentives provided they invest in preferred areas of investment identified annually in the Investment Priorities Plan (IPP). The electronics industry was listed as a preferred area of investment from 1988 to 1994 and from 2006 to 2007. For 1993 and 1994, semiconductors and telecommunications were amongst the preferred activities in the IPP. Most electronics firms are located in PEZA, Subic and Clark, which offer non-fiscal incentives in the form of streamlined government procedures, infrastructure services, and good transport links to ports and seaports that are not available to firms operating outside the zones. These paved the way for the clustering of Japanese semiconductor and disk-drive assemblers in industrial parks located in the Laguna area.

2.2.3 *New industrial policy*

In line with the government's new industrial policy, the electronics industry is formulating its roadmap covering the period 2016–2030. The goal is to move up the value chain and strengthen semiconductor and electronics manufacturing in the country by increasing government–industry–academe cooperation and collaboration in the following areas: talent development, a conducive business environment, industry/country promotion, SME development, and R&D capability development.

Recently, the Department of Science and Technology launched three projects: namely, the Advance Device and Materials Testing Laboratory (ADMATEL), the Electronics Product Development Center (EPDC), and the Philippine Institute

Table 6.1 Investment incentives by type of investment regime

	Investment regime	*BOI OIC*	*PEZA*	*Subic and Clark*
Incentives	Income	48 years income tax holiday (ITH)	4–8 years ITH	No ITH
	Others	After ITH, payment of the regular corporate tax rate of 35% of taxable income	After ITH, exemption from national and local taxes, in lieu of this special rate of 5% tax on gross income	5% tax on gross income in lieu of all local and national taxes
	Importation of raw materials and supplies	Tax credit	Tax and duty exemption	Tax and duty exemption
	Purchase of breeding stocks and genetic materials	Tax exemption within 10 years from registration	Tax and duty exemption	Tax and duty exemption
	Imported capital equipment, spare parts, materials, and supplies	Tax and duty exemption on spare parts (duty and tax free importation of capital equipment expired in 1997)[1]	Tax and duty exemption	Tax and duty exemption

Source: Author's compilation based on existing legislation on investment incentives.

[1]Executive Order 313 (2004) restored these incentives.

for Integrated Circuits (PIIC). These aim to increase value-added capabilities in the industry with a focus on the establishment of facilities for device and materials testing, product development and integrated circuit (IC) design.

3 Performance and current state of the Philippine electronic industry

The electrical machinery sector, considered one of the main drivers of the Philippine economy, is currently the second largest manufacturing subsector in terms of value added (after food manufacturing). Since the 1980s, electrical machinery's average annual contribution to total manufacturing increased substantially, from 3 per cent in the 1980s to 6 per cent in the 1990s and 12 per cent in the 2000s. In 2013, it registered a share of 21 per cent, although this dropped slightly to 20 per cent in 2014. In terms of average annual growth rate, the sector grew by 7.3 per cent in the 1980s and 13.2 per cent in the 1990s, but this slowed down to 5.7 per cent in the 2000s. In 2013 and 2014, the sector grew by 7.3 per cent and 5.5 per cent, respectively.

As of 2010, the industry employed a total of 500,000 workers. Following the trend of average annual value-added growth, average employment growth maintained its double-digit rate, which increased from 14 per cent in the 1980s to 17 per cent in the 1990s, but then slowed in the 2000s with average employment growing by 6 per cent.

In terms of FDI flows, cumulative FDI flows to the electronics and electrical industry for the 1990s reached US$769.3 million, which accounted for about 9 per cent of total FDI flows and 20.2 per cent of total FDI flows to the manufacturing industry. For the period 2000–2009, cumulative flows declined to $526.97 million, which represented 1.4 per cent of total FDI flows into the country and 12.8 per cent of total manufacturing FDI.

The electronics industry has been the country's largest foreign exchange earner, with export value reaching $31 billion in 2007 and accounting for 67 per cent of total Philippine exports (Figure 6.1). Its export contribution increased substantially from 23 per cent in 1991 to about 71 per cent in 1999. However, the pattern was reversed in the late 2000s as its export value fell to $23.9 billion in 2013, triggering a large drop in its export share from 70 per cent in 2000 to 42 per cent in 2013.

The country's electronics exports are highly concentrated in semiconductor components/devices, accounting for an average share of 77 per cent of total electronics exports in the period 1996–1999. In the 2000s, this dropped to about 73.2 per cent. In terms of growth, the sector grew at an average annual rate of

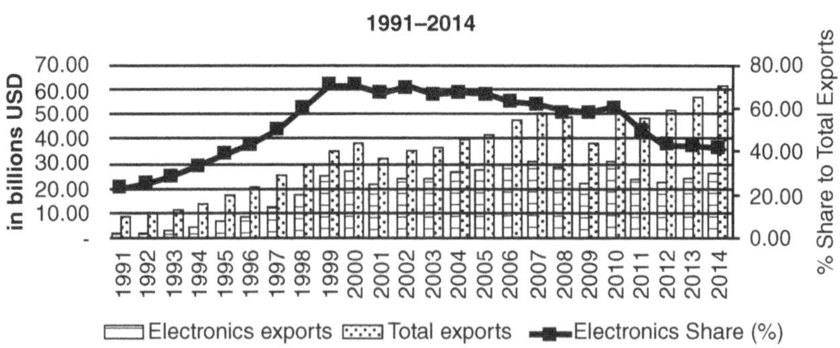

Figure 6.1 Electronics exports share to total Philippine exports

Source: Export Marketing Bureau–Department of Trade and Industry.

almost 9 per cent in the period 2000–2007, but in more recent years, in 2008–2014, the sector registered average annual negative growth of 5 per cent.

Table 6.2 shows how the Philippine's international specialization evolved in the electronics industry using revealed comparative advantage[2] (RCA) measures from 1991 to 2011. The country remained relatively competitive in automatic data processing equipment, parts for office or ADP machines, electricity distributing equipment, nes, and transistors, valves, and so forth. Emerging competitiveness in electric power machinery and parts is also evident.

Table 6.3 presents a summary of the RCA measures for 2010, based on the 1996 six-digit HS classification for the ASEAN-5 countries (Philippines, Thailand, Indonesia, Malaysia, and Singapore). Malaysia and Singapore are leading the ASEAN-5 countries in terms of export competitiveness in the manufacture of electronics products distributed across the different electronics product groups.

Table 6.2 Export competitiveness in the Philippine electronics industry, 1991–2011

SITC 3 description	*91–95*	*96–00*	*01–05*	*06–10*	*2011*
Automatic data-processing equipment	0.95	2.92	4.14	6.36	3.79
Parts for office or ADP machines	0.52	3.47	4.27	3.83	4.22
Radio broadcast receivers	2.33	1.05	1.32	1.12	1.26
Electrical switching relay circuit	0.63	2.06	1.63	1.26	1.09
Electricity distributing equipment, NES	6.84	2.90	2.84	2.96	4.61
Transistors, valves, etc.	4.62	10.12	11.39	11.75	7.50
Telecommunications equipment and parts, NES	2.23	1.21	0.77	0.67	0.51
Office machines	0.01	0.39	1.27	0.63	0.09
Electrical machinery and apparatus, NES	0.28	0.22	1.83	2.21	0.49
Electric power machinery and parts	0.75	0.47	0.57	4.43	6.20
Sound recorders, phonograph	0.54	0.21	0.04	0.05	0.20
Electro-medical, x-ray equipment	0.00	0.00	0.01	0.09	0.26
Domestic electrical, non-electrical equipment	0.46	0.18	0.09	0.06	0.07
Optical instruments, NES	0.68	0.65	0.27	0.07	0.08
Medical instruments, NES	0.10	0.12	0.11	0.27	0.71

Source: Calculations were based on WITS Data (SITC three-digit level Revision 3).

Table 6.3 Electronics revealed comparative advantage in ASEAN-5 (2010)

Country	*PH*	*TH*	*INDO*	*MAL*	*SIN*
Components/devices semiconductors	16	24	12	27	30
Electronic data processing	4	3	1	7	6
Communication and radar	2	0	2	1	2
Telecommunications	4	5	2	5	5
Office equipment	3	3	2	5	5
Medical/industrial instrumentation	1	2	1	5	2
Control and instrumentation	3	1	0	13	13
Auto electronics	2	3	2	2	0
Consumer electronics	7	11	10	21	9
Total	42	52	32	86	72

PH = Philippines; TH = Thailand; INDO = Indonesia; MAL = Malaysia; SIN = Singapore.

Source: Calculations based on WITS Data (HS 1996).

The comparative advantage of both countries lies mainly in semiconductor components and devices.

4. Electronics GVC and Philippine participation

4.1 Philippine position in the electronics GVC

The electronics industry is composed of a wide range of component, intermediate, and final products and characterized by rapid technological change and huge R&D investments. Manufacturing processes in electronics are highly automated, quality standards are demanding, and many of its manufacturing and business processes have been formalized, codified, standardized, and computerized (Frederick and Gereffi 2013).

The segments of the industry consist of *intermediate products* such as electronic components, parts, or subassemblies and *final products* covering consumer and industrial electronics. The value chain also includes activities that add value to final products outside the manufacturing process composed of *research, product and process development, design, marketing, and after sales services.* The most important activities include new product development, circuitry and semiconductor design, software integration, and overall product architecture development, which are the most profitable in the chain and controlled primarily by original equipment manufacturers (OEM) or leading component suppliers (Frederick and Gereffi 2013).

In the Philippines, the semiconductor and electronics industry is divided into two categories: semiconductor manufacturing service (SMS) and electronics manufacturing service (EMS). Semiconductors and other components is the largest subsector of the electronics industry, consisting of companies that manufacture integrated circuits (ICs), transistors, diodes, resistors, capacitors, coils, transformers, printed circuit boards (PCBs), and other components. These are mostly subsidiaries of the world's largest semiconductor companies, such as Texas Instruments, ON Semiconductor, Amkor, Analog Devices, and NXP.

SMS has four major segments: semiconductor design, foundry operations, packaging/product design, and outsourced SMS. Philippine participation is confined largely at the labour-intensive, back-end component of the value chain covering assembly, testing, cleaning, and packaging activities. EMS has three major segments: OEM R&D/marketing, outsourced EMS operations, and customer service. As with semiconductors, Philippine participation in the value chain is limited to labour-intensive, high-capital activities of logistics, assembly, testing, and shipping.

4.2 Trade-in-value-added (TiVA) analysis

The OECD–WTO trade-in-value-added (TiVA) database links national input–output (I–O) tables with bilateral trade data to develop inter-country I–O tables that provide a wide range of indicators on GVCs (OECD 2013). One indicator is the foreign value-added content of exports, presented in the decomposition of electronics exports in Table 6.4.

The foreign value added embedded in Philippine electronics exports has generally increased, indicating the reliance of the industry on foreign imports of intermediates during the past 15 years. Foreign value added increased from 51 per cent in 1995 to 61 per cent in 2000, but declined to 53 per cent in 2008 and 50 per cent in 2009. Domestic value-added content remained the same at 30 per cent in 2000 and 2005 with a slight increase to 31 per cent in 2008 and 2009. Indirect domestic content from domestic intermediates declined from 21 per cent in 1995 to 9 per cent in 2000, but went up to 16 per cent in 2008 and 18 per cent in 2009, suggesting that exporters are now sourcing more intermediate inputs domestically and creating more value added.

The same trend is observed in Malaysia, where the foreign value-added content of exports remained high at 59 per cent in 2000–2005, but this declined to 57 per cent in 2008 and 56 per cent in 2009. Domestic value-added content was 36 per cent in 2008 and 2009, while indirect domestic content was 7 per cent in 2009. Similarly, Thailand's foreign value-added content of exports increased from 47 per cent in 1995 to 59 per cent in 2005 and 2008, although this dropped to 55 per cent in 2009. Direct domestic value-added content remained at 31 per cent from 2000 to 2009, while indirect domestic content declined from 15 per cent in 1995 to 13 per cent in 2009.

China, which has the largest gross exports in terms of value amongst the four countries, had the lowest foreign value-added content at 43 per cent in 2009 and 45 per cent in 2008. Its direct domestic value-added content is also the lowest, at 18 per cent in 2009 and 19 per cent in 2008. However, its indirect domestic content is the highest, at 37 per cent in 2009 and 33 per cent in 2008. This suggests that China is increasingly sourcing its intermediate goods domestically and is less reliant on foreign imports of intermediates as indicated by the increasing

Table 6.4 Decomposition of electronics gross exports

Philippines	*1995*	*2000*	*2005*	*2008*	*2009*
Gross Exports (US$ million)	4,039.5	20,879.7	25,901.9	39,382.6	33,288.6
Direct domestic industry value-added content of gross exports	28	30	30	31	31
Indirect domestic content of gross exports originating from domestic intermediates	21	9	11	16	18
Re-imported domestic value-added content of gross exports	0	0	0	0	0
Foreign value-added content of gross exports	51	61	59	53	50
Domestic value added embodied in gross exports					
Malaysia	**1995**	**2000**	**2005**	**2008**	**2009**
Gross Exports (US$ million)	19,333.1	25,047.8	35,658.2	49,703.6	41,587.3
Direct domestic industry value-added content of gross exports	36	35	35	36	36
Indirect domestic content of gross exports originating from domestic intermediates	7	5	5	6	7
Re-imported domestic value-added content of gross exports	0	1	1	1	1
Foreign value-added content of gross export	57	59	59	57	56
Domestic value added embodied in gross exports					
Thailand	**1995**	**2000**	**2005**	**2008**	**2009**
Gross Exports (US$ million)	21,670.2	24,677.6	37,429.1	58,871.1	54,365.1
Direct domestic industry value-added content of gross exports	38	31	31	31	31
Indirect domestic content of gross exports originating from domestic intermediates	15	14	10	10	13

Philippines	*1995*	*2000*	*2005*	*2008*	*2009*
Re-imported domestic value-added content of gross exports	0	0	0	0	0
Foreign value-added content of gross export	47	55	59	59	55
Domestic value added embodied in gross exports					
China	**1995**	**2000**	**2005**	**2008**	**2009**
Gross Exports (US$ million)	21,964.5	72,859.5	254,334.2	480,368.7	431,446.9
Direct domestic industry value-added content of gross exports	37	25	18	19	18
Indirect domestic content of gross exports originating from domestic intermediates	50	45	29	33	37
Re-imported domestic value-added content of gross exports	0	0	2	2	2
Foreign value-added content of gross export	13	29	51	45	43
Domestic value added embodied in gross exports					

Source: TiVA Database. Available at: http://stats.oecd.org/Index.aspx?DataSetCode=GVC_INDICATORS (accessed 17 August 2014).

indirect domestic activities. As Pilat et al. (2012) indicated, Chinese firms have increasingly moved from simple contract assembly to 'full-package' manufacturing with Chinese firms controlling all stages from material procurement to product design.

Using the TiVA database, De Backer and Miradout (2013) developed other GVC indicators: participation index and length of GVC. The *GVC Participation Index* measures the depth of a country's participation in the GVC indicated by the share of foreign inputs (backward participation) and domestically produced inputs used in third countries' exports (forward participation) expressed as a percentage of gross exports. Figure 6.2a shows that the share of foreign inputs in Philippine electronics exports (looking backwards along the value chain) increased substantially from 8.5 per cent in 1995 to 32.5 per cent in 2000 to 34.4 per cent in 2008, although a decline is evident in 2009 as the share dropped to 32.4 per cent. Similarly, the share of domestically produced inputs used in third countries' exports (looking forward along the value chain) rose from 2.2 per cent in 1995 to 8.4 per cent in 2000 to 16.2 per cent in 2008, but decreased to 14.3 per cent in 2009. The level of Philippine participation in the electronics GVC increased dramatically from 10.7 per cent in 1995 to 50.55 per cent in 2008, but this dropped to 46.7 per cent in 2009. Note that due to the global economic crisis, electronics' gross exports declined by 15.5 per cent between 2008 and 2009. Figure 6.2b compares the participation index of selected countries in Southeast Asia. The Philippines had the highest participation index in 2009, followed by Taiwan.

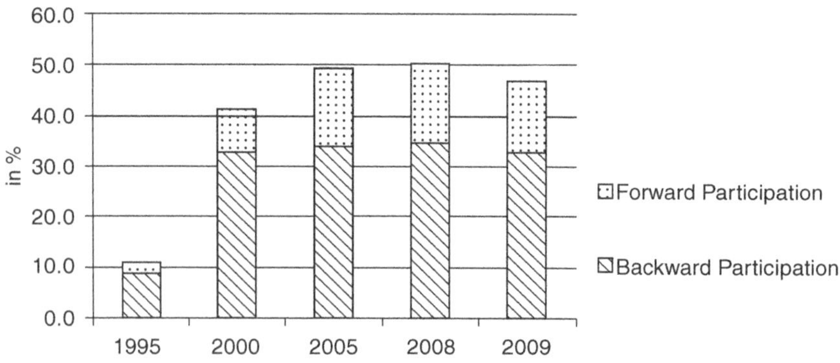

Figure 6.2a Participation index

Source: OECD Global Value Indicators May 2013.

http://stats.oecd.org/Index.aspx?DataSetCode=GVC_INDICATORS (accessed 17 August 2014).

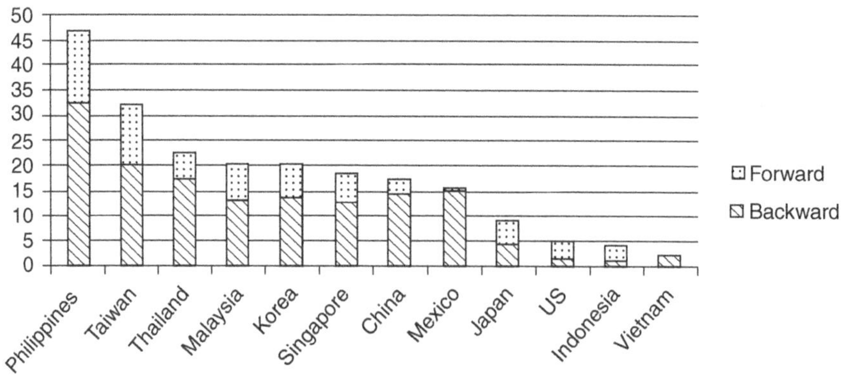

Figure 6.2b Participation index in selected countries, 2009

Source: OECD Global Value Indicators May 2013.

http://stats.oecd.org/Index.aspx?DataSetCode=GVC_INDICATORS (accessed 17 August 2014).

The *Length of GVC* measures the number of production stages involved in the GVC. The index takes the value of one if there is a single production stage in the final industry, and its value increases when inputs from the same industry or other industries are used, with a weighted average of the length of the production involved in these sectors. Figure 6.3a shows very little change in terms of the number of production stages in the Philippines for the period 1995 to 2009. Figure 6.3b compares the index of the number of stages with selected countries, with China having the highest index.

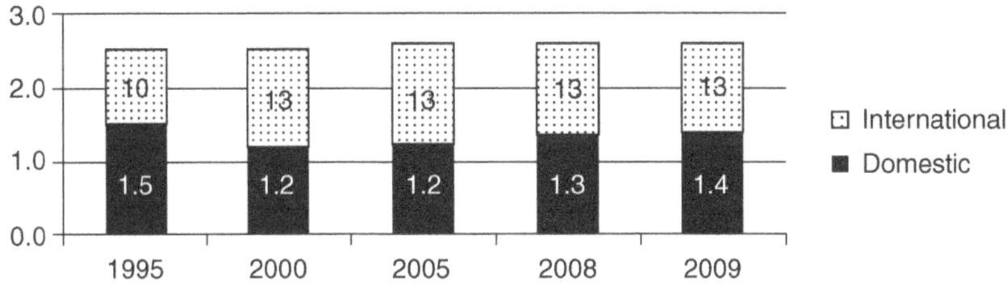

Figure 6.3a Number of production stages

Source: OECD Global Value Indicators May 2013.

http://stats.oecd.org/Index.aspx?DataSetCode=GVC_INDICATORS (accessed 17 August 2014).

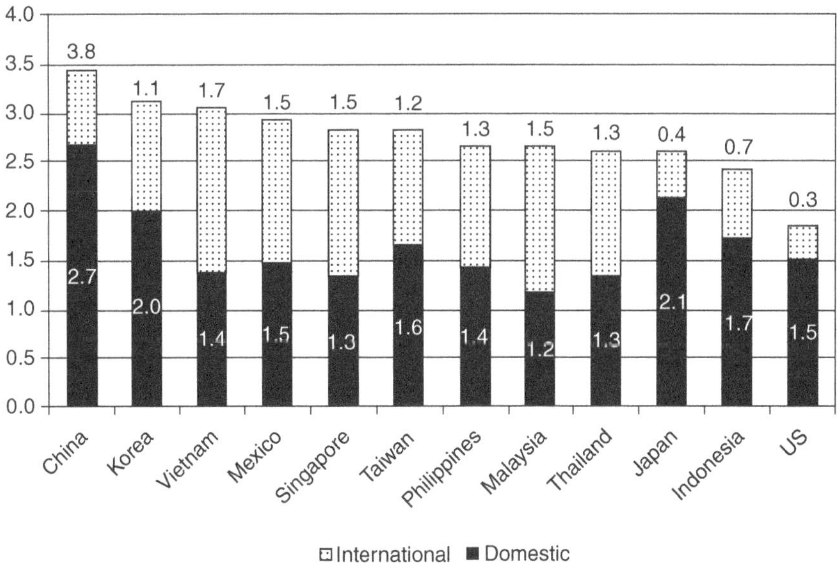

Figure 6.3b Number of production stages in selected countries, 2009

Source: OECD Global Value Indicators May 2013.

http://stats.oecd.org/Index.aspx?DataSetCode=GVC_INDICATORS (accessed August 17, 2014).

Given the high level of industry aggregation of the TiVA statistics and limited number of years covered, the foregoing analysis is supplemented with extensive margins of trade to measure Philippine GVC participation in more specific electronics products. The analysis is further extended by examining the value chain to identify gaps in the value chain and possible ways on how to close these gaps.

4.3 Extensive margins of trade

Following the approach of Ando and Kimura (2013a, 2013b), the trade patterns in the electronics industry are examined by origin/destination countries and regions distinguishing between final, intermediates, and total goods.[3] The analysis covers the years 2000, 2007, and 2013 and selected countries in East and Southeast Asia, chosen based on their importance in the electronics regional/global production network.

As Table 6.5 shows, total electronics exports and imports increased from 2000 to 2007, but this was not sustained as total electronics exports decreased from $31 billion in 2007 to $21 billion in 2013, and electronics imports fell from $25.8 billion to $16.4 billion during the same period. Using 2007 as base year (2007 = 1), the import value index dropped to 0.63, and the export value index declined to 0.69 from 2007 to 2013.

Table 6.5 also shows the regional composition of the Philippine electronics trade, which changed markedly from 2000 to 2013. In terms of the regional distribution of imports, the share of Japan continued to decline from 21 per cent in 2000 to 16 per cent in 2007 and to 11 per cent in 2013. Similarly, Korea's share dropped from 10 per cent in 2000 to 5 per cent in 2013. Imports from the ASEAN 4 countries (Indonesia, Malaysia, Singapore, and Thailand) increased from 15 per cent in 2000 to 19 per cent from 2007 to 2013. Imports from China also rose significantly, from 0.8 per cent in 2000 to almost 11 per cent in 2013.

In terms of exports, Japan's share dropped slightly, from 14 per cent in 2000 to 12 per cent in 2013, and Korea's share of 4 per cent remained unchanged in the period under study. Similar to their import performance, the share of the ASEAN-4 rose from 9 per cent in 2000 to 18 per cent in 2013, as China's share increased markedly, from 1 per cent in 2000 to 14 per cent in 2007, although it shrank to 12 per cent in 2013.

To further assess the changes in trade patterns and the extent or depth of Philippine participation in electronics production networks, extensive margins (the number of traded products and number of trading partners) are calculated. Figures 6.4a and 6.4b present the number of products exported to and imported from the world regardless of partner countries. The number of products exported to the world dropped in 2007 and remained largely unchanged up to 2011, although some recovery is evident in 2012–2013. The number of parts imported from the world has fluctuated greatly and since 2007 a declining trend has been evident.

Figures 6.5a and 6.5b show the number of exported/imported product country pairs with East Asia (Japan, China, and Korea) and the ASEAN region in 2000, 2007, and 2013. As can be observed from Figure 6.5a, for imports, the

Table 6.5 Trade patterns in the electronics industry

			Imports			*Exports*		
	Year	Destination	Total	Parts	Final	Total	Parts	Final
Value in US$B	2000	World	15.88	14.78	1.09	23.64	18.02	5.62
	2007	World	25.83	24.71	1.12	31.08	21.71	9.37
	2013	World	16.36	14.69	1.67	21.39	15.60	5.79
Value index (2007 = 1)	2000	World	0.61	0.60	0.97	0.76	0.83	0.60
	2000	Japan	0.81	0.80	1.22	0.98	1.38	0.67
	2000	Korea	0.96	0.81	14.30	0.79	0.83	0.62
	2000	China	0.08	0.07	0.09	0.08	0.10	0.01
	2000	ASEAN4	0.48	0.44	0.89	0.44	0.54	0.03
	2000	CLMV	0.01	0.01	0.00	0.47	0.41	0.65
	2013	World	0.63	0.59	1.49	0.69	0.72	0.62
	2013	Japan	0.45	0.44	1.22	0.78	1.38	0.31
	2013	Korea	0.56	0.52	3.69	0.82	0.83	0.78
	2013	China	1.05	0.91	1.62	0.60	0.40	1.26
	2013	ASEAN4	0.64	0.59	1.33	0.82	0.94	0.37
	2013	CLMV	0.36	0.31	0.99	16.84	19.19	9.63
Share (in %)	2000	Japan	21.01	21.99	7.74	13.81	11.12	22.44
	2000	Korea	9.68	8.76	22.14	3.57	3.97	2.28
	2000	China	0.79	0.66	2.58	1.41	1.80	0.18
	2000	ASEAN4	14.97	13.91	29.36	8.83	11.40	0.58
	2000	CLMV	0.01	0.01	0	0.02	0.02	0.03
	2007	Japan	16.03	16.48	6.16	11.00	6.70	20.07
	2007	Korea	6.22	6.43	1.51	3.00	3.99	2.20
	2007	China	6.35	5.33	28.71	14.00	15.48	11.12
	2007	ASEAN4	19.3	18.71	32.19	15.00	17.50	10.20
	2007	CLMV	0.75	0.72	1.28	0.03	0.04	0.03
	2013	Japan	11.38	12.1	5.03	12.00	13.00	10.00
	2013	Korea	5.46	5.66	3.74	4.00	5.00	3.00
	2013	China	10.5	8.13	31.29	12.00	9.00	23.00
	2013	ASEAN4	19.49	18.44	28.76	18.00	23.00	6.00
	2013	CLMV	0.43	0.38	0.85	1.00	1.00	0.40

ASEAN4 = Indonesia, Malaysia, Singapore, and Thailand; CLMV = Cambodia, Laos, Malaysia, and Viet Nam.

Source: UN Comtrade http://comtrade.un.org/data/ (accessed on 30 March 2015).

number of product–country pairs or relationships the Philippines has with Japan, China, Korea, and the ASEAN region increased from 1,056 in 2000 to 1,127 in 2007, but this dropped to 1,103 in 2013. For exports, Figure 6.5b shows a decline from 573 in 2000 to 460 in 2007, but the number of product country pairs rose to 751 in 2013.

Figures 6.6a to 6.6c present the number of imported product–country pairs by origin from 2000 to 2013. As can be observed, the index has remained largely unchanged during this period except for the CLMV (Cambodia–Lao PDR–Myanmar–Viet Nam) countries that have started to participate in the electronics

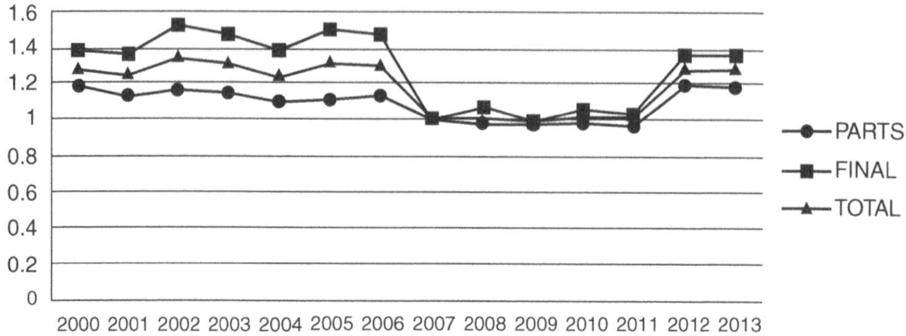

Figure 6.4a Number of products exported to world (2007 = 1)

Source: Author's calculations.

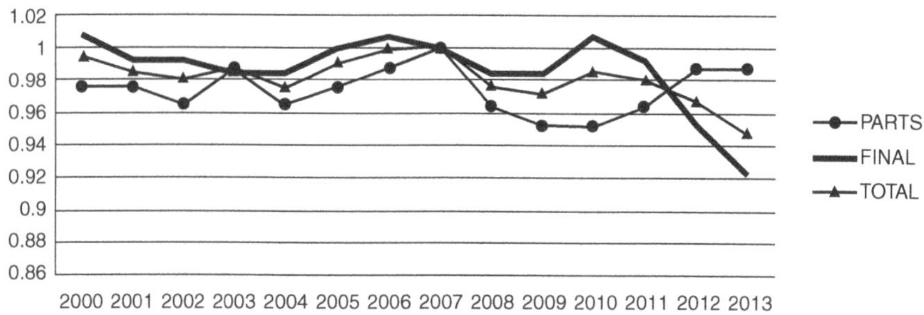

Figure 6.4b Number of products imported from world (2007 = 1)

Source: Author's calculations.

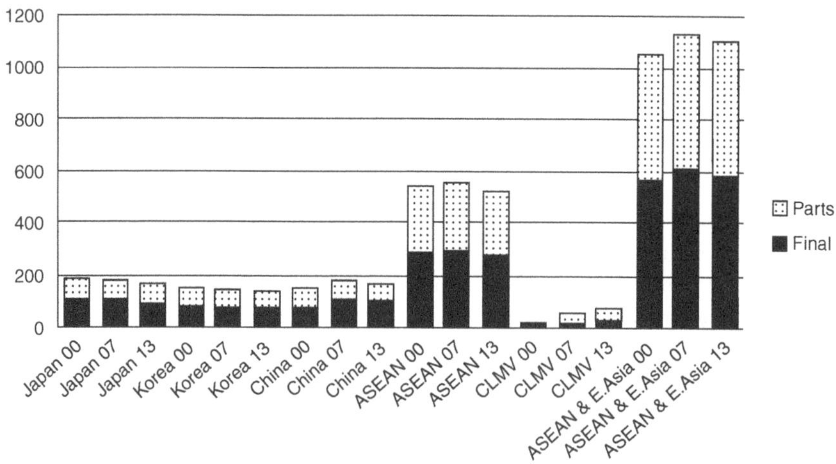

Figure 6.5a Number of product–country pairs: imports

Source: Author's calculations.

production network. The number of imported product–country pairs from the CMLV increased from 15 in 2000 to 56 in 2007 and 80 in 2013 (Figure 6.4a). Except for CMLV, the indexes remained just below one after 2007 for Japan, Korea (except in 2012), and ASEAN–4. China's index also declined, to 0.98 in 2013. On the whole, for 2000–2013, the index was low and remained largely stagnant, with a gradual decline from 2007.

Figures 6.6d–6.6f present the number of exported product–country pairs by destination during the same period. With some fluctuations, the indexes rose from 2000 to 2006, but except for the CLMV, they started to decline in 2007–2011. In examining extensive margins in East Asia for the period 2007–2011, the same declining pattern was also observed by Ando and Kimura (2013a) for the Philippines. The study indicated that whereas exports of China, Korea, and

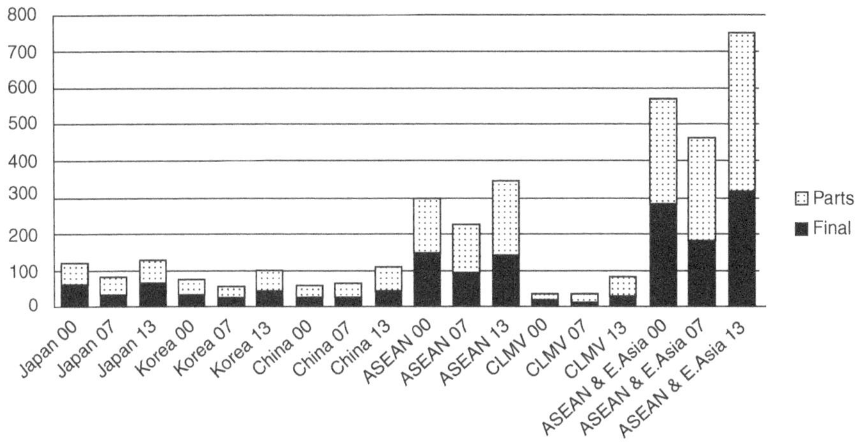

Figure 6.5b Number of product–country pairs: exports

Source: Author's calculations.

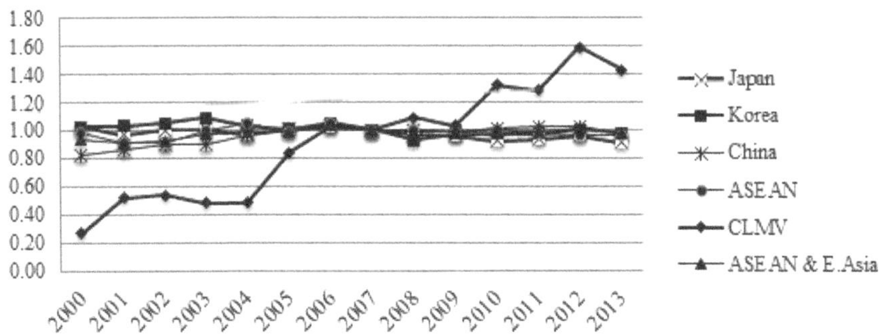

Figure 6.6a Number of product–country pairs: total electronics imports (2007 = 1)

Source: Author's calculations.

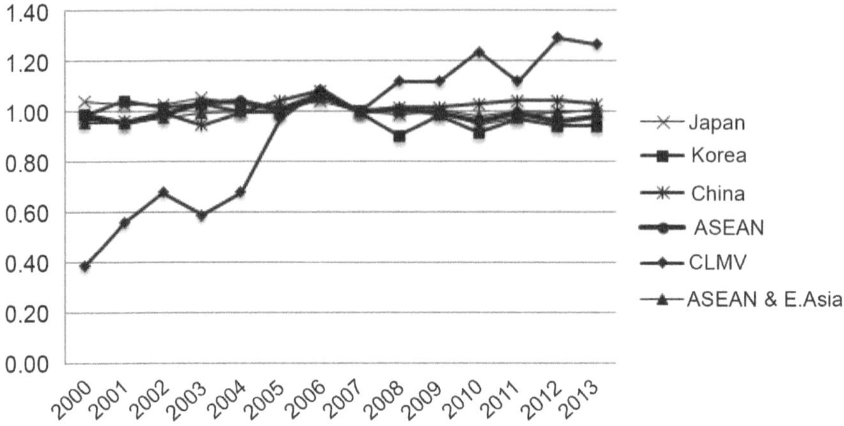

Figure 6.6b Number of product–country pairs: electronics parts imports (2007 = 1)

Source: Author's calculations.

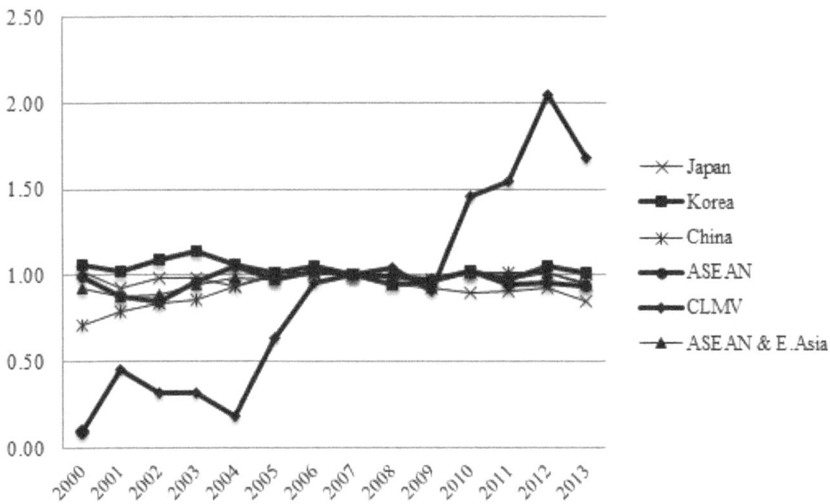

Figure 6.6c Number of product–country pairs: electronics final goods imports (2007 = 1)

Source: Author's calculations.

Viet Nam steadily increased in both machinery parts and components, machinery exports of the Philippines fell, suggesting a loss of its position in the regional production networks in the machinery sectors.

From 2012 to 2013, some recovery in exports can be seen, possibly due to the entry of new electronics firms and new product exports. The number of imported

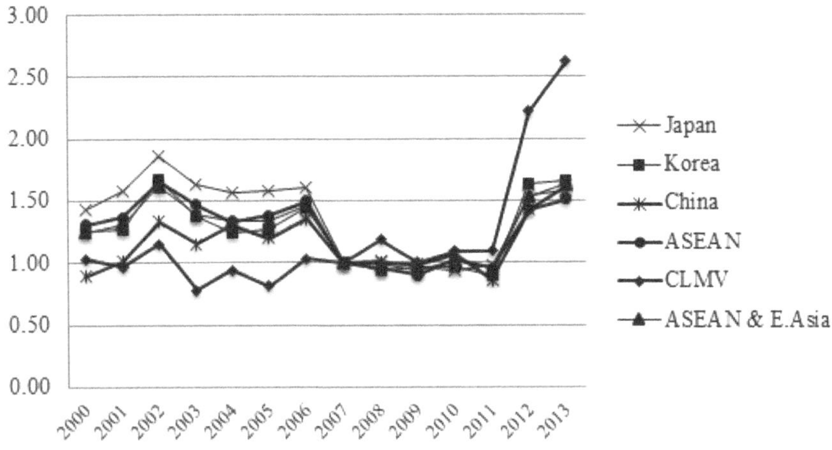

Figure 6.6d Number of product–country pairs: total electronics exports (2007 = 1)

Source: Author's calculations.

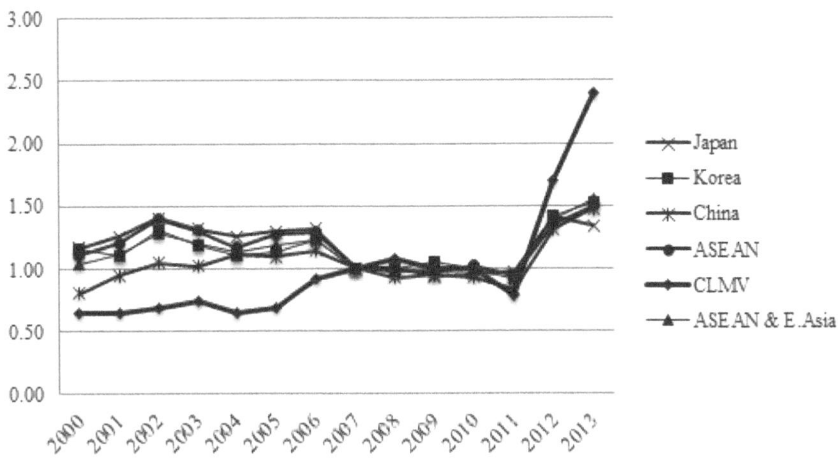

Figure 6.6e Number of product–country pairs: electronics parts exports (2007 = 1)

Source: Author's calculations.

parts remained relatively low and was characterized by a gradual declining trend. This may be attributed to the high concentration and specialization of the Philippines in the assembly and test of components imported from East Asia and ASEAN in the past two decades, highlighting the need to diversify and upgrade its position in the GVC.

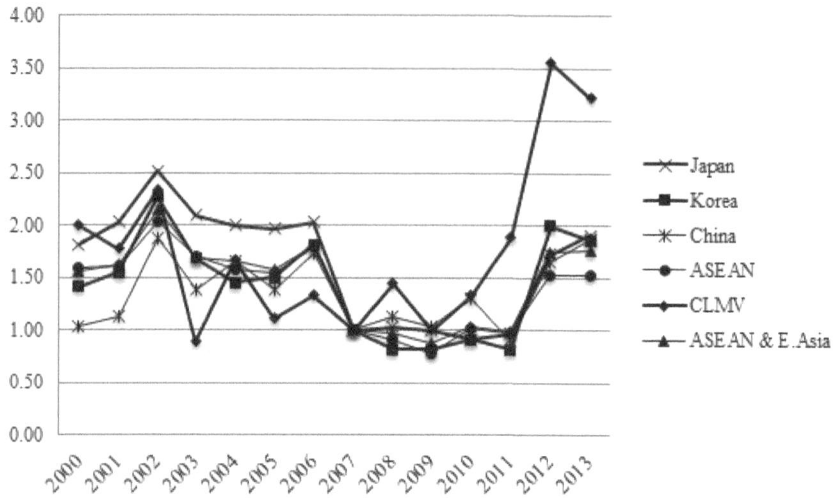

Figure 6.6f Number of product–country pairs: electronics final foods exports (2007 = 1)
Source: Author's calculations.

4.4 *Value chain analysis and challenges in deepening GVC participation*

The Philippine electronics industry has been largely dominated by the semiconductor sector. Its participation in global/regional production networks has relied mainly on mature and legacy products and processes focusing on semiconductor assembly and test services (SATS), which is the back-end part of the semiconductor manufacturing services (SMS). The electronics manufacturing services (EMS) and original design manufacturing (ODM) are currently underrepresented in the country. This is the opposite of the global pattern, which shows that the semiconductor and electronics industry is dominated by EMS. Growing at 30 per cent in the period 2012–2013, the sector was valued at $1.4 trillion in 2013, with a market share of 78 per cent (Gartner, Elcina as cited in EU–TRTA 2014).

As the RCAs indicated, the Philippines has remained strong in semiconductors but has been lagging behind Singapore, Malaysia, and Thailand. The extensive margin analysis shows that the Philippines has been regaining its position in regional production networks as indicated by the rising number of exported products to the region. Attracting more EMS companies will be crucial in sustaining the position of the Philippines in regional production networks. The gradually declining trend in the number of imported parts indicates the need to diversify and upgrade the industry's GVC participation by moving from semiconductors to EMS.

The global supply chain is complex with less clear dividing lines between the different segments. Hence, players have been forced to remain open to new types of collaboration to enable them to secure or capture their position in the GVC. The major segments of the global integrated circuit (IC) supply chain consist of: (1) IC design, (2) IC front-end (silicon/semiconductor compounds wafer fabrication), (3) IC middle-end (advanced packaging new technologies performed in a front-end environment, wafer-level processing for example), (4) IC back-end (IC assembly and test), (5) subsystem module (a derivative from EMS but also includes microelectronics and micro-assembly processes), and (6) electronics manufacturing services or EMS (printed circuit board assembly, devices, systems).

Both IC front-end and IC middle-end are not present in the Philippines. Activities focused on the back-end side either integrated device manufacturing or IDM; assembly and test houses; or outsourced semiconductor assembly and test services or OSAT.

On the EMS side, original design manufacturers or ODM, original equipment manufacturers or OEM, and EMS companies are present in the country. In terms of the subsystem/module segment, there is a base of local companies with EMS capabilities that are looking more at the subsystem/module segment, although their activities are still quite fragmented.

In addressing these major gaps in the value/supply chain, the development of new capabilities will be crucial. Although the absence of wafer fabrication capability needs to be addressed, the country might not be able to establish its own infrastructure due to critical constraints such as electricity cost, power instability, and lack of access to high-quality water and other critical raw materials.

Given the current pool of engineering talent and the emerging IC design activities in the country, one possible way to upgrade and diversify is to move from semiconductor to electronics design and service with a focus on developing IC design as an outsourced service (EU TRTA 2014). To realize this, innovation and research will be crucial, developing linkages with major global foundries from Singapore or Malaysia, and creating linkages with ODM/OEM business focusing on high-potential sectors, such as auto electronics, consumer electronics, electronic data processing, and renewable energy. Talent development for IC design services is equally necessary. At present, the country is still not able to develop PhD-level engineers in microelectronics, and layout engineering competencies are limited due to the worldwide shortage of layout engineers in the industry (EU TRTA 2014). Moreover, support industries such as materials and equipment along with telecommunications and logistics must be developed.

On EMS, in the short term, it will be important to study how to position the industry given the potential in auto electronics, EDP, consumer electronics, and renewable energy. In identifying long-run actions, it will be necessary to conduct technology scanning and assessment of the semiconductor and electronics value chain to help in identifying areas of specialization for the future growth opportunities of the country. There is also a need to upgrade the IC back-end sector to more middle-end advanced packaging technologies, as well as to develop niche ODM and subassembly/module manufacturing service industry.

5. Conclusions and recommendations

The preceding analysis has shown that the Philippines has remained competitive in the highly commoditized, labour-intensive, low-value-added assembly, testing, cleaning, and packaging segment of the semiconductor and electronics industry. However, in terms of overall export competitiveness in the manufacture of electronics products, it has been lagging behind Singapore, Malaysia, and Thailand. Lower costs for power and wages in countries such as Viet Nam and Indonesia have made them attractive investment destinations for similar activities. Viet Nam seems to be an aggressive player, given that it has established its own wafer fabrication plant to serve as a backbone for its electronics industry and is emerging as one of the Philippines' strongest competitors. Malaysia and Thailand are similar to the Philippines in terms of having a strong legacy of semiconductor assembly and test services (back-end and middle-end), but both are now focusing more on advanced types of technology. Within ASEAN, Singapore is the most mature and experienced country in terms of the semiconductor industry, with state-of-the-art infrastructure and end-to-end R&D capabilities from IC design to system development.

It is within this context that the Philippines must discover how to reposition the activities of the industry in the global/regional production networks and upgrade and move up the GVC. A structural transformation that shifts its production of low-value-added goods and services to more diverse, complex and high-value-added production is needed. As the extensive margin analysis shows, the Philippines must attract more EMS companies and create industry agglomeration to sustain its position in regional production networks. The gradually declining trend in the number of imported parts indicates the need to diversify and upgrade the industry's GVC participation by moving from semiconductors to EMS.

Strengthening its competitiveness is necessary to transform and deepen its position in the GVC, not only in semiconductor devices, but also in electronics manufacturing services. The value-chain analysis shows major gaps in both the IC front-end and IC middle-end segments of the supply chain. The nascent activities in the IC design segment and highly fragmented activities in the subsystem/module segment must be supported and unified. Upgrading options for the government include the following:

- Functional upgrading: shift to semiconductor and electronics design and service with a focus on developing IC design;
- Market upgrading: move to the EMS segment and continue to attract investments in areas with high growth potential, such as auto electronics, power electronics, EDP, and consumer electronics; and
- Linkages upgrading: upgrade the IC back-end sector to more middle-end advanced packaging technologies and develop niche ODM and subassembly module manufacturing service industry.

The policy challenge is how to ensure that FDI creates spillover effects to the domestic economy and the country remains plugged into GVCs, particularly as export activities of the electronics GVCs create value and employment that tend to 'stick' to the country. The upgrading process will require the development of new capabilities and human resources to ensure the availability of manpower and talents, establishment of an innovation ecosystem, better protection and enforcement of intellectual property rights (IPR) to promote innovation, design, and creativity, creation of incubation centres for SMEs and start-ups, efficient logistics and infrastructure, and development of parts, supplies, and materials sectors to support the industry. Investing in education and human capital, particularly R&D scientists and engineers, is one of the most important factors in attracting and developing higher-value added activities. Logistics forms the backbone of an effective and efficient supply chain management and would entail the development of both hard (physical transport and storage facilities) and soft (policies, rules, and regulations) infrastructure.

Strengthening domestic entrepreneurship and creation of start-ups will also be crucial in upgrading and diversifying industry activities. Investment in advanced infrastructure, particularly in high-speed communication networks and logistics, will help, along with policies to strengthen the domestic business environment. In particular, policy focus should be on developing the country's innovation system through efficient and homogeneous standards, testing, and quality assurance institutions; high-tech and R&D infrastructures and investments in scientific and technological research; and incentives for the establishment of science and technology parks to facilitate inter-firm and academe–industry collaboration in high-tech activities, as well as incentives for local R&D and training.

Table 6.A Trade patterns in major electronics subsectors

1. Components/Devices semiconductors			*Imports*			*Exports*		
	Year	*Destination*	*Total*	*Parts*	*Final*	*Total*	*Parts*	*Final*
Value in US$ billion	2000	World	11.11	10.99	0.12	14.87	14.77	0.09
	2007	World	18.86	18.70	0.16	19.36	18.34	1.01
	2013	World	11.65	11.40	0.24	15.23	13.63	1.61
Value index (2007 = 1)	2000	World	0.59	0.59	0.73	0.77	0.81	0.09
	2000	Japan	0.75	0.76	0.47	1.17	1.46	0.10
	2000	Korea	0.91	0.91	3.28	0.87	0.87	0.79
	2000	China	0.08	0.08	0.09	0.07	0.07	0.01
	2000	ASEAN4	0.39	0.39	0.91	0.42	0.42	0.00
	2000	CLMV	0.01	0.01	0.04	1.55	6.03	0.05
	2013	World	0.62	0.61	1.51	0.79	0.74	1.59
	2013	Japan	0.49	0.48	1.21	1.48	1.73	0.56
	2013	Korea	0.53	0.52	4.18	1.05	0.90	46.56
	2013	China	0.64	0.60	1.55	0.41	0.36	10.44
	2013	ASEAN4	0.56	0.55	1.81	1.01	0.99	2.66
	2013	CLMV	0.15	0.08	6.64	308.93	1133.7	33.98

(*Continued*)

Table 6.A (Continued)

1. Components/Devices semiconductors			*Imports*			*Exports*		
	Year	*Destination*	*Total*	*Parts*	*Final*	*Total*	*Parts*	*Final*
Share	2000	Japan	13.94	13.96	12.74	9.86	9.74	28.79
	2000	Korea	10.94	10.95	10.78	4.24	4.25	1.86
(in %)	2000	China	0.61	0.59	3.01	1.44	1.44	0.22
	2000	ASEAN4	12.64	12.64	12.20	9.24	9.29	0.15
	2000	CLMV	0.01	0.01	0.03	0.00	0.00	0.02
	2007	Japan	10.89	10.81	19.55	6.48	5.38	26.37
	2007	Korea	7.06	7.10	2.39	3.72	3.91	0.22
	2007	China	4.59	4.42	23.60	15.37	16.14	1.56
	2007	ASEAN4	18.87	18.94	9.74	16.98	17.70	3.96
	2007	CLMV	0.35	0.34	0.46	0.00	0.00	0.03
	2013	Japan	8.63	8.47	15.74	12.17	12.50	9.36
	2013	Korea	6.09	6.08	6.63	4.94	4.77	6.43
	2013	China	4.78	4.36	24.35	8.02	7.75	10.25
	2013	ASEAN4	17.11	17.22	11.70	21.79	23.58	6.63
	2013	CLMV	0.09	0.04	2.05	0.95	0.98	0.75

2. Electronic data processing			*Imports*			*Exports*		
	Year	*Destination*	*Total*	*Parts*	*Final*	*Total*	*Parts*	*Final*
Value in US$ billion	2000	World	2.45	2.26	0.20	7.12	2.48	4.64
	2007	World	5.10	4.85	0.26	9.87	1.93	7.94
	2013	World	2.50	2.14	0.36	4.20	0.66	3.53
Value index	2000	World	0.48	0.47	0.77	0.72	1.28	0.58
(2007 = 1)	2000	Japan	0.67	0.67	2.31	0.81	1.43	0.69
	2000	Korea	0.06	0.05	0.84	0.57	0.50	0.61
	2000	China	0.09	0.06	0.23	0.07	0.25	0.01
	2000	ASEAN4	0.57	0.53	0.78	0.49	1.36	0.02
	2000	CLMV	0.01	0.01	0.59	0.34	0.38	0.13
	2013	World	0.49	0.44	1.41	0.43	0.34	0.44
	2013	Japan	0.38	0.38	0.61	0.20	0.25	0.19
	2013	Korea	0.45	0.41	4.05	0.17	0.12	0.20
	2013	China	1.10	0.82	2.58	0.91	0.41	1.09
	2013	ASEAN4	0.68	0.64	0.92	0.29	0.39	0.24
	2013	CLMV	0.38	0.36	3438	2.46	0.90	9.02
Share	2000	Japan	51.11	55.11	5.36	20.95	17.19	22.96
(in %)	2000	Korea	0.51	0.46	1.06	2.64	2.61	2.66
	2000	China	1.13	0.70	6.08	1.42	3.73	0.19
	2000	ASEAN4	24.71	21.70	59.09	9.63	26.77	0.48
	2000	CLMV	0.03	0.03	0.00	0.03	0.07	0.00
	2007	Japan	36.71	38.56	1.79	18.74	15.47	19.54
	2007	Korea	4.18	4.35	0.97	3.36	6.77	2.53
	2007	China	6.27	5.54	20.16	13.93	18.77	12.75
	2007	ASEAN4	21.00	19.02	58.56	14.05	25.33	11.31
	2007	CLMV	2.20	2.32	0.00	0.05	0.23	0.01
	2013	Japan	28.60	33.31	0.77	8.98	11.47	8.51
	2013	Korea	3.82	4.00	2.77	1.34	2.34	1.15
	2013	China	14.12	10.28	36.85	29.88	22.53	31.26
	2013	ASEAN4	29.17	27.66	38.08	9.70	28.55	6.16
	2013	CLMV	1.71	1.87	0.77	0.32	0.60	0.27

3. Communication, radar, and telecommunications			*Imports*			*Exports*		
	Year	*Destination*	*Total*	*Parts*	*Final*	*Total*	*Parts*	*Final*
Value in US$ billion	2000	World	1.44	0.95	0.49	0.76	0.21	0.55
	2007	World	1.05	0.81	0.23	0.27	0.06	0.21
	2013	World	1.37	0.80	0.57	0.57	0.31	0.26
Value index (2007 = 1)	2000	World	1.37	1.17	2.10	2.80	3.46	2.61
	2000	Japan	2.21	2.11	9.45	2.35	2.06	3.39
	2000	Korea	8.00	1.05	49.24	1.45	1.59	1.00
	2000	China	0.03	0.05	0.01	1.11	1.34	0.37
	2000	ASEAN4	1.25	0.89	2.34	0.65	0.37	0.94
	2000	CLMV	0.04	0.09	0.00	0.46	0.01	22.51
	2013	World	1.31	0.98	2.45	2.10	5.10	1.25
	2013	Japan	0.93	0.84	7.60	2.16	2.39	1.34
	2013	Korea	1.49	1.50	1.43	2.10	2.11	2.06
	2013	China	1.95	1.86	2.13	4.85	4.12	7.10
	2013	ASEAN4	1.98	1.13	4.67	5.56	8.86	2.02
	2013	CLMV	7.57	9.41	6.24	1.89	1.68	12.04
Share (in %)	2000	Japan	11.18	15.99	1.83	10.24	25.59	4.43
	2000	Korea	16.99	2.88	44.42	0.35	1.09	0.08
	2000	China	0.69	0.99	0.11	0.65	2.18	0.07
	2000	ASEAN4	11.37	9.38	15.26	0.97	1.05	0.93
	2000	CLMV	0.00	0.01	0.00	0.17	0.02	0.23
	2007	Japan	6.95	8.81	0.41	12.17	42.93	3.41
	2007	Korea	2.91	3.20	1.89	0.68	2.37	0.20
	2007	China	27.33	23.52	40.68	1.64	5.60	0.51
	2007	ASEAN4	12.53	12.21	13.68	4.18	9.77	2.58
	2007	CLMV	0.14	0.07	0.36	1.04	4.61	0.03
	2013	Japan	4.93	7.54	1.26	12.51	20.12	3.66
	2013	Korea	3.33	4.90	1.11	0.68	0.98	0.32
	2013	China	40.72	44.52	35.37	3.78	4.52	2.92
	2013	ASEAN4	19.01	13.99	26.08	11.05	16.96	4.17
	2013	CLMV	0.79	0.70	0.91	0.94	1.52	0.26

4. Office equipment			*Imports*			*Exports*		
	Year	*Destination*	*Total*	*Parts*	*Final*	*Total*	*Parts*	*Final*
Value in US$ billion	2000	World	0.06	0.03	0.03	0.08	0.03	0.05
	2007	World	0.07	0.04	0.03	0.13	0.05	0.09
	2013	World	0.07	0.03	0.05	0.17	0.13	0.03
Value index (2007 = 1)	2000	World	0.90	0.88	0.91	0.60	0.74	0.53
	2000	Japan	2.08	1.76	4.40	0.68	1.03	0.53
	2000	Korea	1.87	0.67	2.14	2.48	2.61	0.49
	2000	China	0.46	0.59	0.43	0.41	2.53	0.01
	2000	ASEAN4	0.78	0.94	0.71	0.04	0.02	0.10
	2000	CLMV	0.00	0.00		7.01	0.00	
	2013	World	1.04	0.73	1.40	1.25	2.99	0.37
	2013	Japan	0.20	0.08	1.07	0.66	1.50	0.27
	2013	Korea	2.07	4.07	1.64	11.62	5.19	104.70
	2013	China	2.76	2.21	2.89	3.97	19.28	1.10
	2013	ASEAN4	0.88	0.63	0.99	1.67	2.27	0.12
	2013	CLMV	637.86	608.1		1451.8	1283	

(*Continued*)

Table 6.A (Continued)

4. Office equipment			*Imports*			*Exports*		
	Year	*Destination*	*Total*	*Parts*	*Final*	*Total*	*Parts*	*Final*
Share (in %)	2000	Japan	43.20	60.96	23.47	16.76	18.84	15.29
	2000	Korea	1.03	0.12	2.04	0.67	1.59	0.01
	2000	China	6.00	2.86	9.49	6.09	14.41	0.19
	2000	ASEAN4	24.30	17.41	31.96	0.53	0.43	0.60
	2000	CLMV	0.00	0.00	0.00	0.00	0.00	0.00
	2007	Japan	18.64	30.61	4.87	14.72	13.55	15.31
	2007	Korea	0.49	0.17	0.87	0.16	0.45	0.02
	2007	China	11.64	4.27	20.11	9.03	4.23	11.46
	2007	ASEAN4	27.90	16.42	41.11	7.75	16.65	3.25
	2007	CLMV	0.00	0.01	0.00	0.00	0.00	0.00
	2013	Japan	3.55	3.27	3.73	7.69	6.82	11.22
	2013	Korea	0.98	0.92	1.02	1.51	0.79	4.43
	2013	China	30.87	12.99	41.52	28.55	27.29	33.67
	2013	ASEAN4	23.63	14.31	29.17	10.33	12.64	1.00
	2013	CLMV	2.96	7.56	0.22	0.18	0.19	0.10

5. Medical/Industrial and control instrumentation			*Imports*			*Exports*		
	Year	*Destination*	*Total*	*Parts*	*Final*	*Total*	*Parts*	*Final*
Value in US$ billion	2000	World	0.12	0.04	0.08	0.04	0.01	0.03
	2007	World	0.33	0.10	0.22	0.08	0.05	0.03
	2013	World	0.24	0.08	0.16	0.42	0.27	0.15
Value index (2007 = 1)	2000	World	0.37	0.43	0.34	0.51	0.23	1.02
	2000	Japan	0.73	0.61	0.90	0.53	0.50	0.58
	2000	Korea	0.46	0.92	0.31	0.11	0.10	0.15
	2000	China	0.01	0.17	0.01	0.02	0.00	0.07
	2000	ASEAN4	0.56	0.48	0.59	0.46	0.19	0.53
	2000	CLMV	0.00	0.00	0.00	69.98	0.00	
	2013	World	0.73	0.73	0.73	5.44	5.41	5.49
	2013	Japan	0.79	0.50	1.17	2.32	2.84	1.53
	2013	Korea	6.44	2.01	7.87	8.87	10.36	4.74
	2013	China	0.23	2.59	0.17	9.79	8.74	12.70
	2013	ASEAN4	0.94	0.88	0.97	10.64	41.60	1.82
	2013	CLMV	0.30	2.58	0.07	388.78	161.32	
Share (in %)	2000	Japan	28.79	36.99	23.96	45.97	87.33	28.41
	2000	Korea	1.58	2.10	1.28	1.29	2.78	0.66
	2000	China	1.11	1.05	1.15	0.25	0.00	0.35
	2000	ASEAN4	16.88	12.40	19.52	5.71	1.75	7.40
	2000	CLMV	0.00	0.00	0.00	0.46	0.00	0.66
	2007	Japan	14.56	25.95	9.17	44.06	41.00	49.73
	2007	Korea	1.28	0.97	1.43	5.90	6.68	4.46
	2007	China	33.10	2.65	47.50	6.92	7.82	5.23
	2007	ASEAN4	11.23	11.07	11.31	6.35	2.17	14.11
	2007	CLMV	2.06	0.59	2.76	0.00	0.01	0.00
	2013	Japan	15.69	17.81	14.69	18.79	21.49	13.85
	2013	Korea	11.30	2.68	15.37	9.62	12.78	3.86
	2013	China	10.57	9.37	11.13	12.44	12.63	12.10
	2013	ASEAN4	14.43	13.28	14.98	12.42	16.66	4.68
	2013	CLMV	0.85	2.09	0.27	0.24	0.15	0.40

6. *Auto Electronics*			*Imports*			*Exports*		
	Year	*Destination*	*Total*	*Parts*	*Final*	*Total*	*Parts*	*Final*
Value in US$billion	2000	World	0.01	0.01	0.00	0.33	0.20	0.12
	2007	World	0.02	0.01	0.01	0.61	0.56	0.05
	2013	World	0.01	0.01	0.00	0.48	0.33	0.15
Value index (2007 = 1)	2000	World	0.63	0.73	0.41	0.54	0.36	2.62
	2000	Japan	2.29	2.30	1.55	1.56	0.14	3.67
	2000	Korea	0.74	0.73		0.14	0.00	14.08
	2000	China	0.41	0.41	0.38	2.00	2.33	0.00
	2000	ASEAN4	0.35	0.33	0.36	0.25	0.63	0.00
	2000	CLMV	0.00	0.00		1049.6	1049.6	
	2013	World	0.53	0.51	0.57	0.79	0.59	3.26
	2013	Japan	0.60	0.61	0.12	2.35	1.67	3.35
	2013	Korea	1.16	1.15		3.18	2.98	22.71
	2013	China	0.81	0.80	0.90	10.36	10.95	6.77
	2013	ASEAN4	0.68	0.85	0.59	2.44	1.47	3.09
	2013	CLMV	1.64	1.64		131.81	131.81	
Share (in %)	2000	Japan	43.55	54.01	1.83	31.04	2.64	78.18
	2000	Korea	3.21	3.98	0.11	0.06	0.00	0.15
	2000	China	4.36	5.21	0.99	1.20	1.92	0.00
	2000	ASEAN4	24.69	9.94	83.51	0.39	0.63	0.01
	2000	CLMV	0.00	0.00	0.00	0.17	0.27	0.00
	2007	Japan	11.96	17.06	0.48	10.75	6.96	55.92
	2007	Korea	2.74	3.95	0.00	0.22	0.23	0.03
	2007	China	6.67	9.16	1.07	0.32	0.30	0.58
	2007	ASEAN4	44.20	21.78	94.70	0.83	0.36	6.45
	2007	CLMV	0.04	0.06	0.00	0.00	0.00	0.00
	2013	Japan	13.58	20.25	0.10	31.85	19.87	57.52
	2013	Korea	5.97	8.89	0.07	0.88	1.19	0.19
	2013	China	10.11	14.29	1.67	4.23	5.64	1.21
	2013	ASEAN4	56.42	35.97	97.68	2.57	0.91	6.13
	2013	CLMV	0.13	0.19	0.00	0.01	0.02	0.00

7. *Consumer electronics*			*Imports*			*Exports*		
	Year	*Destination*	*Total*	*Parts*	*Final*	*Total*	*Parts*	*Final*
Value in US$ billion	2000	World	0.68	0.50	0.18	0.44	0.31	0.13
	2007	World	0.41	0.19	0.21	0.76	0.72	0.04
	2013	World	0.53	0.24	0.29	0.32	0.27	0.05
Value index (2007 = 1)	2000	World	1.68	2.57	0.85	0.58	0.43	3.41
	2000	Japan	3.89	4.10	2.46	1.23	0.83	51.37
	2000	Korea	2.38	2.28	3.23	1.68	1.71	1.45
	2000	China	0.28	0.25	0.32	0.19	0.17	1.16
	2000	ASEAN4	1.03	1.90	0.75	0.36	0.31	1.37
	2000	CLMV	0.07	1.00	0.00	0.08	0.34	0.00
	2013	World	1.30	1.25	1.35	0.42	0.38	1.20
	2013	Japan	0.44	0.35	1.02	0.69	0.51	22.51
	2013	Korea	0.61	0.48	1.69	1.54	0.95	6.15
	2013	China	3.97	4.45	3.53	1.01	1.00	1.74
	2013	ASEAN4	0.94	0.63	1.05	0.75	0.71	1.74
	2013	CLMV	0.30	3.10	0.10	3.31	9.87	1.19

(*Continued*)

Table 6.A (Continued)

7. Consumer electronics			*Imports*			*Exports*		
	Year	*Destination*	*Total*	*Parts*	*Final*	*Total*	*Parts*	*Final*
Share (in %)	2000	Japan	44.66	55.92	13.68	21.53	20.26	24.61
	2000	Korea	9.13	10.68	4.87	4.76	6.07	1.59
	2000	China	2.12	1.21	4.62	1.25	1.55	0.51
	2000	ASEAN4	24.20	15.16	49.10	3.89	4.64	2.08
	2000	CLMV	0.07	0.10	0.00	0.03	0.04	0.00
	2007	Japan	19.25	35.09	4.75	10.22	10.67	1.64
	2007	Korea	6.43	12.05	1.29	1.65	1.54	3.73
	2007	China	12.53	12.70	12.36	3.75	3.87	1.49
	2007	ASEAN4	39.24	20.56	56.32	6.39	6.45	5.19
	2007	CLMV	1.75	0.25	3.13	0.18	0.05	2.71
	2013	Japan	6.50	9.95	3.58	16.79	14.46	30.78
	2013	Korea	3.01	4.66	1.61	6.06	3.88	19.18
	2013	China	38.17	45.18	32.23	9.07	10.22	2.16
	2013	ASEAN4	28.39	10.30	43.70	11.50	12.15	7.55
	2013	CLMV	0.41	0.62	0.22	1.42	1.20	2.69

Based on HS 1996 product codes and following Ando and Kimura (2013b) parts and final goods were defined as follows:

Intermediate Parts: 850490; 851790; 851840; 851850; 851890; 853090; 900990; 901090; 901190; 901290; 902790; 903090; 903190; 903290; 847310; 847321; 847329; 847330; 847350; 851220; 851230; 852290; 852910; 852990; 853110; 853180; 853190; 853210; 853221; 853222; 853223; 853224; 853225; 853229; 853230; 853290; 853310; 853321; 853329; 853331; 853339; 853340; 853390; 853400; 853610; 853641; 853649; 853650; 853669; 854011; 854012;854020;854040; 854050; 854071; 854072; 854079; 854081; 854089; 854091; 854099;854110; 854121;854129; 854130; 854140; 854150; 854160; 854190; 854212; 854213; 854230; 854240; 854250; 854290; 854411; 854419; 854420; 854441; 854449; 854451; 854459; 854470; and 870839

Final Goods: 846911; 846912; 847010; 847021; 847029; 847030; 847040; 847050; 847090; 847110; 847130; 847141; 847149; 847150; 847160; 847170; 847180; 847190; 847290; 847310; 847321; 847329; 847330; 847350; 850410; 850421; 850422; 850423; 850431; 850432; 850433; 850434; 850440; 850450; 851220; 851230; 851711; 851719; 851721; 851730; 851750; 851780; 851810; 851821; 851822; 851829; 851830; 851910; 851921; 851929; 851931; 851939; 851992; 851999; 852020; 852033; 852039; 852090; 852110; 852190; 852290; 852311; 852312; 852313; 852320; 852410; 852431; 852432; 852439; 852440; 852451; 852453; 852460; 852491; 852499; 852510; 852520; 852530; 852540; 852610; 852691; 852692; 852712; 852713; 852719; 852721; 852729; 852731; 852732; 852739; 852790; 852812; 852813; 852821; 852822; 852830; 852910; 852990; 853010; 853080; 853110; 853180; 853190; 853210; 853221; 853222; 853223; 853224; 853225; 853229; 853230; 853290; 853310; 853321; 853329; 853331; 853339; 853340; 853390; 853400; 853610; 853641; 853649; 853650; 853669; 854011; 854012; 854020; 854040; 854050; 854071; 854072; 854079; 854081; 854089; 854091; 854099; 854110; 854121; 854129; 854130; 854140; 854150; 854160; 854190; 854212; 854213; 854230; 854240; 854250; 854290; 854320; 854411; 854419; 854420; 854441; 854449; 854451; 854459; 854470; 870839; 900911; 900912; 901010; 901110; 901120; 901210; 901600; 901720; 901811; 901812; 901813;901814; 901819; 901820; 902140; 902150; 902212; 902213; 902219; 902221; 902229; 902230; 902290; 902300; 902610; 902620; 902680; 902720; 902730; 902740; 902750; 902780; 903010; 903020; 903031; 903039; 903040; 903089; 903220; 903281; and 903289

Notes

1 The ITA was signed during the Ministerial Conference of the World Trade Organization in Singapore in December 1996. The agreement covers a wide range of products, including computer hardware and software, semiconductors, telecommunications equipment, electronic office equipment, and manufacturing equipment particularly for use in semiconductor production. Excluded are consumer electronics.

2 The RCA compares how much a country is exporting a given product relative to its total trade, in comparison to the share of that product in world trade. A country is said to have a revealed comparative advantage when its share of export of a given product exceeds the equivalent share of export of the world. This is captured when the RCA is above one. An RCA below one suggests that the country does not have a revealed comparative advantage in a given product.

3 See Table 6.6 for the specific HS 1996 product coverage.

References

Ando, M and Kimura, F 2013a, 'What are the opportunities and challenges for ASEAN?', *ERIA Discussion Paper Series* 2013–31, ERIA, Jakarta.

Ando, M and Kimura, F 2013b, 'Production linkage of Asia and Europe via Central and Eastern Europe', *Journal of Economic Integration* (June), vol. 28, pp. 204–240.

De Backer, K and Miradout, S 2013, 'Mapping global value chains', *OECD Trade Policy Papers* No. 159, OECD, Paris.

European Union–Trade Related Technical Assistance Project 2014, *Mission Report: Philippine IC Design Industry*. Department of Trade & Industry Philippines and European Union.

Frederick, S and Gereffi, G 2013, *Costa Rica in the Electronics Global Value Chain: Opportunities for Upgrading*, Duke Center on Globalization, Governance & Competitiveness, Duke University, Durham, NC.

Maurer, A and Degain, C 2010, 'Globalization and trade flows: what you see is not what you get', *WTO Staff Working Paper* No. ERSD 2010-12, WTO, Geneva, Switzerland.

OECD 2014, Measuring Trade in Value Added: An OECD-WTO joint initiative. www.oecd.org/sti/ind/measuringtradeinvalue-addedanoecd-wtojointinitiative.htm.

Organisation for Economic Cooperation and Development [OECD] 2013, *Interconnected Economies: Benefiting from Global Value Chains*, OECD, Paris.

Pietrobelli, C and Rabellotti, R 2011, 'Global value chains meet innovation systems: are there learning opportunities for developing countries?', *World Development*, vol. 39, no. 7, pp. 1261–1269.

Pilat, D, Yamano, N, and Yashiro, N 2012, 'Moving Up the Value Chain: China's Experience and Future Prospects', in *China in Focus: Lessons and Challenges*, ed. OECD, OECD, Paris, pp. 72–88.

7 Singapore's participation in global value chains

Perspectives of trade in value added

Mun-Heng Toh

1. Overview of Singapore's economy

Singapore's economy has evolved from an *entrepôt* economy in the early 1960s into one that is powered by modern industries, such as electronics, chemicals and pharmaceuticals, and sophisticated service industries in the areas of finance, business consultancies, and medical and education services. It is now a hub for many types of economic activities: financial services, information technology (IT) services, medical services, electronics, aviation, and education services.

1.1 International trade

The remarkable growth performance of Singapore's economy over the past five decades is attributed to the twin strategies of trade expansion and powering its industrialization driven by foreign capital, technology, and international market access.

Following the recommendation of a United Nations Industrial Mission (1960–1961) study taskforce, import substitution industrialization was selected as a solution to move Singapore away from its overdependence on *entrepôt* trade. But this policy proved ineffective as it tended to develop inefficient domestic manufacturing industries, especially at a time when the domestic market was limited and lacked sophistication. Inflows of foreign capital were unimpressive despite various fiscal incentives and concessions the government provided through its investment promotion agency, the Economic Development Board (EDB), established in 1961. The separation of Singapore from Malaysia in 1965 spelt the end of the import substitution phase.

The policy that gave Singapore a head start in attracting foreign capital was the government's highly liberal stance on ownership, at a time when foreign investment was viewed with suspicion by other developing countries following the experience of the Latin American economies. Transnational corporations (TNCs) were seen as footloose and exploitative. After 1965, the government consistently maintained an open policy towards foreign ownership and operations. There were no restrictions on equity ownership, no foreign exchange controls, and no limits on the repatriation of capital, dividends, interest, or royalties. There were no restrictions on foreign borrowings from the domestic capital market and no regulations governing the transfer of technology. Furthermore, the government was willing to co-invest with foreign companies if there was a need for risk sharing and the nurturing of business confidence.

Nonetheless, the transition from an import substitution mindset towards an export promotion strategy was far from easy. Several structural adjustments had to be made. In Singapore's case, the domestic market was small and unable to absorb the goods manufactured by foreign enterprises. It was therefore vital to gain access to foreign markets, which in turn required Singapore to become an open economy. Free trade zones that were previously restricted to areas around the ports expanded their coverage to include the whole island and virtually all tariffs were rapidly removed.

Singapore's role as an *entrepôt* port for the region did not really decline when the country embarked on its industrialization strategy. Its strategic geographical location has made Singapore a key dissemination centre for the region. Goods are imported into Singapore in bulk before being broken into smaller consignments to be re-exported to other destinations. Re-exports as a proportion of total exports were high, at 62 per cent in 1970, but this declined to 34 per cent in 1990, before rising to 47 per cent in 2013.

Singapore's domestic exports are exports of Singaporean origin and comprise (1) primary commodities grown or produced in Singapore; and (2) goods that have been transformed, that is, manufactured, assembled, or processed in Singapore, including those with imported materials or parts. In recent years, almost half of the total exports are domestic exports, compared with about 40 per cent in 1970. The changing composition of goods exported from 1970 to 2013 is shown in Figure 7.1.

Figure 7.1 also shows the share of services exports as a percentage share of total merchandise exports. Although the share varies from year to year, it remains relatively stable within a band of 20 per cent to 40 per cent since 1983. Participation in global value chains (GVCs), as some analysts have claimed, allows the contribution of services exports to be maintained.

1.2 Foreign direct investment

From the late 1960s, with job creation a priority and local entrepreneurship and expertise essential for kick-starting Singapore's industrialization campaign lacking, the government invited foreign capital and entrepreneurs to come to Singapore to establish their production bases and to create employment opportunities. Singapore became an outsourcing centre for the foreign multinational companies (MNCs). Fiscal incentives such as tax holidays and the restriction-free repatriation of profits, as well as incentives and grants for export expansion, were put in place. Complementing these fiscal incentives were industrial sites with prefabricated factory buildings and physical infrastructure such as communications, telecommunications, and transportation, all of which contributed directly to efficiency and competitiveness.

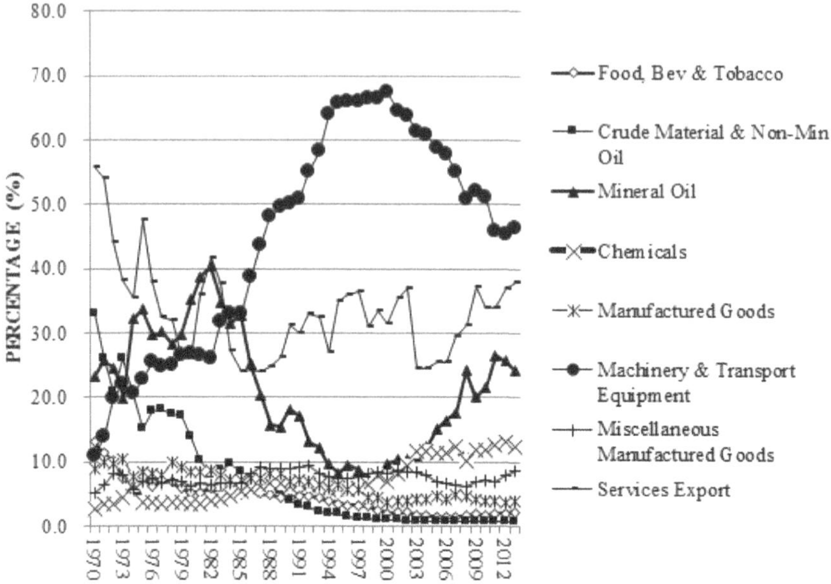

Figure 7.1 Percentage composition of merchandise exports and services exports as percentage of total merchandise exports, 1970–2013

Source: Yearbook of Statistics, Singapore Department of Statistics.

The many efforts made to attract foreign capital to Singapore soon bore fruit. FDI not only supplemented the limited savings and capital formation in Singapore. Foreign enterprises also brought along their technological know-how, managerial skills, and market accessibility in developed economies, in addition to the many jobs created for local workers. Healthy industrial relations and favourable labour market conditions formed a virtuous circle and created a conducive environment for FDI.

Inflows of FDI enabled Singapore to plug into GVCs and global production networks (GPNs)[1] relatively quickly. It is not surprising that Singapore's industrialization drive attracted predominantly FDI in the manufacturing sector in the initial years. However, as the manufacturing sector grew, so did demand for supporting services in logistics, finance, and professional business services, and FDI in these sectors increased in tandem.

2. Economic development philosophy and strategy

In 1985–1986, the city-state was hit by a severe recession and a new approach to generate growth was needed. The publication of the Strategic Economic Plan (SEP) by the Ministry of Trade and Industry in 1991 marked the beginning of a new development philosophy. It incorporated concepts such as competitive advantage, value chains, and agglomeration economies in industrial development and strategic business management.

In the SEP, it was recognized that an industrial policy that takes into account the relative strengths of Singapore in specific areas and that intelligently supports those cluster enterprises with the best chance of becoming world class would counter the limitations of small size. Fourteen clusters were identified comprising commodity trading, shipping, precision engineering, electronics, information technology, petroleum and petrochemical, construction, heavy engineering, finance, insurance, general supporting industries, tourism, international hub, and domestic industries. Each of these clusters included enterprises that had some common features or core capabilities in the form of natural advantages, created competitive advantages, or industry structures or attributes.

Manufacturing is a key engine of Singapore's economy, accounting for some 20 per cent to 25 per cent of GDP. Since 1990, the manufacturing sector has been reorganized into six major industrial clusters: electronics, chemicals, precision engineering, biomedical, transport engineering, and general manufacturing.

The EDB considered how the cluster approach could enable the electronics industry to achieve further expansion and growth. With rising wages and rentals, Singapore risked losing its manufacturing base and the threat of industrial hollowing out could not be ignored. The solution came in the form of two transitions that Singapore had to navigate: first, an internal transition for TNCs, from labour-intensive to automation; and second, from automation to integrated manufacturing, and from vertically integrated TNCs to dynamic clusters.

The threat of de-industrialization was addressed by relocation and reorganizing value chains. TNCs were enlightened enough to take a cluster and value-chain view of their production decisions. TNCs did not relocate their entire operations to lower-wage, labour-surplus destinations. Instead, they maintained non-labour-intensive manufacturing and higher value-added-services activities in Singapore, and relocated only the labour-intensive activities off-shore.

Singapore has managed to thrive on GVCs. However, it is not an easy task to maintain Singapore's relevance in GVCs. From a long-term perspective, the government adheres to the plan of developing Singapore into a knowledge-based economy. It decided to move upstream in GVCs.

3. Global value chains and trade in value added: importance and implications for development policies in Singapore

Trade data based on gross flows are increasingly inadequate as a basis for understanding modern trade, as the value of a final good now comes from many countries (Grossman 2010). Around 80 per cent of all trade takes place within the international production networks of TNCs, including contractual relationships between firms. But around one-third of global trade is now estimated to be intra-firm trade, occurring within the ownership structure of a single firm or TNC

(UNCTAD 2013). Current trade data, based on gross flows, are failing to capture this shift, hampering a thorough understanding of modern trade within GVCs.

Policy makers need to better understand where production is taking place and how value is being added. This can only be known through understanding the proportion of subcontracting components made elsewhere. The input–output technique pioneered by Nobel Laureate Leontief (1951) offers one way of estimating the source(s) of value (by country and industry) that is added in producing goods and services for export (and import). It recognizes that growing GVCs mean that a country's exports increasingly rely on significant intermediate imports (and so, value added by industries in upstream countries). The availability of global I/O matrices[2] has led to the development of methodological contributions suggesting more general metrics of GVCs. In particular, it helps to quantify the value added embodied in the goods and services traded internationally. Several recent articles generalize the vertical specialization concept of Hummels et al. (2001) and capture different dimensions of value added embedded in trade. Essentially, these new databases measure the extent to which countries are involved in vertically fragmented production.[3] This is approximated by the sum of the value of imported inputs in the overall exports of a country (the backward linkage) plus the percentage of exported goods and services used as imported inputs to produce other countries' exports (the forward linkage). The value-added shares describe the participation of a country in GVCs, both as a user of foreign inputs and as a supplier of intermediate goods and services used in other countries' exports.

3.1 Singapore participation in GVCs from the perspective of trade in value added

The Organisation for Economic Co-operation and Development (OECD) – World Trade Organization (WTO) Trade-in-Value-Added (TiVA) database was made public recently and has been used in many policy-oriented studies. OECD (2013) summarized the main evidence and policy implications of the OECD's work on GVCs, including trade and investment policies targeted to GVCs. In addition, the OECD has produced several comparable country notes, including indicators on the relevance of value-added trade and the participation in GVCs.[4]

According to information gathered from the OECD–WTO TiVA database, in 2009, world gross exports amounted to US$17.05 trillion. However, world value-added exports amounted to $13.7 trillion (around 19 per cent lower than gross exports), emphasizing the extent of double-counting in total trade due to trade in intermediate inputs related to production network spanning across countries. Whilst world gross exports as a proportion of GDP increased from 19 per cent in 1995 to 25 per cent in 2005, and then to 29 per cent in 2009, world value-added exports were much lower and increased from 16 per cent to 18 per cent in 2005, and then to 24 per cent in 2009. In this chapter, we use the information in the OECD–WTO TiVA database to consider Singapore's participation in GVCs in the light of the new TiVA indicators available.

Composition of value added in gross exports

The global input–output table enables users and policy makers to decompose the entire value of any good, exported by industry I, into the following components:

1. Direct domestic value added from industry I;
2. Indirect domestic value added generated via purely domestic transactions, disaggregated by all domestic industries;
3. Indirect domestic value added embodied in imports (broken down by all domestic industries);
4. Indirect imported (foreign) value added (broken down by producing country and industry)

In Table 7.1, at the aggregate level, the gross exports of Singapore for selected years between 1995 and 2009 are decomposed into the four VA components. Foreign value added (FVA) made up about 50 per cent of gross exports in 2009, and this proportion is higher than that in 1995 and slightly lower than in 2005.

The global I/O tables help in estimating the 'domestic value-added' content in gross exports of a country. 'Domestic value-added exports' will therefore differ from 'gross exports' and can be estimated by subtracting foreign value added (FVA), that is, value added created in other countries that is imported and enters exports of the country. In Figure 7.2, the gross exports and domestic value-added exports for each of the 21 selected countries are shown. The figures in per cent for each country indicate the excess of gross exports over domestic value-added export (i.e., foreign value-added export) expressed as a percentage of gross exports.

The extent of the difference between gross exports and domestic value-added exports (which equals the FVA in gross exports), varies across countries depending

Table 7.1 Decomposition of Singapore gross exports into VA components

	Gross export (US$ m)	*Direct domestic VA*	*Indirect domestic VA*	*Reimported domestic VA*	*Total domestic VA*	*Foreign VA in gross export*
2009	212,449	72,616	33,143	643	106,401	106,048
2005	160,821	51,422	24,695	519	76,636	84,185
2000	91,860	28,321	16,571	374	45,265	46,595
1995	80,306	24,843	17,757	205	42,804	37,501
Percentage Distribution						
2009	100.0	34.2	15.6	0.3	50.1	49.9
2005	100.0	32.0	15.4	0.3	47.7	52.3
2000	100.0	30.8	18.0	0.4	49.3	50.7
1995	100.0	30.9	22.1	0.3	53.3	46.7

Source: OECD–WTO Trade-in-value-added (TiVA) database, May 2013.

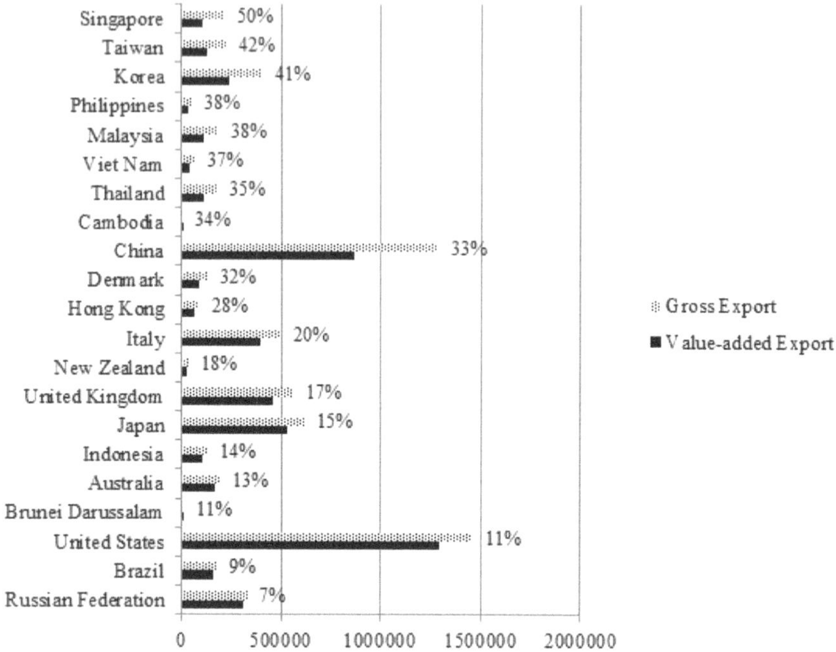

Figure 7.2 Domestic value-added exports and gross exports, 2009

Source: OECD WTO Trade-in-value-added (TIVA), May 2013.

on a country's engagement in network trade. The difference in gross exports and value-added exports is most prominent for Newly Industrialized Countries tier 1 (NICs 1), such as Singapore (50 per cent); Taiwan (42 per cent); Republic of Korea (henceforth, Korea) (41 per cent); followed by NICs 2 – Malaysia (38 per cent); the Philippines (38 per cent); Thailand (35 per cent); China (33 per cent); and Hong Kong (28 per cent). For most developed countries, FVA in gross exports is less than 30 per cent, with the United Kingdom at 17 per cent, the United States at 11 per cent, and Germany at 27 per cent.

The share of domestic value added in gross exports indicates the value-added gains for a country from exports. The information from the OECD–WTO TiVA database enables the comparison of the ratio over time. In Figure 7.3, we plot the percentage point changes of the ratio between 2009 and 1995 for the 21 selected countries.

Domestic value added in gross exports has declined substantially for many developing countries, indicating a rise of foreign value addition in their gross exports. However, for some countries, domestic value added increased in this period. These are the United Kingdom, Italy, Malaysia, the Russian Federation, and Hong Kong. The decline in the United States has been marginal

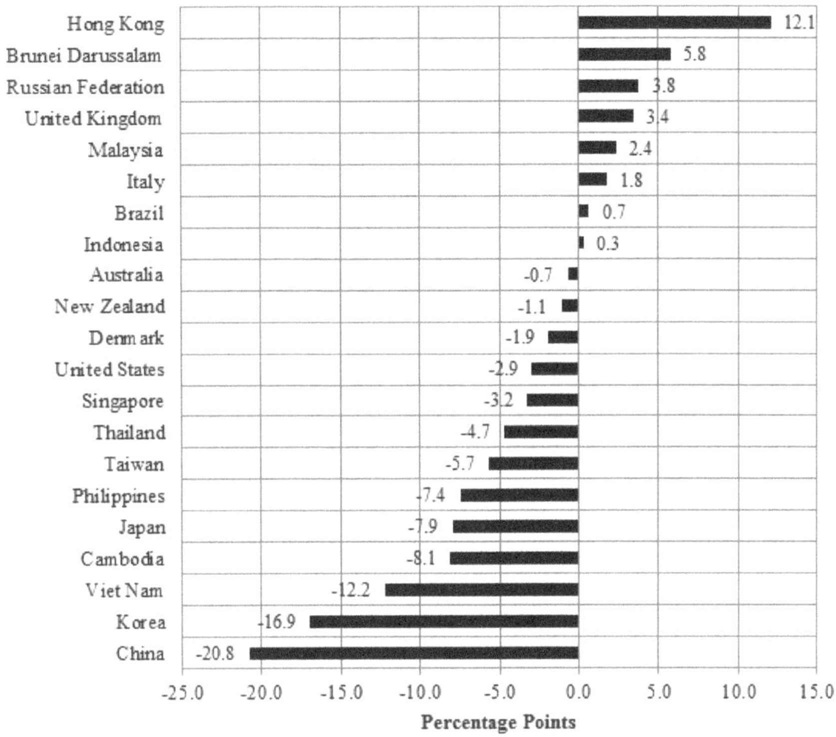

Figure 7.3 Percentage point change in share of domestic value added in gross exports: 2009 over 1995

Source: OECD–WTO Trade-in-value-added (TiVA), May 2013.

(3 percentage points) but very high for countries such as China (21 percentage points), and Korea (17 percentage points). Singapore, Taiwan, and Korea have lower domestic value-added shares in gross exports between 1995 and 2009, shaving off 3, 6, and 17 percentage points, respectively.

Typically, the share of re-imported domestic value added in gross exports is small. However, it is an indicator that measures the intensity of value added crossing borders and returning to its originating source. It is a rather intuitive measure of participation in the global production network. Table 7.2 presents reimported domestic value-added (VA) share for the top 20 countries in each of the four benchmark years. Germany and the United States have consistently ranked amongst the top three in the list. The shares for Singapore were 0.26 per cent in 1995, 0.41 in 2000, 0.32 per cent in 2005, and 0.30 per cent in 2009. China's share increased steadily, from only 0.13 per cent in 1995 to 1.1 per cent in 2009, displacing Germany as the top performer. A similar rate of improvement

Table 7.2 Top 20 countries in terms of the share of reimported domestic VA in gross exports

Rank	*2009*	*ReIm* **DVA**	*2005*	*ReIm* **DVA**	*2000*	*ReIm* **DVA**	*1995*	*ReIm* **DVA**
1	*China*	1.10	Germany	1.20	Germany	1.01	Germany	0.81
2	Germany	1.03	*China*	0.86	USA	0.91	USA	0.45
3	USA	0.58	USA	0.74	France	0.54	France	0.40
4	Taiwan	0.53	Taiwan	0.59	UK	0.49	UK	0.39
5	Korea	0.48	France	0.50	Malaysia	0.46	Netherlands	0.36
6	Netherlands	0.40	Korea	0.50	Canada	0.41	Belgium	0.35
7	France	0.39	Malaysia	0.47	**Singapore**	0.41	Canada	0.28
8	Malaysia	0.38	Netherlands	0.45	Netherlands	0.37	Sweden	0.26
9	Japan	0.37	Japan	0.43	Japan	0.36	Italy	0.26
10	**Singapore**	0.30	UK	0.41	Italy	0.33	Malaysia	0.26
11	Belgium	0.28	Belgium	0.38	Belgium	0.32	**Singapore**	0.26
12	Switzerland	0.28	Italy	0.36	Sweden	0.27	Czech Rep	0.26
13	UK	0.26	Canada	0.34	Spain	0.25	Japan	0.22
14	Norway	0.25	**Singapore**	0.32	Norway	0.24	Norway	0.21
15	Sweden	0.24	Denmark	0.30	*China*	0.23	Slovak Rep	0.19
16	Italy	0.24	Sweden	0.30	Austria	0.23	Austria	0.18
17	Czech Rep	0.23	Spain	0.26	Czech Rep	0.19	Spain	0.17
18	Austria	0.23	Czech Rep	0.25	Taiwan	0.19	Switzerland	0.16
19	Canada	0.22	Austria	0.25	Korea	0.18	Taiwan	0.15
20	Philippines	0.22	Norway	0.24	Switzerland	0.17	*China*	0.13

Source: OECD–WTO Trade-in-value-added (TiVA) database, May 2013.

can be seen in the two Asian newly industrializing economies (NIEs): Taiwan and Korea.

FVA in gross exports of a country reflects the total value added created in other countries that enters the exports of a country. This measure is considered a better indicator than the 'import content of exports' measure. Banga (2013) highlighted three aspects. First, FVA will not double-count, as it includes FVA in all inputs of the products only once and the number of times the inputs cross borders will not affect its calculation. It also includes the services component that enters the value addition. Second, in the GVCs, the FVA will contain not just the FVA content in bilateral trade, but also FVA included in exports of the country's bilateral trading partner. For example, if Singapore imports intermediate products from Malaysia, FVA content in Singapore's exports will be the sum of value added created in Malaysia, as well as value added created in other countries from where Malaysia imported its inputs for producing its intermediate product. It therefore includes all direct imports and indirect imports (from countries where there is no direct trade). This can have important implications for the bilateral trade balance. Third, the re-imported domestic value added will be netted out.

Table 7.3 reports FVA in gross exports (per cent) in 21 selected countries. The Asian tiger economies, that is, Singapore, Taiwan, and Korea, had relatively

Table 7.3 Total FVA in gross exports (%), 1995–2009

	2009	*2005*	*2000*	*1995*
Singapore	49.9	52.4	50.7	46.7
Taiwan	41.5	42.2	35.4	35.8
Korea	40.6	37.7	32.9	23.7
Philippines	38.4	45.6	45.9	30.9
Malaysia	37.9	41.5	43.0	40.3
Viet Nam	36.7	35.0	29.6	24.4
Thailand	34.5	38.5	34.8	29.9
Cambodia	34.1	37.9	34.5	26.0
China	32.6	36.4	18.8	11.9
Denmark	32.0	32.0	26.2	30.1
Hong Kong	28.5	28.3	32.6	40.6
Italy	20.1	27.1	25.3	21.9
New Zealand	18.4	19.6	20.2	17.4
United Kingdom	17.3	20.3	18.4	20.7
Japan	14.8	13.8	9.9	6.9
Indonesia	14.4	17.8	19.3	14.7
Australia	12.5	13.0	13.5	11.8
Brunei Darussalam	11.3	6.8	10.4	17.1
United States	11.3	11.1	8.9	8.4
Brazil	9.0	13.0	11.5	9.7
Russian Federation	6.9	8.2	12.5	10.7

Source: OECD–WTO Trade-in-value-added (TiVA) database, May 2013.

high shares of FVA in gross exports. Indeed, other than Luxembourg, Singapore had the highest share in 2009. Korea and Taiwan were ranked fifth and sixth, respectively, in the same year. Other Asian economies, such as the Philippines, Malaysia, Thailand, and Viet Nam, have also experienced a steady rise in their share of FVA in gross exports, whereas there has been a steady decline in the case of Hong Kong.

Shares of FVA in gross exports for most developed countries have remained at less than 30 per cent, and those of Japan, the United States, the United Kingdom, and Australia have remained at less than 20 per cent. This brings us to the question of whether FVA in gross exports is an appropriate indicator for measuring the extent of a country's participation in GVCs. We examine this issue in detail in the next section.

Participation in GVCs confers considerable benefits. It may allow suppliers in developing countries to meet standards and regulations that give them access to developed country markets; it may allow imports under privileged tariff treatment for intra-firm trade; it may permit the utilization of network technology that would not otherwise be available; and it may open up new sources of capital. The OECD–WTO database provides an indicator (*participation index*) that measures the extent of participation in GVCs. The measure is based on the share of exports involved in a vertically fragmented production process (Hummels et al. 2001; Koopman et al. 2010). The index is expressed as a percentage of gross exports and indicates the share of foreign inputs (backward participation) and domestically produced inputs used in third countries' exports (forward participation).

The participation index at the country level is represented in Figure 7.4 for selected countries. Open economies such as Singapore and Taiwan are amongst the top 10 economies with a high participation index. Small open economies such as Singapore source more inputs from abroad in GVCs than large countries, such as the United States or Japan. The forward and backward participation indices, which sum to give the participation index, offer additional information on the type of participation. Singapore and Taiwan have the same score in the participation index. However, Singapore's backward participation is larger than that of Taiwan, reflecting Singapore's relatively greater reliance on foreign inputs and value added to support its export activities. If gains are measured in terms of 'net value added' by participation in GVCs, the higher the forward linkages compared with backward linkages, the higher the gains. This would imply that by its participation in GVCs, a country is creating and exporting more domestic value added than the FVA that it is importing. The

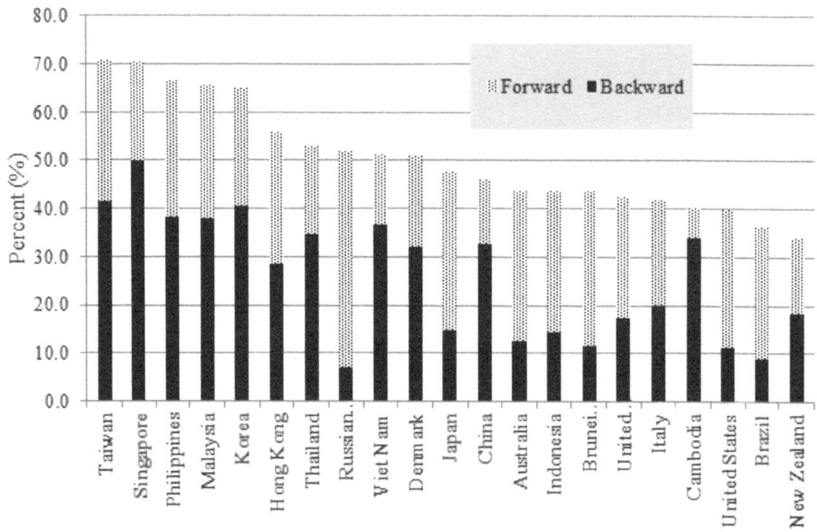

Figure 7.4 GVC participation index for selected economies, 2009

Note: Foreign inputs (backward participation) and domestically produced inputs used in third countries' exports (forward participation), as a share of gross exports (%).

Source: OECD–WTO Trade-in-value-added (TiVA), May 2013.

ratio of forward to backward participation, therefore, can be a measure of the extent of net gains. From this perspective, the Taiwanese economy has a ratio that is higher than Singapore, and hence enjoyed a higher gain in participation in GVCs in 2009.

Another popular indicator of participation in GVCs is the index of the number of production stages. The length of production (LoP) index takes a value of one if there is a single production stage in the final industry, and its value increases when inputs from the same or other industries are used.[5] In Table 7.4, we present the LoP index together with the participation index for Singapore and other top 10 countries shown in Figure 7.4. The LoP indices shown correspond to the highest scores achieved during the period 2005–2009. As would be expected, the number of production stages can be more than three in the case of the transport equipment sector, which includes the automobile manufacturing industry. Other sectors that have a high LoP index include basic metal and fabricated metal product industries, food products and beverages, and construction services industries.

Table 7.4 Participation index in 2009 and highest length index, 2005–2009

Country	*Participation index*	*Length of production index*	*Industry associated with the length of production index*
Taiwan	70.99	3.192	Basic metals and fabricated metal products
Singapore	70.66	3.293	Construction
Philippines	66.65	2.801	Transport equipment
Malaysia	65.57	2.826	Food products and beverages
Korea	65.03	3.370	Basic metals and fabricated metal products
Hong Kong	55.79	3.119	Construction
Thailand	52.82	2.619	Wood, paper, paper products, printing and publishing
Russian Federation	51.83	2.328	Transport equipment
Viet Nam	51.35	3.239	Basic metals and fabricated metal products
Denmark	50.98	2.482	Food products and beverages
Japan	47.75	3.091	Transport equipment
China	46.06	3.543	Transport equipment
Australia	43.81	2.542	Transport equipment

Source: OECD–WTO Trade-in-value-added (TiVA) database, May 2013.

3.2 Trade in value added and Singapore's exports

The composition of Singapore's domestic exports of merchandise and services is shown in Table 7.5. Exports of merchandise constitute about 60 per cent of total exports of goods and services. The major merchandise export items are mineral fuels and lubricants, electronics and equipment, machine and equipment, and

Table 7.5 Export of merchandise and services in 2009 and 2013

	2009 *S$ m*	*2013* *S$ m*	*2009* *%*	*2013* *%*
Total Export (Merchandise and Services)	**319,023**	**445,996**	**100.0**	**100.0**
Domestic Exports – Merchandise	**200,003**	**274,192**	62.7	61.5
Food	3,138	4,948	1.0	1.1
Beverages and tobacco	403	579	0.1	0.1
Crude materials	1,115	1,798	0.3	0.4
Mineral fuels and lubricants	58,655	106,476	18.4	23.9
Animal and vegetable oils	246	228	0.1	0.1
Chemicals and chemical products	36,821	46,397	11.5	10.4
Manufactured goods	5,427	6,243	1.7	1.4
Electronics and equipment	51,404	48,872	16.1	11.0
Machine and equipment (non-electronics)	24,062	29,992	7.5	6.7
Miscellaneous manufactures	15,484	25,349	4.9	5.7
Miscellaneous	3,248	3,310	1.0	0.7
Services Exports	**119,020**	**171,803**	37.3	38.5
Transport	43,552	56,041	13.7	12.6
Travel	13,418	24,151	4.2	5.4
Telecommunication	3,804	6,117	1.2	1.4
Construction	1,545	2,198	0.5	0.5
Finance and insurance	19,447	27,683	6.1	6.2
Business services	25,916	40,520	8.1	9.1
Others	11,340	15,093	3.6	3.4

Source: International Enterprise Singapore and Department of Statistics.

chemical products. In terms of services, transportation services, business services, and financial services are the key sectors.

The main export markets for Singapore are shown in Table 7.6. The ASEAN countries absorb more than a quarter of Singapore's domestic exports of merchandise. Malaysia and Indonesia account for more than 70 per cent of Singapore's merchandise exports to ASEAN. China is the second-largest export market, followed closely by the European Union (EU)–27. In the case of services export, ASEAN is an important market but has a smaller share than that of the United States or the EU–27. All other destinations in the world market take up a larger share of services exports than that of merchandise exports.

Figures 7.5a and 7.5b depict the gross exports of nine commodities[6] in 2009 and 1995, respectively. The gross exports are decomposed into direct domestic value added, indirect domestic value added, and FVA. The reimported domestic value added is miniscule, ranging from 0.1 to 0.3 per cent of gross exports, and is omitted from the charts.

We can see from the charts that the share of FVA increased between 1995 and 2009 for every commodity except chemical and non-metallic mineral products. Generally, the increase is matched by a decrease in the share of indirect domestic value added. This is another indication of rising involvement in outsourcing activities and increasing participation in GVCs and is corroborated by the comparison of foreign value-added export ratios (FVAX) in Figure 7.6.

The shares of FVA in gross exports (FVAX) for each of the commodities reported in the TiVA database are shown in Figure 7.6 for 1995 and 2009. The

Table 7.6 Major export markets for Singapore

	Domestic export of goods				*Export of services*			
	2009 S$ m	*2013 S$ m*	*2009 %*	*2013 %*	*2009 S$ m*	*2013P S$ m*	*2009 %*	*2013 %*
Total	200,003	274,192	100.0	100.0	119,020	171,803	100.0	100.0
ASEAN	48,232	76,768	24.1	28.0	12,483	17,202	10.5	10.0
Indonesia	*13,462*	*22,964*	*6.7*	*8.4*	*3,821*	*4,513*	*3.2*	*2.6*
Malaysia	*18,923*	*31,474*	*9.5*	*11.5*	*4,096*	*4,758*	*3.4*	*2.8*
China	18,026	30,568	9.0	11.1	5,706	8,825	4.8	5.1
Hong Kong	20,781	25,863	10.4	9.4	4,371	5,029	3.7	2.9
Taiwan	6,997	11,222	3.5	4.1	2,054	2,182	1.7	1.3
Korea	6,882	8,785	3.4	3.2	2,303	2,673	1.9	1.6
Japan	9,677	10,614	4.8	3.9	5,629	8,495	4.7	4.9
US	15,755	17,330	7.9	6.3	13,628	20,353	11.5	11.8
EU–27	24,841	25,397	12.4	9.3	19,471	22,932	16.4	13.3
Others	48,811	67,645	24.4	24.7	53,375	84,114	44.8	49.0

Source: Yearbook of Statistics 2013, Singapore Department of Statistics.

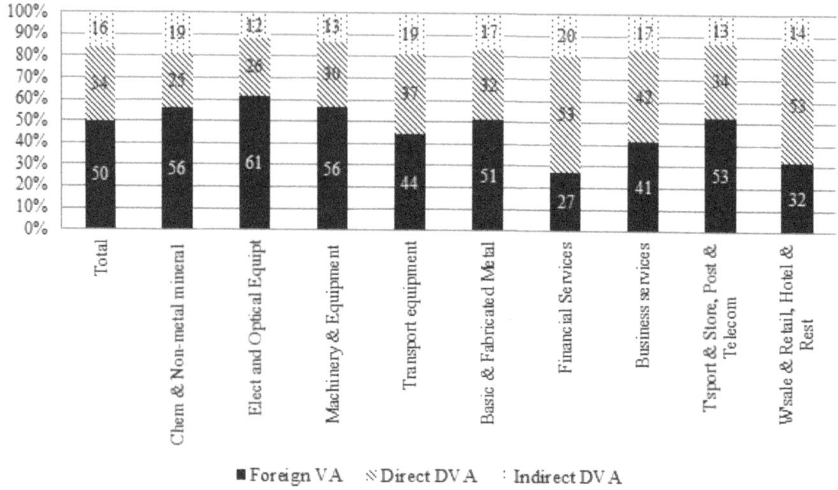

Figure 7.5a Composition of gross exports in 2009

Source: OECD WTO Trade-in-value-added (TIVA), May 2013.

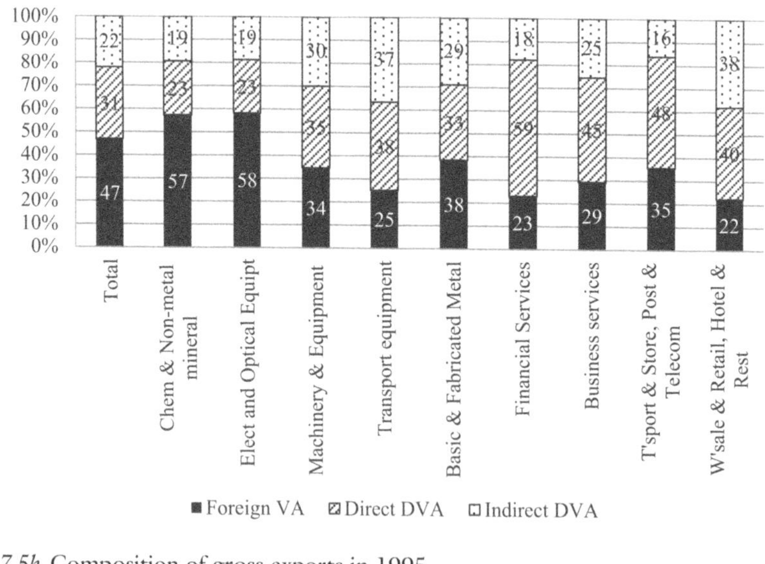

Figure 7.5b Composition of gross exports in 1995

Source: OECD WTO Trade-in-value-added (TIVA), May 2013.

commodities are arranged in descending order according to the difference of the FVAX in the two years.

Every commodity, except for chemicals and non-metallic mineral products, had a larger FVA share in 2009 relative to that in 1995. The machinery and equipment sector had the largest increase in the FVA ratio. The electrical and optical equipment sector, which includes electronic peripheral and components and already had a high FVA share of 58 per cent in 1995, managed to increase its share to 61 per cent in 2009.

Of interest is to determine those countries/regions that are contributing to FVA in gross exports. The tables in the TiVA database enable that information to be extracted and this is tabulated in Table 7.7. In 2009, half of the value added embodied in the aggregate exports was foreign. The EU–27 contributed 11.4 per cent of the total VA, 8.3 per cent of total VA originated from the United States, and 5 per cent from the Association of Southeast Asian Nations (ASEAN) region.

Electronic products are included in the category of electrical and optical equipment. The export of commodities from this category had the lowest share of domestic value-added amongst all the categories listed in the table. Slightly more than 60 per cent of the value added originated from abroad. The EU–27, the United States, and the NIEs contributed between 10 per cent and 13 per cent each to the total VA embodied in gross exports of this category.

As expected, exports from the services sector had higher domestic value-added content than the non-services sectors. In particular, the financial services sector had the highest domestic VA share, of 73 per cent, and the bulk of its foreign VA originated from the EU–27 region.

3.3 Services export and service value added in exports

The share of services exports in total exports rose from 37.3 per cent in 2009 to 38.5 per cent in 2013 (Table 7.5). Based on TiVA database information for Singapore, the value-added contribution of services exports to the economy increased considerably between 2000 and 2009, rising from 24 per cent of GDP to 31 per cent of GDP.

According to the OECD research brief,[7] accounting for the value added by services in the production of goods shows that the services sector in 2009 contributed more than 50 per cent of total exports in the United States, the United Kingdom, France, Germany, and Italy, and nearly one-third in China, with a significant contribution (typically one-third in 2009) across all manufactured goods provided by both foreign and domestic service providers, with the contribution rising between 5 per cent and 10 per cent in many countries since 1995.

In Table 7.8, the services content of Singapore's gross exports and the sources of services value added are extracted from the information available in the TiVA database.

The services VA content of Singapore gross exports increased from 49 per cent in 1995 to 57 per cent in 2009. The bulk of the services VA originated from

Table 7.7 Sources of value added for Singapore sectoral gross exports by country/region in 2009

	Total	*Singapore*	*ASEAN-S*	*China*	*NIEs*	*Japan*	*US*	*EU–27*	*Row*
Sectoral Commodities									
Total	**212,449**	**106,401**	**10,687**	**6,324**	**7,991**	**8,246**	**18,243**	**24,242**	**30,315**
Chemicals and non-metal mineral products	54,972	24,373	3,134	1,124	994	1,430	4,101	5,220	14,596
Electrical and optical equipment	44,867	17,392	3,969	2,733	4,390	2,129	4,701	5,710	3,843
Machinery and equipment	10,897	4,745	510	385	290	696	1,586	1,603	1,083
Transport equipment	7,152	3,976	220	157	149	239	1,314	635	462
Basic and fabricated metal product	3,890	1,913	345	143	119	249	203	321	598
Financial services	12,775	9,291	78	121	118	124	1,027	1,559	457
Business services	11,531	6,764	242	308	216	286	992	1,786	937
Transport and store, post and telecom	30,890	14,639	1,032	664	1,016	1,906	2,358	4,161	5,115
Wholesale and retail, hotel and restaurant	28,805	19,579	511	438	544	957	1,558	2,696	2,521
Others OEC	6,670	3,730	648	251	155	231	402	551	702
Percentage Distribution									
Total	**100.0**	**50.1**	**5.0**	**3.0**	**3.8**	**3.9**	**8.6**	**11.4**	**14.3**
Chemicals and non-metal mineral product	100.0	44.3	5.7	2.0	1.8	2.6	7.5	9.5	26.6
Electrical and optical equipment	100.0	38.8	8.8	6.1	9.8	4.7	10.5	12.7	8.6
Machinery and equipment	100.0	43.5	4.7	3.5	2.7	6.4	14.6	14.7	9.9
Transport equipment	100.0	55.6	3.1	2.2	2.1	3.3	18.4	8.9	6.5
Basic and fabricated metal product	100.0	49.2	8.9	3.7	3.1	6.4	5.2	8.2	15.4
Financial services	100.0	72.7	0.6	0.9	0.9	1.0	8.0	12.2	3.6
Business services	100.0	58.7	2.1	2.7	1.9	2.5	8.6	15.5	8.1
Transport and store, post and telecom	100.0	47.4	3.3	2.2	3.3	6.2	7.6	13.5	16.6
Wholesale and retail, hotel and restaurant	100.0	68.0	1.8	1.5	1.9	3.3	5.4	9.4	8.8
Others	100.0	55.9	9.7	3.8	2.3	3.5	6.0	8.3	10.5

Source: OECD–WTO Trade-in-value-added (TiVA) database, May 2013.

Table 7.8 Services value added embodied in gross exports by source country, as percentage of gross exports

	2009	*2005*	*2000*	*1995*
Total	**56.5**	**51.1**	**47.6**	**49.3**
Singapore	30.0	25.7	24.9	30.2
ASEAN	1.3	1.1	1.6	2.0
China	1.0	0.6	0.4	0.6
NIEs	1.8	1.7	1.7	1.4
Japan	2.3	2.3	2.5	3.3
US	5.6	6.0	5.7	3.4
EU–27	8.9	9.2	6.9	6.1
Others	5.6	4.6	4.0	2.3

Source: OECD–WTO Trade-in-value-added (TiVA) database, May 3013.

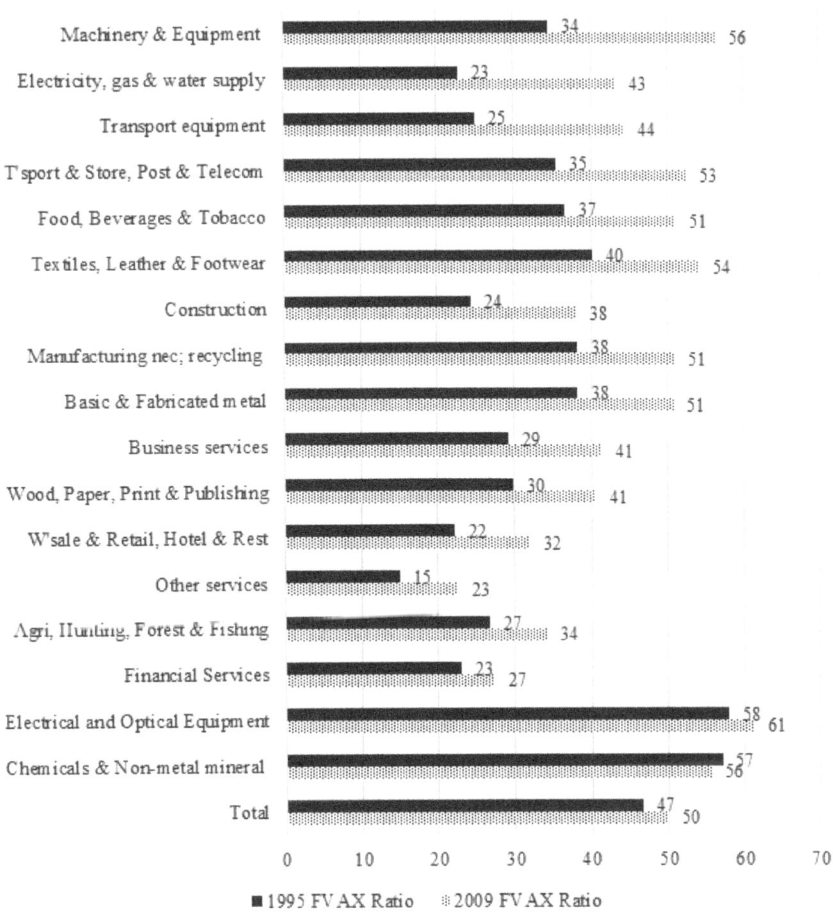

Figure 7.6 Foreign value-added share in gross exports, 1995 vs. 2009

Source: OECD–WTO Trade-in-value-added (TiVA) database, May 2013.

Singapore, whilst the triad, namely, the EU–27, the United States, and Japan, are the main contributors of the foreign services VA content in Singapore's gross exports.

The service VA content by sectors is shown in Figure 7.7. Amongst the exports of goods, the average service VA content was 34 per cent in 2009. In the electrical and optical equipment sector, the services content of exports was more than 35 per cent. This could possibly reflect the increased knowledge intensity (e.g., design, R&D, software) of electrical and optical equipment.

The importance of services exports to Singapore's economy is expected to continue to increase, supported by three broad trends.[8] First, the demarcation between manufacturing and services is becoming more obscure. Increasing the services content in commoditized manufacturing products has become a way to maintain product differentiation and competitiveness. A well-known example is the Rolls Royce company, famous for its aircraft engine manufacturing, which now earns more than 60 per cent of its revenue by undertaking a wide spectrum of activities, including R&D, testing, repair, overhaul (MRO), and overall services and parts management in many countries, including Singapore. Second, the trend towards fragmentation is expanding. This will engender more business for services relating to transport and logistics, financing, legal services, business consultancy, and management. Singapore is well placed and well equipped to provide such services. Third, the successful implementation of the various regional trade

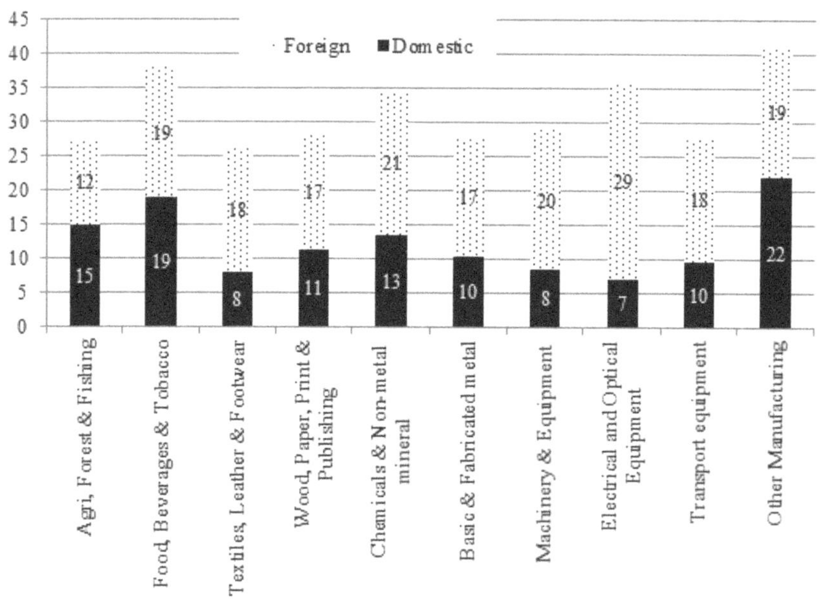

Figure 7.7 Service VA content of export of goods, 2009

Source: OECD–WTO Trade-in-value-added (TiVA) database, May 3013.

agreements – such as the Trans-Pacific Partnership (TPP), the Regional Comprehensive Economic Partnership (RCEP), the ASEAN Free Trade Area (AFTA), and the ASEAN Framework Agreement on Services (AFAS) – will provide greater opportunities to boost trade in services in tandem with rising incomes and the removal of barriers to the movement of goods.

4. Conclusion

What new insights and implications do the analysis of trade in terms of value added have on Singapore development policies and strategies? In fact, the studies done by researchers at OECD and other leading research institutions generally affirm Singapore has been following the right approach in recognizing the importance and benefits that can be derived from active participation in the global value chain and embedding itself into the global production network (GPN). Singapore has provided an example that success in pursuing a strategy of trade-led economic growth is translated into successful trading in integrated global markets and upgrading within the GVCs.

In the pursuit of upgrading, it is not confined to simply moving up the value chain, shifting to upstream activities to create and capture more value. Singapore practitioners recognize that it is also about strengthening technological capabilities such that process upgrading, product upgrading, functional upgrading (e.g., from production to design and research activities), and chain upgrading (e.g., shifting to other higher value chains) can be achieved. To sustain Singapore as an important node in the GPN, resources have to be invested in R&D and encourage innovation and entrepreneurialism. Recently, the Singapore government has announced the commitment of S$19 billion over a five-year period (2016–2020) to boost research, innovation, and enterprise and to sustain the competitive edge of the economy.

The focus on value added in trade and export has deepened the understanding on how the economy has contributed to value creation. It has provided a better understanding of the direct link between trade, income, and jobs. It delineates more lucidly the specific functions performed by the workers and enterprises, rather than by sectors, in generating the values embedded in goods and services produced in the GVCs and traded across borders. GVCs approach has helped to bring gain to Singapore in terms of improved competitiveness, and better access to global markets and expansion of production and jobs. It has also helped Singapore to increase productivity and avoid the middle-income trap. Economies in a middle-income trap cannot compete with low-income economies that retain low labour-cost advantages. At the same time, they are not yet able to compete effectively with advanced economies in high-tech products and services. Moving up the value chain or upgrading provides a remedy to avoid the trap (ADB 2010).

Singapore recognized the importance of GVCs and GPNs at an early stage in its development. Instead of blanket subsidies for exports and FDI, efforts were made to attract MNCs to produce key inputs or to bring specific knowledge needed by clusters with the ability to absorb them. Without policies to develop local capabilities, MNC-led exports will likely remain technologically stagnant, leaving developing countries unable to progress beyond the assembly of imported components (Chandra and Kolavalli 2006)

The new trade-in-value-added (TiVA) database launched by the OECD and the WTO in January 2013 reveals that services play a far more important role in global trade than suggested by the standard measurement of gross flows of exports and imports. The value created by services as intermediate inputs represents, on average, more than 30 per cent of the total value added in manufactured goods. Liberalization of the services trade would allow for more efficient and higher-quality services, thus enhancing the competitiveness of manufacturing firms and allowing them to better participate in global production networks. In this regard, Singapore concerted efforts in promoting trade liberalization in services relating to transport, logistics, finance, and communication, via the regional and bilateral FTAs are steps in the right direction. It is an integral part of the strategy in transforming Singapore into a knowledge-based services economy. Its prospects are not diminished with expansion and proliferation of GVCs.

The decomposition of the trade (export) into domestic and foreign value-added should not evolve into a debate between analysts advocating one type of value-added export over the other. Increasing the share of foreign value added is considered by some analysts as a favourable deepening participation in GVC. However, some consider that development may be at the expense of diminished competitiveness of local industries and industrial development. There could be no optimal share that is applicable at all times. Adverse impacts from participation in the GVCs can be mitigated by having adequate absorptive capacity in terms of skilled manpower, entrepreneurial capabilities, and infrastructure and regulatory governance. From the perspective of Singapore's experience, it is better to consider foreign value added as complementary rather than a substitute to domestic value added. Local and foreign assets and resources work in concert to produce goods and service that consumers in the world demand.

The Singapore government has been actively involved in prodding and pushing enterprises to participate in GVCs via cluster-based development policies. Though the latter is undeniably a form of industrial policy, but is generally regarded as a pro-market intervention grounded on informational externalities (Hausmann and Rodrik 2003), it is now widely accepted. However, risk of failure is not eliminated. The risk can be mitigated by close surveillance of trends and performance, together with flexibility and courage to modify and change policies along the way.

Singapore has adapted its policies to meet the needs of international investors and has proved able to retain their presence in the economy to generate employment and income. Trade and investment with active participation in GVCs will remain the key pillars to sustain Singapore's economy for a long time to come.

Notes

1 The value chain describes the full range of activities that firms and workers do to bring a product from its conception to its end use and beyond. This includes activities such as design, production, marketing, distribution, and support to the final consumer. The activities are often carried out in different parts of the world, hence the term *global value chain*. The fragmentation of production along GVCs takes the form of global production networks (GPNs). More detail discussion of the concepts can be found in Kaplinsky (2000) and Wood (2001).

2 A recent special issue of the *Economic Systems Research* provides a very useful and detailed description of several global multiregional I/O databases currently available (see Tukker and Dietzenbacher 2013).

3 The first studies of the measurement of the value added of trade in an international I/O framework were those of Daudin et al. (2011), Johnson and Noguera (2012), and Koopman et al. (2012), using the Global Trade Analysis Project (GTAP) database.

4 Description of the database and the methodologies used in the computation of the various TiVA indicators are available in Ahmad (2013) and Backer and Miroudot (2013).

5 The formula for computation is found in Backer and Miroudot (2013, p. 45), and is included in the OECD–WTO TiVA database.

6 The nine export commodities constituted more than 97 per cent of gross exports in 2009.

7 OECD (2013), measuring trade-in-value-added.

8 Ministry of Trade and Industry, Singapore: Economic Survey of Singapore, Second Quarter 2014.

References

Ahmad, N 2013, 'Measuring trade in value added and beyond', paper presented at the *Conference on Measuring the Effects of Globalization*, Washington, DC. Available from www.upjohn.org/MEG/papers/Ahmad_Trade_in_VA.pdf [20 August 2014].

Asian Development Bank (ADB) 2010, *Asian Development Bank Outlook 2010 Update: The Future of Growth in Asia*, ADB, Manila.

Backer, KD and Miroudot, S 2013, 'Mapping global value chains', *OECD Trade Policy Papers*, No. 159, OECD, Paris. Available from: www.oecd-ilibrary.org/trade/mapping-global-value-chains_5k3v1trgnbr4-en [20 August 2014].

Banga, R 2013, 'Measuring value in global value chains', *UNCTAD Regional Value Chain Background Paper* No. RVC-8, UNCTAD, Geneva, Switzerland.

Chandra, V and Kolavalli, S 2006, 'Technology, Adaptation, and Exports: How Some Countries Got It Right', in *Technology, Adaptation, and Exports: How Some Countries Got It Right*, ed. V Chandra, The World Bank, Washington, DC, pp. 1–48.

Daudin, G, Rifflart, C, and Schweisguth, D 2011, 'Who produces for whom in the world economy?', *Canadian Journal of Economics*, vol. 44, no. 4, pp. 1403–1437.

Hausmann, R and Rodrik, D 2003, 'Economic development as self-discovery', *Journal of Development Economics*, vol. 72, pp. 603–633.

Hummels, D, Ishii, J, and Yi, K-M 2001, 'The nature and growth of vertical specialization in world trade', *Journal of International Economics*, vol. 54, no. 1, pp. 75–96.

Johnson, RC and Noguera, G 2012, 'Accounting for intermediates: production sharing and trade in value added', *Journal of International Economics*, vol. 86, no. 2, pp. 224–236.

Kaplinsky, R 2000, 'Globalization and unequalisation: what can be learned from value chain analysis?', *Journal of Development Studies*, vol. 37, no. 2, pp. 117–146.

Koopman, R, Wang, Z, and Wei, S-J 2012, 'Estimating domestic content in exports when processing trade is pervasive', *Journal of Development Economics*, vol. 99, no. 1, pp. 178–189.

Leontief, WW 1951, 'Input–output economics,' *Scientific American* (October), pp. 15–21.

Ministry of Trade and Industry 2014, 'Rising contribution of services exports to the Singapore economy', *Economic Survey of Singapore*, Second Quarter 2014.

OECD 2013, *Measuring Trade in Value Added: An OECD–WTO Joint Initiative.* Available from: www.oecd.org/sti/ind/measuringtradeinvalue-addedanoecd-wto-jointinitiative.htm.

OECD and WTO 2013, *OECD–WTO Database on Trade in Value-Added First Estimates: 16 January 2013*, OECD and WTO, Paris.

Tukker, A and Dietzenbacher, E 2013, 'Global multiregional input–output frameworks: an introduction and outlook', *Economic Systems Research*, vol. 25, no. 1, pp. 1–19.

UNCTAD 2013, *World Investment Report: Global Value Chains: Investment and Trade for Development*, United Nations Conference on Trade and Development. Geneva, Switzerland. Available from: http://unctad.org/en/publicationslibrary/wir2013_en.pdf.

Wood, A 2001, 'Value chains: an economist's perspective', *IDS Bulletin*, vol. 32, no. 3, pp. 41–5.

8 Thai automotive industry

International trade, production networks, and technological capability development

Patarapong Intarakumnerd[1]

1. An overview of the automotive industry in Thailand

The automotive industry in Thailand started in the early 1960s under an import substitution policy and a revision of the investment promotion law to encourage automotive assembly. The automotive industry has contributed significantly and increasingly to the Thai economy in terms of value added and employment, especially since 2001. Data from the National Economic and Social Development Board indicate that the auto industry employed 310,000 persons and accounted for 7 per cent of the country's total value added in 2011.

Currently, firms in the industry can be classified into three groups: 17 car assemblers, 648 first-tier suppliers, and approximately 1,700 second- and third-tier suppliers, including supporting companies. Most of these are small and medium-sized companies (Figure 8.1). Most assemblers are subsidiaries of transnational corporations (TNCs). They are dominated by Japanese TNCs as well as the big three car companies in the United States, namely, DaimlerChrysler, General Motors, and Ford, whose prime objective is to produce and export one-ton pickups from Thailand. Due to a sufficient pool of qualified engineers and technicians and an extensive supplier network enabling integrated production, Thailand is clearly the strongest automotive production base in Southeast Asia.

Nonetheless, indigenous Thai suppliers are mainly involved in 'non-functional' parts such as body parts, accessories, and other similar parts while foreign suppliers are focused on 'functional' parts that require higher manufacturing and design capabilities to produce, such as engine, electrical transmission, and suspension parts (Table 8.1).

The outline of this chapter is as follows. Section 2 reviews the literature on the Thai automotive industry. Section 3 analyses trade in value added and the production network of the Thai automotive industry. Section 4 investigates the industry by using the concept of the sectoral innovation system. Section 5 provides concluding remarks.

2. Literature review on the Thai automotive industry

There have been several studies on the industry from different perspectives, including those concerning intra-industry trade (IIT), international product

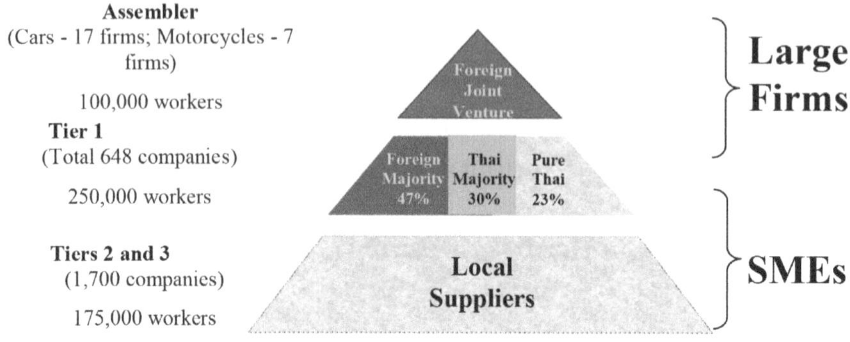

Figure 8.1 Structure of manufacturers in the automotive industry in Thailand
Source: Thai Automotive Institute (TAI) (2014).

Table 8.1 Number of automotive suppliers classified by part

Group of part	*Thai*	*Thai majority*	*Foreign majority*	*Total*
Engine parts	20	8	35	63
Electrical parts	15	10	27	52
Drive / Transmission	17	6	29	52
Suspension / Brake	13	1	21	35
Body parts	57	17	47	119
Accessories	18	2	19	39
Others	214	24	111	349
Total	**354**	**68**	**287**	**709**

Source: Thai Automotive Association (2011).

fragmentation, and technological capability development. Techakanont (2011) discussed the experience of Thailand's development in the auto parts industry and intra-regional trade integration. Although the main focus was on Thailand's auto parts industry, the paper also touched on the integration of Thailand with the Association of Southeast Asian Nations (ASEAN) and other Asian countries. Product fragmentation increased in the past decades as shown by the expansion of trade in parts and components. The Thai government, in particular, had clear policies and strategies and has always been flexible and aligned with the interests of multinational firms as will be illustrated in Section 4. Japanese assemblers were crucial in bridging productive resources among countries via their international production network.

Regarding the roles of TNCs and their spillover impacts on the upgrading of the technological capabilities of local firms, Intarakumnerd and Charoenporn (2010) reported that since 1986, foreign direct investment (FDI) has played an

Table 8.2 Thailand's automotive industry trade, by Country (value in millions of US$, share in %), 1992–2011

Country/ Region	*Exports*				*Imports*			
	1992	*1996*	*2002*	*2011*	*1992*	*1996*	*2002*	*2011*
ASEAN	5.42	20.14	14.73	24.42	0.54	1.73	10.05	11.88
Indonesia	0.70	2.20	5.06	10.67	0.06	0.13	2.82	5.89
Malaysia	2.30	3.85	4.63	6.32	0.16	0.48	2.19	1.86
Philippines	1.14	4.34	2.52	3.84	0.32	1.12	5.00	3.33
Viet Nam	1.28	9.75	2.52	3.59	0.00	0.00	0.05	0.80
ASEAN+3	18.31	29.29	28.25	34.29	78.99	79.64	76.83	81.25
Japan	11.85	8.03	12.08	8.25	76.54	75.65	64.02	58.98
China	0.37	0.50	0.92	1.24	0.37	0.51	1.65	7.02
Korea	0.67	0.62	0.52	0.38	1.54	1.76	1.10	3.36
ASEAN+6	21.09	32.39	38.55	47.85	79.24	80.26	77.23	83.86
India	0.03	0.09	1.11	2.75	0.08	0.10	0.15	2.31
Australia	2.74	3.02	9.19	10.82	0.17	0.51	0.25	0.31
World	100.00	100.00	100.00	100.00	100.00	100.00	100.00	100.00
US$ million	468.00	1,239.04	4,305.95	27,413.44	3,170.63	6,131.11	3,554.11	13,452.74

Source: Authors' own calculations based on UN Comtrade database.

important role in Thailand's industrialisation. Before that, compared with investment in neighbouring countries like Singapore and Malaysia, private investment in Thailand was based more on domestic resource mobilisation and less on FDI. As a percentage of gross domestic product, FDI was 0.2 per cent in 1978, and it increased to 6.54 per cent in 1998. Furthermore, foreign investments in the manufacturing industry during that period mainly took the form of joint ventures with an emphasis on assembly of final goods for the domestic market, whereas both capital-goods industries and parts-producing industries remained relatively underdeveloped. From 1986 onwards, FDI became more significant and Thailand has almost doubled its share of net foreign investments in ASEAN. One major change during this period was that a significant amount of FDI went to export industries and to industries that provide intermediate inputs such as electronic parts and components. As a result, Thailand's export performance looked quite impressive during the 1986–1996 period, before a major economic crisis hit in 1997 (Lauridsen 2004).

Similar to the experience of East Asian newly industrialised economies, there was also a significant change in the structure of Thai manufacturing exports. The share of once-dominant resource-based and labour-intensive exports has gone down while that of science-based and differentiated exports has gone up, especially in the 1990s. A range of new 'high-tech' products (especially computers and parts, electrical equipment, and integrated products) emerged as key export goods. However, unlike East Asian newly industrialised economies, much of Thailand's high-tech export was, in reality, manufactured through rather simple, labour-intensive assembly of high-tech components imported from advanced industrialised countries and East Asian newly industrialised economies. As a result, in terms of level of technological capability, there was no remarkable movement away from the period of resource-based, labour-intensive, light- and final-goods-assembly industries before 1986.

A powerful explanation for this is that, whereas FDI in Thailand has been significant in terms of volume, and the role of TNCs in the technological development of local firms and subsidiaries of TNCs is supposed to be substantial, the links for technological development between TNCs and their subsidiaries and Thai suppliers, in general, were rather limited and trivial (Sribunruang 1986; Kaosa-Ard 1991; Intarakumnerd et al. 2002). Although foreign investment brings in product management and process technology, the transfer of technology has tended to be limited to the operational level. There was no design- or product-specific technology transfer. Machinery was imported with minimal instructions on operational procedures given by suppliers, resulting in inefficiently operated and inadequately maintained equipment. TNCs themselves were not active in developing subcontractors or giving technical assistance to local suppliers. This was due to the inefficiency and backwardness of local supporting industries. Equally important, TNCs lacked the willingness and effort to devote resources and time to upgrading the capability of local suppliers (Dahlman et al. 1991). The weakness in knowledge linkages between local suppliers and TNCs is noted in the comprehensive studies of Dahlman and Brimble (1990) and Kaosa-Ard (1991).

As a result, there was insufficient transfer of technology at higher levels, such as design and engineering. TNCs in Thailand made minimal investment in research and development (R&D). In 2000, only 39 TNCs, or 1.4 per cent of the total number of TNCs operating in Thailand in the manufacturing sector, established R&D centres in Thailand. Most of the R&D centres of these TNCs were small and involved primarily in adapting their products to local market needs. Nonetheless, the situation has started to change in the past few years. There have been some indications of investment made by TNCs in higher-technology activities (e.g., R&D, engineering, and design) in certain sectors such as the automotive and hard disk drive industries (Intarakumnerd et al. 2002).

3 Analysis on trade in value added and production network

Thailand's automotive sector has become part of the global production network of many car manufacturers. Completely built-up vehicles and completely knocked-down kits are produced by locally based suppliers and have been a major export product since 2000. Automobile production in Thailand surpassed 1 million units in 2005 and reached a new record high of 1.6 million units in 2010. Annual production of one-ton pickup trucks also exceeded 1 million units for the first time in 2010. In 2011, domestic production and exports dropped because of two natural disasters: the tsunami in Japan and the floods in Thailand. Nevertheless, production and sales in Thailand recovered quickly. Domestic production reached 2.4 million units; sales, 1.4 million units, and exports, 1.02 million units in 2012 (Figure 8.2), before the figures dropped later due to a slump in domestic and exported demand. According to the International Organization of Motor Vehicle Manufacturers (the Organisation Internationale des Constructeurs d'Automobiles, or OICA), Thailand is the largest producer car in ASEAN, followed by Indonesia (1.2 million units) and Malaysia (600,000 units). This marks another important milestone for the Thai automotive industry. The success can be attributed to the strategic investments of all Japanese carmakers and the investment promotion by the Board of Investment of the small eco-car programme.

We used secondary data from the UN Comtrade spanning from 1992 to 2011 to calculate the extent of IIT, which is commonly measured by the Grubel–Lloyd Type Trade Decomposition, as follows:

$$IIT_{jkt} = \frac{\sum_{i=1}^{n}\left(X_{ijkt} + M_{ijkt}\right) - \sum_{i=1}^{n}\left|X_{ijkt} - M_{ijkt}\right|}{\sum_{i=1}^{n}\left(X_{ijkt} + M_{ijkt}\right)}$$

where X_{ijkt} and M_{ijkt} are Thailand's exports and imports of product i in industry j with country k at time t.

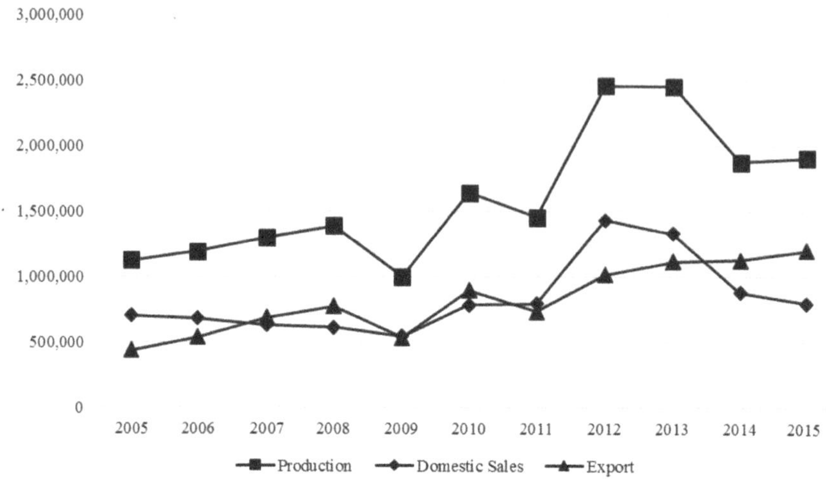

Figure 8.2 Thailand's production, sales, and exports (2005–2015)

Source: Automotive Intelligence Unit, Thailand Automotive Association (2015), Available at: http://data.thaiauto.or.th/iu3/.

We computed an aggregate Grubel–Lloyd index after which we summed up the export and import value of all commodities in a particular industry and calculated the Grubel–Lloyd index using this formula:

$$GL_{jkt} = 100\left\{1 - \frac{\left|\sum_{i=1}^{n} X_{ikt} - \sum_{i=1}^{n} M_{ikt}\right|}{\sum_{i=1}^{n} X_{ikt} + \sum_{i=1}^{n} M_{ikt}}\right\}$$

In this chapter, we attempted to examine the trade pattern and the IIT in the automotive industry in detail. This involved breaking down bilateral trade flows into two types: (a) inter-industry trade or one-way trade if $\frac{Min\left(X_{ijkt}, M_{ijkt}\right)}{Max\left(X_{ijkt}, M_{ijkt}\right)} \leq 0.1$ and (b) IIT for automobile products (or completely built-up vehicles) and auto parts, otherwise.

To be able to see the evolution of the trade pattern between Thailand and selected trade partners over the past two decades, we had to break down the bilateral trade flows in each detailed vehicles and parts into two types: (a) inter-industry trade (one-way trade) and (b) IIT. Based on data obtained from the UN Comtrade database, we found that the degree of IIT between Thailand and selected trade partners has been increasing. Figure 8.3 indicates that the degree of IIT in the industry lies in the auto parts trade. For automobile products

a. Automobile Products

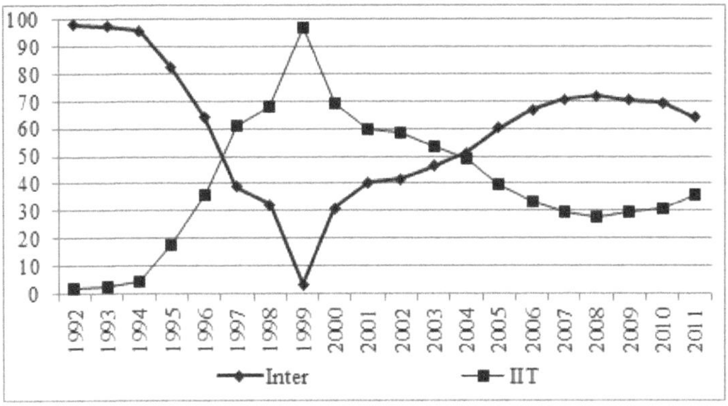

b. Auto Parts

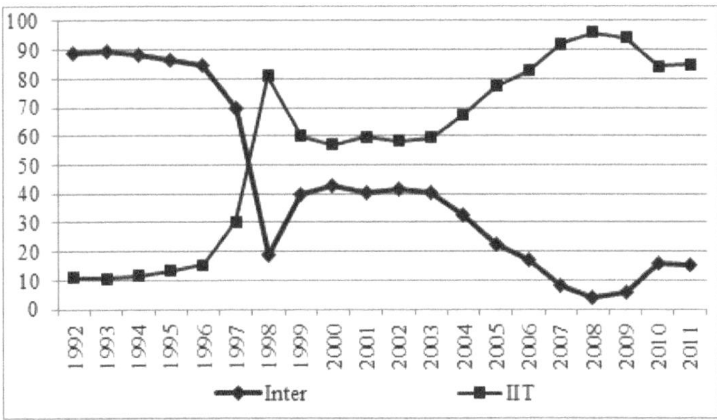

Figure 8.3 Development of intra-industry trade in Thailand's automotive Industry G–L Index (%), 2000–2011

a. Automobile Products
b. Auto Parts

Note: Thailand's automotive industry trade with selected countries, including Australia, China, India, Indonesia, Japan, the Republic of Korea, Malaysia, Philippines, and Viet Nam.

Source: Authors' own calculations.

(completely built-up units), we found that the IIT pattern was mainly one-way trade before 1997. However, during the recovery process from 1997 to 2000, IIT surged. This shift in trend was due to the fact that carmakers in Thailand were forced to utilise their enormous excess capacity, especially for incumbent firms that had previously focused only on the domestic market. Some orders were transferred from Japan to Thailand, resulting in the surge in IIT in 1999 (Poapongsakorn and Techakanont 2008; Techakanont and Charoenporn 2011).[2] Later, several carmakers decided to use Thailand as a production base for export, as indicated by a significant increase in export volume (Figure 8.2). The degree of IIT index in automobile products declined to 36 in 2011 whereas one-way trade has been increased. This is not surprising. Thailand's production consists mainly of one-ton pickup trucks and small to medium-sized passenger cars. The majority of the makers are Japanese. Hence, Thailand could be a major exporter of these products to corresponding trade partners, explaining the surge in one-way trade.

With regard to auto parts trade, we found that the degree of IIT increased significantly in 1996 and 2011. Similar to automobile products, the transition in inter- and intra-industry trade occurred during the financial crisis in 1997 when many subsidiaries of multinational corporations changed their status from joint ventures to wholly foreign-owned firms. As a result, they were more confident in increasing their production volume and using Thailand as their export hub. Product fragmentation in parts and component trade became an even clearer phenomenon after global car manufacturers made Thailand their export base for one-ton pickups starting 2000 and for eco-cars starting 2007. Automobile production requires physical proximity to parts suppliers to minimise cost, so many Japanese parts suppliers invested in Thailand. These firms supplied local carmakers and exported to other locations in the production network of carmakers. Therefore, the G–L index for auto parts was almost 90 per cent, indicating that Thailand's automotive industry has been linked to the automotive industries in other countries.

In Table 8.3, we provide more detail of the inter- and intra-industry trade with nine trade partners. Looking at the G–L Index in four periods (1992, 1996, 2002, and 2011), we can see different patterns amongst trade partners and products (automobile products and automotive parts). In general, Thailand tends to have a higher degree of IIT in automobile products with Japan and Indonesia. This is because both Thailand and Indonesia became important bases for Japanese carmakers. As there are more Thai automotive parts suppliers and these parts suppliers are relatively more technologically capable, Thailand exports parts and components to Japanese car manufacturers in Indonesia. IIT with Malaysia and the Philippines is lower because Malaysia has a strategy of developing its own national car and the Philippines is not an important production base. Interestingly, IIT with China increased (as evidenced by data from 2002 and 2011) because China significantly increased its production of cars, especially those with their own brands. Nonetheless, IIT increased significantly for almost all trade partners in 1992 and 2011, especially with China, India, and Viet Nam. Clearly, Thailand has become a hub in the production network of Japanese carmakers.[3]

There is no available public data on the proportion of value added of Thailand's domestic suppliers vis-a-vis that of foreign car manufacturers and parts makers, with most firms perceiving such data as confidential. However, as Thai auto parts makers produce mostly 'non-functional' parts, their proportion of total value added is likely to be low. In addition, the profit margin

Table 8.3 Development of intra-industry Trade (IIT) in Thailand's automotive industry G–L Index (1992, 1996, 2002, and 2011)

Countries	*1992*		*1996*		*2002*		*2011*	
	Inter	*IIT*	*Inter*	*IIT*	*Inter*	*IIT*	*Inter*	*IIT*
a. Automobile products								
Australia	95.88	4.12	0.89	99.11	99.96	0.04	99.82	0.18
China	42.04	57.96	83.29	16.71	91.54	8.46	12.98	87.02
India	100.00	0.00	67.60	32.40	98.59	1.41	91.95	8.05
Indonesia	100.00	0.00	86.36	13.64	70.29	29.71	72.18	27.82
Japan	99.85	0.15	98.42	1.58	33.76	66.24	10.49	89.51
Korea	100.00	0.00	98.02	1.98	94.27	5.73	84.80	15.20
Malaysia	66.19	33.81	33.34	66.66	99.90	0.10	80.67	19.33
The Philippines	100.00	0.00	99.95	0.05	48.38	51.62	80.71	19.29
Viet Nam	100.00	0.00	100.00	0.00	99.46	0.54	85.02	14.98
Total	98.10	1.90	64.42	35.58	41.71	58.29	63.91	36.09
b. Automobile parts								
Australia	50.53	49.47	30.32	69.68	76.23	23.77	87.85	12.15
China	95.87	4.13	94.67	5.33	37.82	62.18	54.10	45.90
India	89.21	10.79	77.64	22.36	79.65	20.35	49.24	50.76
Indonesia	7.64	92.36	54.81	45.19	27.10	72.90	46.58	53.42
Japan	93.19	6.81	95.07	4.93	65.78	34.22	60.56	39.44
Korea	67.01	32.99	52.73	47.27	0.19	99.81	55.22	44.78
Malaysia	37.87	62.13	20.36	79.64	36.67	63.33	72.19	27.81
Philippines	33.39	66.61	27.14	72.86	30.77	69.23	4.74	95.26
Viet Nam	99.84	0.16	99.85	0.15	95.04	4.96	78.43	21.57
Total	88.95	11.05	84.46	15.54	41.81	58.19	15.00	85.00

Source: Author's own calculations based on UN Comtrade database.

of parts makers has been continuously squeezed by car manufacturers who have requested them to reduce their selling prices by 2 per cent to 5 per cent every year (Atthawit Techawiboonwong, Director of Autoparts Manufacturer Department, Thailand Automotive Institute, personal communication, 20 December 2014).

4. Sectoral innovation system of Thailand's automotive industry

Franco Malerba, an evolutionary economist, developed a *sectoral system* framework focusing on the nature, structure, organisation, and dynamics of innovation and production in sectors. Based on evolutionary theory and the innovation system approach, he defines a sector as a set of activities unified by some linked product groups for a given or emerging demand and that share some common knowledge. Firms in a sector have some commonalities and, at the same time, are heterogeneous in terms of learning processes and capabilities. There are key elements in a sectoral system of innovation:

1 Firms in the sector. These are the key actors in a sectoral system and have specific learning processes, capabilities and organisational structures, beliefs, expectations, and goals.
2 Other actors. These can be organisations or individuals. Organisations may be suppliers, users, universities, financial institutions, government agencies, trade unions, or technical associations. Individuals may be consumers, entrepreneurs, or scientists. They also have their own specific learning processes, competencies, beliefs, objectives, organisational structures, and behaviours.
3 Networks. Firms are connected in various ways through market and nonmarket relationships. Networks integrate agents' complementarities in knowledge, capabilities, and specialisation. In addition, networks formed between firms and non-firm organisations, for example, universities and public research institutes, can be sources of innovation. The types and structures of networks vary from one sectoral system to another.
4 Demand. Demand can be both domestic and international. Demand comprises individual consumers, firms, and public agencies which are different in size, knowledge, learning processes, and competencies, and are influenced by different underlying institutions.
5 Institutions. Agents' actions and interactions are somewhat determined by institutions. These include norms, routines, common habits, established practices, rules, laws, standards, and so forth. Institutions may be formal (such as laws and regulations) or informal (such as norms). Institutions play a major role in affecting the rate of technological change, the organisation of innovative activity, and performance.
6 The knowledge base. Any sector has a specific knowledge base, technologies, and inputs. Knowledge plays a central role in innovation and affects the types of learning and the capabilities of firms. Knowledge is highly eccentric at the

firm level, does not diffuse automatically and freely among firms, and has to be absorbed by firms through their differential abilities accumulated over time. The boundaries of sectoral systems may change more or less rapidly over time as a consequence of dynamic processes related to the transformation of knowledge; the evolution of, and convergence in, demand; and changes in competition amongst, and learning by, firms.

7 The main processes and coevolution. Innovation is considered a process that involves systematic interactions among a wide variety of actors for the generation and exchange of knowledge relevant to innovation and its commercialisation. Interactions include market and nonmarket relations. Over time, a sectoral system undergoes processes of change and transformation through the coevolution of its various elements. This process involves technology, demand, knowledge base, learning processes, firms, non-firm organisations, and institutions.

Based on the concept of sectoral innovation system, we first examine the roles and capability development of key actors in the Thai automotive sector and subsequently investigate linkages between those actors and the concomitant learning processes. Finally, important underlying institutions such as entrepreneurship and trust will be explored.

4.1 Roles and capabilities of key actors

Key actors in the Thai automotive sector include carmakers and their component suppliers, government policy and promotion agencies, research institutes, universities (with specialised automotive industry programmes), automotive-sector industrial associations, private-sector technology promotion agencies, and so on (Figure 8.4).

4.1.1 Carmakers and suppliers

Automotive firms in Thailand can be classified into three groups: assemblers, first-tier automotive parts suppliers, and second- and third-tier automotive subparts suppliers. Figure 8.4 shows the connection between upstream and downstream manufacturers. Second- and third-tier manufacturers supply raw materials and automotive subparts to first-tier suppliers who produce major automotive parts and modules. The first-tier suppliers then supply their output to automotive assemblers who manufacture for domestic consumption and exports.

GROUP 1: AUTOMOTIVE ASSEMBLERS

There are currently 17 assemblers, all of which are either subsidiaries of TNCs or joint ventures. With the presence of most major carmakers, Thailand has already become one of the key production bases for global players from Japan, the United States, and Europe. Before the 2000s, these carmakers carried out

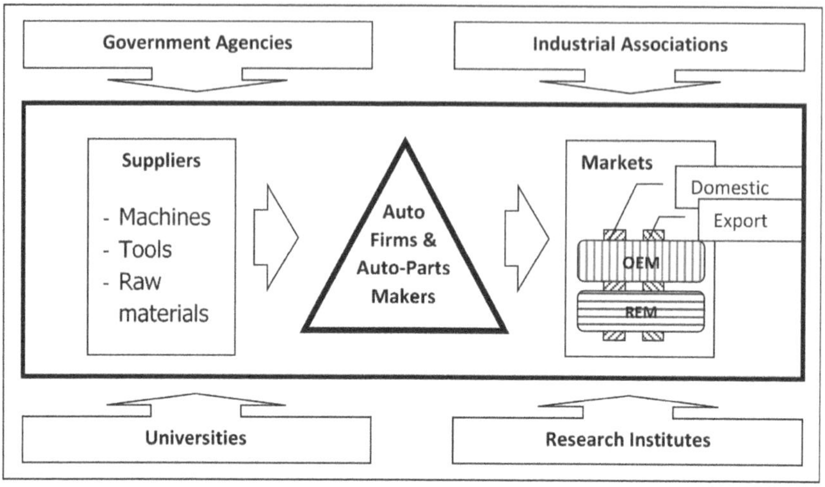

Figure 8.4 Thai automotive sectoral innovation system

OEM = Original Manufacturing Equipment (market); REM = replacement market.

Source: Author's own conceptualisation.

only production activities in Thailand, whereas more sophisticated activities like design and R&D were undertaken in their home countries. The investment strategies of TNCs in the automotive industry started changing in the 2000s. They began investing in more technologically sophisticated activities in Thailand, such as advanced engineering, process and product design, and advanced testing and validation. Several major automotive TNCs (mostly Japanese) have set up technical centres in Thailand, separate from their normal production plants. These include Toyota Motor Asia Pacific Engineering and Manufacturing Co., Ltd; Nissan Technical Center Southeast Asia Co., Ltd; Isuzu Technical Center Asia Co., Ltd; and Honda R&D Asia Pacific Co., Ltd. The R&D activities of these companies started with them modifying their already-designed/existing products to make these products fit local demands and to exploit local advantages resulting from an analysis of appropriate local skills, natural raw materials, and parts to meet international standards or the standards and regulations of importing countries such as members of the European Union (EU). Nonetheless, the trend has been towards more advanced product design being undertaken in Thailand. There are 2,289 automotive parts manufacturers that can be classified into two groups (Groups 2 and 3).

GROUP 2: DIRECT SUPPLIERS OR OEM SUPPLIERS (FIRST TIER)

This group consists of 648 auto parts manufacturers who produce and directly supply auto parts to auto assemblers. Of these manufacturers, 458 produce parts for car assemblers and the remaining 190 make parts for motorcycle assemblers. Figure 8.1 shows that first-tier suppliers can be classified into six groups according to the types of automotive parts they produce. Most Thai-owned firms concentrate on the non-functional parts: body parts, accessories, and similar parts. Supplying first-tier parts requires a rather high technological capability and proactivity about learning. There is little evidence of local firms moving up from the lower tiers to the first tier. On the other hand, several local first-tier suppliers actually dropped to the second tier because they could not cope with constant demands from carmakers to design and manufacture technologically demanding parts by themselves without the aid of a blueprint at reduced prices.

GROUP 3: INDIRECT SUPPLIERS OR RAW MATERIAL SUPPLIERS (SECOND- / THIRD-TIER)

This group includes raw material suppliers and subcontractors for the first-tier suppliers. There are 1,641 indirect suppliers which can be classified as follows:

1 Suppliers supporting the industry for raw materials and subassembly parts (e.g., leather, plastic, rubber, steel and iron, electrical and electronic parts, glass, painting, and surface treatment)
2 Suppliers supporting the industry for manufacturing or equipment (e.g., mould and die, jigs and fixtures, forging, casting, tooling, cutting, surface treatment, heat treatment, precision, electronic connectors, engineering plastic)

These indigenous Thai parts suppliers have low technological capability and largely depend on the technology joint-venture partners or licensers provide, but they also cannot absorb whatever technology is transferred due to a lack of skilled labour (Brooker Group 2002, pp. 2–14).

4.1.2 Government: policy and sector-specific promotion agency

The policies of the Thai government on the automotive industry can be divided into three phases.

PHASE I: IMPORT SUBSTITUTION REGIME

The automotive industry in Thailand started in the early 1960s under an import substitution policy and a revision of the investment promotion law to encourage local automotive assembly. From 1961 to 1969, nine assembly plants were set up as joint ventures between Thai and foreign carmakers. To boost investments in the domestic production of automotive parts, the Thai government imposed a minimum local content requirement of 25 per cent on automotive assembly in 1969. After the requirement was enacted, carmakers had to start purchasing locally. However, Japanese carmakers could not rely on Thai locally owned firms, and they requested affiliated Japanese automotive parts suppliers to build plants locally and supply to them.

In the late 1970s, a localisation policy was formulated to reduce the trade deficit and boost the industry. In addition to implementing import bans and raising tariff rates on completely knocked-down units and completely built-up units, the Thai government limited the number of automotive models and increased the local content requirement from 25 per cent to 50 per cent for passenger cars. Since the local automotive industry suffered from low demand for cars in the early 1980s, carmakers preferred to produce in house, subcontract activities requiring the use of casting machines, and produce the automobiles themselves to use their excess production capacity. To further boost the development of the parts industry, the government raised the local content requirement to 54 per cent for passenger cars and to between 60 per cent and 72 per cent for pickup trucks. This policy provided opportunities for local firms to become suppliers of TNCs in global production networks. Nonetheless, note that the policy was initiated in consultation with foreign carmakers and parts suppliers and implemented with a high degree of flexibility.

PHASE II: MASSIVE FDI INFLOW AND EXPORT PROMOTION

Foreign direct investment (FDI) inflows in the automotive industry were more or less unchanged from 1970 to 1985, with annual inflows amounting to less than US$5 million. But from 1986 to 1995, the annual average value of inflows increased dramatically. In the late 1980s, the appreciation of the Japanese yen had pushed up the cost of major automotive parts imported from Japan. The yen appreciation triggered the relocation of Japanese parts producers to Thailand to reduce production costs. As indicated by the huge increase in FDI inflows, the increased degree of TNC involvement in the Thai automotive industry took place for both carmakers and parts suppliers. Following their customers, Japanese parts suppliers established new affiliates for manufacturing new and more sophisticated parts starting in the late 1980s when Japan experienced dramatic currency appreciation. Apart from the strong yen, Thailand was chosen because of its advantages in terms of geography, readiness of logistical infrastructure, availability of technicians and engineers, more liberal investment policies (having no national car projects), and more attractive business and living conditions compared with its neighbours.

Since the mid-1990s, several world-class, non-Japanese, multiple-parts manufacturers such as Dana (1994), TRW Steering & Suspension (1998), Visteon Thailand (1998), Johnson Controls (1999), Delphi Automotive Systems (2000), and Tenneco Automotive (2002) have established their factories in Thailand. From 1994 to 1997, the value of projects promoted by the Board of Investment was four times higher than from 1990 to 1993. Increased foreign investment in the automotive industry brought in new lines of automotive parts not previously produced in Thailand (e.g., transmission systems). During this period, the Thai government also designated one-ton pickup trucks as the 'product champion' for the automotive industry. Tax incentives and other promotions were specially implemented, leading to remarkable investment in, and subsequent exports of, this product. Thailand has become the second-largest production base of pickup trucks after the United States.

Phase III: Liberalisation and technological upgrading

Thailand faced an economic crisis in 1997. To help affected companies improve their liquidity positions, the Board of Investment removed restrictions on foreign shareholding in November 1997. Prior to this, the policy required that majority ownership be held by a Thai national. Many investors, mostly Japanese, took advantage of this new policy. From November 1997 to September 2000, foreign partners in 164 automotive firms changed their shareholding structure from minor shares to majority shares (Charoenporn 2001). FDI inflows in the Thai automotive industry further increased after the 1997 financial crisis and reached a record high by 2007.

In 2004, the Board of Investment substantially changed its policy by paying more attention to the issues underlying the long-term competitiveness of the country; namely, the development of indigenous technological capability and human resources. A special investment package promoting 'skill, technology, and innovation' (STI) was initiated. Firms can enjoy one or two years of extra tax incentives if they perform the following activities within the first three years: spend at least 1 per cent to 2 per cent of their sales/revenues on R&D or design, ensure that scientists or engineers with at least a bachelor's degree make up at least 5 per cent of their workforce, and ensure that at least 1 per cent of their total payroll cost is spent on the training of their employees and at least 1 per cent on training the personnel of their local suppliers.

In the late 2000s, the economical and ecology-friendly car or 'eco-car' was selected as the second product champion after a long consultation and negotiation with the industry. Very preferential incentives together with significant requirements on producing four out of five engine components locally were given to interested carmakers. This new product champion is part of the *Master Plan for Automotive Industry (2012–2016)*, which aims to establish Thailand as a global green production base. As a result, Thailand has become the hub of eco-car production in Asia. Nissan's March and Honda's Brio, for instance, have been produced and exported to the global market from production bases in Thailand.

Alongside the aforementioned government policies, the Thailand Automotive Institute (TAI) was established by Cabinet resolution on 7 July 1998 to strengthen cooperation between the government and private enterprises to enhance the competitiveness of the Thai automotive industry. As a result, TAI is a *sector-specific promotional and intermediary agency* for the automotive industry. Administratively, TAI is not a part of the national bureaucracy but comes under the Industry Development Foundation set up by the Ministry of Industry.

Therefore, the organisation's administration is rather flexible. It is not subject to the rules and regulations of ordinary government agencies and state enterprises. TAI's governing committee, headed by the permanent secretary of the Ministry of Industry, comprises representatives from the government and private sector as well as academics. TAI provides data, training, consulting, and testing and calibration services to firms, especially local parts suppliers.

4.1.3 Universities and research institutes

Key universities with active roles in helping the automotive industry in Thailand include Chulalongkorn University, King Mongkut's Institute of Technology Ladkrabang (KMITL), King Mongkut's University of Technology Thonburi, King Mongkut's University of Technology North Bangkok, and Thai-Nichi Institute of Technology. After Thailand became a production hub, these universities started offering automotive engineering programmes to produce qualified manpower for the thriving industry.

Chulalongkorn University was the first university in Thailand to offer a degree in automotive engineering. The department was established in 1996 and started offering a master's degree in automotive engineering in 2005. King Mongkut's University of Technology Thonburi has two departments related to the automotive industry: the mechanical engineering department and the tools and materials engineering department. King Mongkut's University of Technology North Bangkok offers a master's degree in automotive engineering through the Sirindhorn International Thai–German Graduate School of Engineering. This is a joint programme with the Rheinisch–Westfälische Technische Hochschule (RWTH) Aachen University in Germany. RWTH, the leading German institute in the field of automotive engineering, closely collaborates with companies like BMW, DaimlerChrysler, and Siemens, as well as Asian and American automotive manufacturers.

The Thai–Nichi Institute of Technology was established in 2005 to develop in people the specialised technological knowledge required to work in some Thai industries. It offers a bachelor's degree in engineering with majors in automotive engineering, production engineering, and computer engineering. The largest research institute in Thailand is the NSTDA, which developed the 'Automotive Cluster Program' in 2003 to enhance the industry's competitiveness.

4.1.4 Private-sector industrial associations

There are five industrial associations concerned with the Thai automotive industry, two of which are under the Federation of Thai Industries. The association that specifically aims to strengthen the capabilities of indigenous suppliers is the Thai Auto-Parts Manufacturers Association (TAPMA), created in 1978 as an association of auto parts manufacturing companies. It aims to serve as the central voice for auto parts industrialists in the country in order to protect, support, and develop Thai industries.

4.1.5 Private-sector technology promotion agency

Established in 1973 with financial support from Japan, the Technology Promotion Association (Thailand–Japan) is the most successful private-sector technology promotion agency. It provides consulting services under the name 'Shindansha Program' to medium-sized and large local firms. The majority of consulting projects are done with the Department of Industrial Promotion as the direct customer.

4.2 Linkages and learning by interacting

There are three main types of linkages and interactive learning in the sector: carmakers–suppliers, firms–universities and research institutes, and firms–government agencies.

4.2.1 Carmakers–suppliers linkages and learning

Foreign carmakers have played an important role in disseminating important technology that has enhanced the formation of technological capability and the growth of Thailand's supporting industries (Techakanont and Terdudomtham 2004). TNCs are actively transferring technology to local suppliers through information sharing and advising. Many TNC parts suppliers, especially the Japanese ones, are involved with local parts suppliers through a technology-licensing contract or as minor shareholders. However, they have expressed their intention to become co-owners and/or majority shareholders. Their prime objective was to take full control of the parts manufacturing operation. This tendency towards strengthening their involvement with local parts suppliers has been observed since the late 1980s.

As the Thai automotive industry has become more export oriented, the local content of locally assembled vehicles has naturally increased. To a certain extent, the increased importance of vehicle exports can be regarded as a structural change. In particular, when the restriction on foreign ownership was abolished at the beginning of the crisis in 1997, these foreign, major-shareholder TNCs started bringing in updated and more cutting-edge technology, the use of which foreign technicians closely supervised. This did not occur when these same TNCs were involved only through the technology-licensing channel or were minor and less-active shareholders (Kohpaiboon 2009). The content of the inter-firm technology transfer was gradually enhanced from being simple 'operational technology' to a higher level of 'process engineering technology'. According to Techakanont's research, there has been another shift towards 'product engineering capabilities' since the year 2000 (Techakanont 2002). This coincides with the new investment strategies of foreign carmakers to set up R&D or technical centres in Thailand and use the country as a regional (and increasingly global) hub for their specific products such as one-ton pickup trucks and eco-cars. Nonetheless, only a few, large, local suppliers with long-term relationships with assemblers and their own human resources and technology-development initiatives have been given the opportunity for this higher-level technology transfer.

4.2.2 University–industry linkages and learning

Universities in Thailand play a significant role in generating basic knowledge and in providing relevant education in this field. Relationships have been established and expanded through various forms, mainly through curriculum development, joint programmes, student internships, and co-research projects. However, the role of universities in helping companies improve their technological capability is still limited.

4.2.3 Firms–government agencies linkages and learning

The key linkages here are the links between firms and TAI. Although TAI has different roles and activities as mentioned earlier, its activity that helps private firms the most is the Automotive Human Resource Development Program, which was implemented from 2006 to 2011. This programme was a joint collaboration between Thailand and Japan. Aside from TAI, the Federation of Thai Industries also joined the programme.

4.3 Institutions

Institutions play major roles in shaping the attitudes and behaviours of actors as well as the patterns of their interaction and learning. Here we pinpoint two crucial institutional factors underlying evolution of the Thai automotive industry: entrepreneurship and trust.

In Thailand, even smaller businesses are mostly owned and operated by Sino-Thais (East Asian Analytical Unit 1995). The dominance of family-owned enterprises established by immigrant Chinese entrepreneurs in Thailand has long been rooted in Thai business norms and culture. Cooperation is less likely in interfamily businesses, and in intra-family enterprises, cooperation often brings family complexity and contention into the mix. This partially explains why local parts-supplier firms from different families are less likely to collaborate with one another and with industrial associations like TAPMA whose main members are from these family-owned businesses and are seen mainly as lobbyist organisations, without too much development-oriented collaborative activities going on. The lack of trust and understanding also explains the mostly weak university–industry collaboration.

Note that the slow and passive technological learning of local firms is rooted in their origins. Unlike local firms in Japan and the Republic of Korea, most Thai firms started their businesses from trading activities. There were only a few industrialist business groups (Suehiro 1993). Many Thai family businesses later formed joint ventures with foreign firms, which own proprietary technologies. They left the technology development and engineering aspects of the ventures to their foreign partners because they were happy to focus on the management and marketing side.

Nonetheless, as illustrated earlier, these institutional factors have started to change. One reason is the pressure from foreign carmakers, which adopted global sourcing strategies. Suppliers, especially the first-tier ones, must have a certain level of design and engineering capability to deliver acceptable-quality parts to these makers. This induces behaviour changes among leading local suppliers. Investments in R&D and related activities by foreign makers induce local firms to follow by setting up their own R&D/technical centres and seeking closer and longer-term collaboration with local universities and public research institutes. Therefore, although institutional factors can be obstacles to the positive development of the sectoral innovation system, these factors can change over time if they are induced by changes in the behaviour of the key actors.

5. Conclusion

Thailand's automotive industry has a long history spanning more than 50 years. It has come a long way from being just a small import-substituting industry to a large and vibrant exporting one. It has contributed significantly and increasingly to the country's gross domestic product and employment. In relation to its neighbouring countries, Thailand has become the most important investment destination of Japanese carmakers in Southeast Asia due to its advantages in geographical location, logistic infrastructure, more liberal and flexible industrial development policies, and a more conductive investment climate and living environment. It has been well integrated into global production networks, especially those of Japanese carmakers. As the industry's products become more fragmented, IIT between Thailand and other countries in East Asia has risen markedly, especially in automotive parts. This reflects the importance of the Japanese production network as Japan's FDI in China and Indonesia has increased markedly in the past decade, enabling the increase in trade between Thailand and these countries.

Looking more closely, we also observe the three main evolution patterns of the Thai automotive sectoral innovation system. First, the system is evolving from a passive-learning and fragmented system to a more active-learning and coherent one. Foreign carmakers and first-tier suppliers and several local suppliers have made considerable effort to enhance their technological and innovative capabilities. Universities and research institutes started to offer sector-specific teaching and research programmes and closer collaboration with the industry. The sector-specific government promotion agency (i.e., TAI) has increasingly acted as an 'intermediary' organisation, establishing and strengthening linkages among foreign firms, local suppliers, universities, and other government agencies. The long-standing institutional obstacles, especially Thai suppliers' attitude of passively learning technologies, have also considerably improved.

Second and related to the first, the results of this change enable the system to gradually upgrade from being a 'production' system to being a somewhat 'innovative' and 'R&D-intensive' one. Foreign carmakers and first-tier suppliers together with selected Thai first-tier and leading second-tier suppliers began setting up R&D/technical centres separate from their production departments. As a result, they have been seeking more 'innovation/R&D-intensive' collaboration with local universities and research institutes.

Last, the system has increasingly become more 'product specific'. Rather than producing and innovating on varieties of vehicles as was done in the past, product champions like one-ton pickup trucks and, subsequently, eco-cars have emerged within the past 15 years. These product champions have significantly increased both in terms of absolute numbers and relative to the country's total production and exports. The global car industry market views Thailand as a specialised production base and, increasingly, also as an R&D base of these products. Nonetheless, to realise this goal, a new type of relationship between first-tier suppliers, especially local ones, and foreign carmakers has to be developed. To co-develop a new product with foreign carmakers, these parts suppliers have to increase their engineering, design, and R&D capabilities, especially product design and development. The government should facilitate the development of these capabilities by contributing financially and acting as intermediaries (as in the case of TAI) in bridging the two sides. Universities and public research institutes should also focus on educating and training design engineers and conducting collaborative R&D initiatives, focusing on product development with these parts suppliers. An R&D consortium among Thai parts suppliers, universities, and public research institutes with inputs from foreign carmakers to develop strategic and functional automotive parts/components should be encouraged and partially financed by the government. The government should also upgrade existing testing facilities or establish a new one advanced enough to verify these new products.

Notes

1 Professor, National Graduate Institute for Policy Studies (GRIPS), 7–22–1 Roppongi, Minato-ku, Tokyo 106–8677 Japan, email: prpu6@hotmail.com
2 In 1998 and 1999, domestic sales were 144,000 and 218,000 units, respectively, whereas production capacity was about 1 million units.
3 Based on our calculation, the degree of intra-industry trade in auto parts between Thailand and selected trade partners was relatively high – more than 90 per cent in recent years and 85 per cent in 2011. This is possible because, as Grubel and Lloyd pointed out, 'the measures of bilateral trade with individual countries, with each country's weights determined by its share of the total export plus import trade, is less than the measure for intra-industry trade between the country and all other countries combined' (1975, p. 24). The calculation of bilateral intra-industry trade in Table 8.5 is also consistent with Grubel and Lloyd's observation.

References

Brooker Group 2002, 'International competitiveness of Asian economies: a cross-country study – Thailand paper', report for the Asian Development Bank. Brooker Group.

Charoenporn, P 2001, 'Automotive part procurement system in Thailand: a comparison of American and Japanese companies', unpublished master's thesis, Faculty of Economics, Thammasat University, Bangkok.

Dahlman, CJ and Brimble, P 1990, 'Technology strategy and policy for industrial competitiveness: a case study of Thailand', *Industrial Paper Series* No. 24, The World Bank, Washington, DC.

Dahlman, CJ, Brimble, P, Dollar, D, and Tan, JP 1991, 'Technology Strategy and Policy for Industrial Competitiveness: A Case Study of Thailand', in *Decision and Change in Thailand: Three Studies in Support of the Seventh Plan*, eds. D Dollar, CJ Dahlman, P Brimble and T Jee-Peng. The World Bank, Washington, DC, pp. 57–61.

East Asia Analytical Unit 1995, *Overseas Chinese Business Networks in Asia*, Department of Foreign Affairs and Trade, Canberra.

Grubel, HG and Lloyd, PJ 1975, *Intra-industry Trade: The Theory and Measurement of International Trade in Differentiated Products*, Macmillan, London.

Intarakumnerd, P, Chairatana, P, and Tangchitpiboon, T 2002, 'National innovation systems in less successful developing countries: the case study of Thailand', *Research Policy*, vol. 8, no. 9, pp. 1445–1447.

Intarakumnerd, P and Charoenporn, P 2010, 'The Roles of Intellectual Property Rights Regime on Thailand's Technological Catching Up', in *Intellectual Property Rights, Development, and Catch Up: An International Comparative Study*, eds. H Odagiri, A Goto, A Sunami, and RR Nelson, Oxford University Press, Oxford, UK, pp. 378–411.

Kaosa-Ard, MS 1991, 'A preliminary study of TNCs' hiring and localization policies in Thailand', *TDRI Quarterly Review*, vol. 6, no. 4, pp. 11–18.

Kohpaiboon, A 2009, 'Global integration of Thai automotive industry', *Discussion Paper Series* No. 16, Faculty of Economics, Thammasat University, Bangkok. Available from: www.econ.tu.ac.th/oldweb/doc/content/569/Discussion_Paper_NO.16.pdf.

Lauridsen, L 2004, 'Foreign direct investment, linkage formation and supplier development in Thailand during the 1990s: the role of state governance', *European Journal of Development Research*, vol. 16, no. 3, pp. 561–586.

Malerba, F 2002, 'Sectoral systems of innovation and production', *Research Policy*, vol. 31, no. 2, pp. 247–264.

Malerba, F 2004, *Sectoral Systems of Innovation: Concepts, Issues and Analyses of Six Major Sectors in Europe*, Cambridge University Press, Cambridge, UK.

Poapongsakorn, N and Techakanont, K 2008, 'The Development of Automotive Industry Clusters and Production Networks in Thailand', in *Production Networks and Industrial Clusters*, eds. K Ikuo and TM Heng, ISEAS, Singapore, pp. 196–256.

Sribunruang, A 1986, *Foreign Investment and Manufactured Exports from Thailand*, Chulalongkorn University Social Research Institute, Bangkok.

Suehiro, A 1993, 'Family business reassessed: corporate structure and late-starting industrialisation in Thailand', *The Developing Economies*, vol. 31, no. 4, pp. 378–407.

Techakanont, K 2002, *A study on inter-firm technology transfer in the Thai automotive industry*, PhD dissertation, Hiroshima University.

Techakanont, K 2011, 'Thailand Automotive Parts Industry', in *Intermediate Goods Trade in East Asia: Economic Deepening through FTAs/EPAs*, ed. M Kagami, BRC

Research Report Series, No. 5, Bangkok Research Center/IDE–JETRO, Bangkok, pp. 193–229.

Techakanont, K and Charoenporn, P 2011, 'Evolution of automotive clusters and interactive learning in Thailand', *Science, Technology and Society*, vol. 16, no. 2, pp. 147–176.

Techakanont, K and Terdudomtham, T 2004, 'Evolution of inter-firm technology transfer and technological capability formation of local parts firms in the Thai automobile industry', *Asian Journal of Technology Innovation*, vol. 12, no. 2, pp. 151–183.

Thailand Automotive Association 2011, Available from: www.thaiauto.or.th/Records/eng/vehicleandpartexport_eng.asp [15 January 2011].

Thailand Automotive Association 2015, Available from: http://data.thaiauto.or.th/iu3/.

9 Trade in value added

The case of Viet Nam[1]

Tri Thanh Vo,[2] *Anh Duong Nguyen and Trinh Bui*

1. Introduction

The world economy has been finding its way to robust recovery since the outbreak of the 2007–2009 global financial crisis. Global exports continue to increase and have become one of the key drivers of economic recovery. The volume of global trade has retained modest growth in recent years, and even shows good prospects for the years ahead. A recent forecast by the World Bank (2011) shows that world trade may increase by up to 4.9 per cent until 2017. More important, global trade has widened to encompass almost all countries and territories. The deepening liberalization of trade, which induces further specialization of production based on comparative advantage, has become an almost irreversible process, therefore.

The process of international economic integration has continued, even deepened. The Doha Development Round was revitalized with the hard-earned achievements in Bali in 2013. In parallel, a number of free trade agreements (FTAs) are being negotiated or implemented, covering a growing list of areas such as trade in goods, services, competition, government procurement, and investment. The East Asian and Pacific region is the most active in this FTA trend. Several mega-FTAs involve major economies in the region, such as the Trans-Atlantic Trade and Investment Partnership (TTIP), the Trans-Pacific Partnership (TPP), and the Regional Comprehensive Economic Partnership (RCEP). Meanwhile, various FTAs have been concluded and implemented, such as the Association of Southeast Asian Nations (ASEAN) Free Trade Area/ASEAN Trade in Goods Agreement, the ASEAN–Japan Comprehensive Economic Partnership, the ASEAN–Korea FTA, the ASEAN–China FTA, and the Korea–Australia FTA.

The important characteristic of economic integration in East Asia is that it is accompanied by the development of a regional production network. The flows of foreign direct investment (FDI) into and out of various countries, with investors' efforts to retain existing networks of suppliers and customers, essentially induce this development. This process has been even more noticeable as East Asian countries are widely heterogeneous in various aspects such as market scale and production capacity. Trade in intermediate goods within East Asia has also been increasing. Some estimates[3] even show that intermediate goods account for more than a half of trade in the region, and this share amounts to up to 60 per cent in the case of imports. Such shares surpass the corresponding figures at the global level. Exports of regional countries become increasingly dependent on imports of intermediate goods. The question then arises as to whether the growing trade in intermediate goods, driven by increasing specialization and more developed production networks, led to reasonable shares of participating countries in gross value added of final products. This issue is particularly important for developing economies, as they embark on progressive integration, given their relatively small size.

Since the start of Doi Moi (Renovation), active efforts towards international economic integration have constituted a key pillar of Viet Nam's reform process. Looking back, regional economic integration with the Association of Southeast Asian Nations (ASEAN) as the hub has significantly broadened opportunities for trade and investment activities in Viet Nam. More important, the integration process increased pressures for domestic reforms, as the opportunities that arose had to be grasped. Gradual and continuous efforts have been made to structurally transform Viet Nam into a market economy. The periods that saw bold and comprehensive economic reforms were also those that saw rapid integration attempts and associated economic achievements. Notwithstanding its inherent weaknesses, Viet Nam's economy has benefited significantly from pro-active integration.[4] At the same time, its enterprises also started to become more deeply involved in the regional production network. This deeper involvement has made the country confident about future integration attempts.

The trend of economic integration is expected to continue, particularly as Viet Nam aims to develop a strong and regionally integrated industrial economy. The key ASEAN-centred FTAs, such as the ASEAN+1 FTAs and the RCEP, have shaped and will continue to drive trade and investment liberalization in Viet Nam. The implications of these FTAs for the regional production network – of which Viet Nam is an emerging member – can be profound. Enhancing Viet Nam's export potential, therefore, depends not only on access to foreign markets, but also on the country's capacity to manage the supply of inputs for export-oriented production. Along with this trade expansion again emerges the question of whether Viet Nam generates a reasonable value-added content from trade activities.

This chapter attempts to analyze the issue of how Viet Nam acquires and improves value added in its trade activities. It seeks to answer two questions in particular – (1) whether Viet Nam is depending more on imported products to meet its export growth; and (2) whether export growth brings about sufficient benefits to the domestic economy. In doing so, the authors take a neutral stance on domestic value-added content, that is, the share of domestic content in exports of products may not necessarily require a higher priority than the gross amount of domestic value added from exports. In other words, together with trade expansion and greater integration into the regional production network, the margin for domestic stakeholders can become smaller, yet Viet Nam can still benefit if more value added in exports is sourced domestically in absolute terms. This chapter relies on a combination of analytical methods. It uses an updated version of the input–output table for Viet Nam to calculate the domestic value-added content of exports. And it incorporates a comprehensive review of the literature and qualitative statistics on Viet Nam's trade.

The remainder of this chapter is structured as follows. Section 2 provides a review of the literature on trade in value added in Viet Nam. Section 3 specifies the analytical method used to calculate the domestic value added content of exports. Section 4 discusses the key findings from the analytical work with the input–output table in 2011. Section V concludes with several recommendations as to how Viet Nam can better identify priorities and benefit from engagement in the regional production network.

2. Literature review

Tran et al. (2011) offered the first rigorous quantitative estimation of import contents of Viet Nam's exports. This study basically followed the approach of Hummels et al. (2001), that is, it is based on the concept of vertical specialization. The study utilized the updated input–output table for Viet Nam in 2007, with 138 sectors. Overall import content was found to be 40.8 per cent in 2007, though it varied across sectors. The lowest figure was for paddy (3.1 per cent), and the highest for animal and vegetable oil and fat (75.9 per cent). Viet Nam's key exports also had high import contents, for example, exploited marine products (54.2 per cent) and garments (59.7 per cent). This was also observed in some strategic manufacturing sectors, such as automobiles (65.4 per cent), shipbuilding (61.7 per cent), fibre (60.4 per cent), and electronics components (58.0 per cent).

But the study has several limitations. First, it relies on key assumptions, which reduces the robustness of findings. For example, exported products and locally consumed ones are assumed to be homogenous, which facilitates the calculation of import contents. However, this assumption may not hold strictly in Viet Nam, since export products are often of higher quality and may find themselves significantly differentiated from those for domestic consumption. Moreover, the analytical findings are only meaningful if the imported inputs can be readily substituted with the domestic products. Again, some of the products may not be available in Viet Nam, which makes it necessary to import them. Quality matching presents another challenge, which may weaken the previous assumption. Second, the employed input–output table is rather outdated, which may not reflect structural changes in Viet Nam's national economy and economic sectors in recent years. In particular, various studies assessing structural and policy changes in Viet Nam during 2007–2014 reflect its adaptation to a drastic surge in imports in 2007–2008 and its response to the global financial crisis from 2008 to 2009.[5] Third, the study only considers import contents of different sectors in 2007. Without similar calculations for other years, it is difficult to assess changes in import content over time.

Bui et al. (2008) carried out another notable study on the direct and indirect import contents of some aggregate sectors in Viet Nam. Their research was done on the basis of input–output tables for 1996, 2000, and 2005. They employed the concept of the Keynesian, Leontief, and Miyazawa multiplier to estimate domestic intermediate consumption, domestic final consumption, domestic investment, and exports. However, findings with important implications for policy making are scarce in this study.

Other studies mostly focus on the value chains of various products in Viet Nam. Ha (2012) described the different stages of the global value chains for textiles and garment products, with an emphasis on the linkages between the textiles subsectors and garments. His study adopted an output-based approach and considered the value chains for textiles and garment exports of four groups: (1) international customers; (2) local producers; (3) suppliers of inputs; and (4) intermediaries. On this basis, the study explained why Viet Nam was able to acquire only a modest share of value added from exporting textiles and garment products.

Similarly, Nadvi and Thoburn (2003) used a global value chain framework to analyze Viet Nam's textiles and garments sector. The authors mapped Viet Nam's changing position in the global industry, and looked into ties between Vietnamese firms and global buyers. The study also assessed the links within the Vietnamese garments and textiles sectors, and considered the impact of global challenges on Vietnamese firms and workers. This helped to distinguish between potential 'winners' and 'losers', and the study underlined the wider poverty implications. However, a limitation of this study was that it is only a qualitative description of the issues and challenges confronting the textiles and garment supply chain in Viet Nam, without estimating the value added produced by those chains.

Dao and Nguyen (2013) examined the relationship between FTAs in the Asian region and the agricultural value chain. They found that FTAs had a positive impact on Viet Nam's economy as a whole, but their impact on agriculture was unclear. Prices of Vietnamese products exported to the world markets are lower than those of the same products from other exporting countries and the share of value added in the agricultural value chains for producers is modest. Farmers, therefore, do not have much of an incentive to expand production. In another aspect, fruits and vegetables have potential in terms of export-oriented production. Consequently, to increase the value added for agricultural value chains of these potential products, Viet Nam should follow the requirements under geographical indicators (GI), areas of operations (AO), and other food safety standards. The shortcoming of this study is that it fails to quantitatively document the impacts on value-added content of Viet Nam's agricultural exports.

The World Bank (2011) concentrated on the value chain of rice products in the Mekong River Delta. Apart from analyzing simple statistics on the rice industry, the study concluded that the underdevelopment of the rice export value chain was due to several factors, including the length of the chain; little application of collective actions at the industry level; minimal levels of vertical integration and very modest levels of vertical coordination; and undifferentiated products. Again, this study did not identify the value-added content contributed by each stage of the value chain, which limits its usefulness for informing policy in terms of the extent of intervention into relevant stages.

Various other studies – such as CIEM (2010, 2013); Truong et al. (2011); Vo and Nguyen (2006) – looked into the patterns of trade and investment in Viet Nam from 2007 and the underlying drivers of those patterns. The studies used a range of methods, including qualitative analysis and econometric estimation, to explain the variations in trade and investment in Viet Nam. These studies concluded that failure to account for the reliance of exports on imported materials may lead to misguided macroeconomic policy management. As an example, devaluing the national currency against the US dollar could not help restore current account balance as such a nominal depreciation was not accompanied by a real depreciation. Underlying the absence of real appreciation was that reliance on imports also caused domestic prices to increase (even more rapidly) following the nominal depreciation. But these studies barely analyzed the issues surrounding the value-added contents of Viet Nam's exports, though asserting that these issues explain a significant part of the changes in Viet Nam's current account deficit since its World Trade Organization (WTO) accession in 2007.

3. Approach and methodology

This chapter attempts to fill the current research gap by analyzing the import content of exports in Viet Nam at the sector level. In doing so, the authors adopt an approach to measuring trade in value added which '*focuses on the value added contents of a traded product, and considers each country's contribution to the value added generation in a production process*' (Inomata 2013, p. 3). This concept is also consistent with that of Hummels et al. (2001), namely, that '*The key idea behind fragmentation, outsourcing, slicing-the-value chain, etc. is that countries increasingly link sequentially to produce good*' (Hummels et al. 2001, p. 77). Merely considering official trade data is insufficient as it does not provide a complete understanding of the trade balance and inadequate policy interventions. For instance, Viet Nam has had a trade deficit with ASEAN since its accession to ASEAN in 1995; nonetheless, intermediate products accounted for a major share of Viet Nam's imports from ASEAN[6] and imports of such products enabled Viet Nam to expand its exports to and trade surplus with other major markets such as the United States, the European Union (EU), and Japan.

Adopting this approach, this chapter starts with the Leontief system as follows:

$$X - AX = C + I + E - M \tag{1}$$

where *X, C, I, E, M* denote the vectors of gross output, household consumption, gross capital formation, exports, and imports, respectively.

Equation (1) can be written as follows:

$$X - AX = C + I + E - M^P - M^C \tag{2}$$

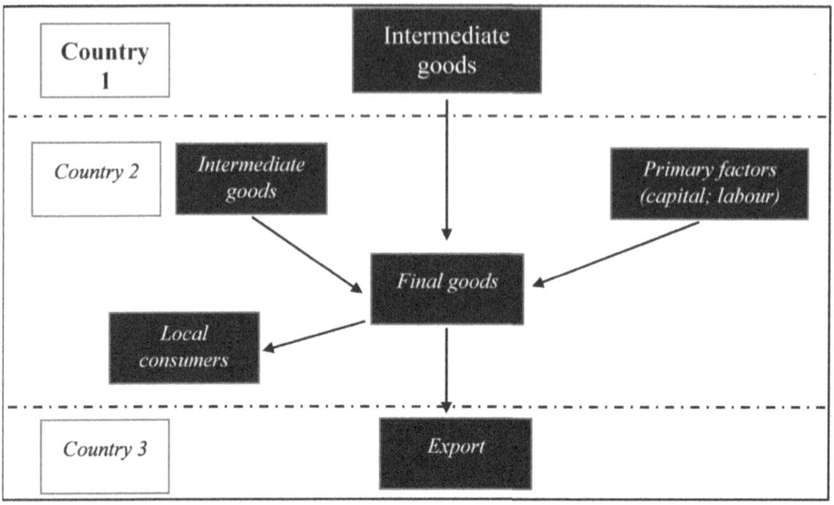

Figure 9.1 Vertical specialization

Source: Tran et al. (2011, p. 30).

where M^P is the import for intermediate input, M^C is the import of final products, $M = M^P + M^C$.

Expanding equation (2) yields:

$$X - A^d X - A^m X = C^d + I^d + E + C^m + I^m - M \quad (3)$$

where $AX = A^d X + A^m X$; $A^m X = M^P$ and $M^c = C^m + I^m$. A^d. A^d is the matrix of intermediate consumption of domestic products, whilst C^m, I^m are the final consumption and gross capital formation vectors of domestic products, respectively.

Using $Y^d = C^d + I^d + E$ where Y^d denotes final domestic demand of products vector, we can rewrite equation (3) as follows:

$$X = (I - A^d)^{-1} . Y^d = (1 + A + A^2 + A^3 + \ldots) Y^d \quad (4)$$

where $(I - A^d)^{-1}$ is the Leontief matrix multiplier that shows domestic product requirements for a unit increase in domestic final demand.

From another perspective, a backward linkage is a measure of the relation between an industry and the suppliers of its inputs from the entire production system. It measures the output increase that will occur in industries that supply inputs to the industry concerned. A backward linkage can be computed as the ratio of the sum of the elements of a column of the Leontief inverse to the average for the whole system. This ratio was described by Rasmussen (1957) as the index of the power of dispersion, μ_j, and is defined mathematically as follows:

$$\mu_j = \frac{\sum_{i=1}^{n} l_{ij}}{\frac{1}{n}\sum_{i=1}^{n}\sum_{j=1}^{n} l_{ij}} \quad (5)$$

where the I_{ij} is the element of the inter-regional Leontief inverse. The higher the value of, μ_j the stronger the influence of production sector *j* as a user of intermediate inputs.

Equation (3) can be re-formulated as follows:

$$X - A^m X = A^d X + C^d + I^d + E + C^m + I^m - M = TDD - M^p + E \quad (6)$$

where total domestic demand includes intermediate expenditure, final consumption, and investment (TDD), that is, $TDD = A^d X + C^d + I^d$

Rearranging (6) gives:

$$X = (I - A^m)^{-1}(TDD + E - M^p) \quad (7)$$
$$\text{or: } X = (I - A^m)^{-1}(TDD + C^m + I^m + E - M) \quad (8)$$

The matrix $(I–A^m)^{-1}$ is the import multiplier matrix. Equations (5) and (6) present the demand of import multiplier by domestic demand. The input–output table is constructed as the non-competitive import type in which intermediate demand and final demand have been separated into domestic products and imports. Accordingly, the input–output table for Viet Nam should be mathematically changed to the non-competitive import type. A^m and A^d are calculated as follows:

$$A^mX = \Phi.A.X \text{ and } A^dX = (I–\Phi).A.X \quad (9)$$

where $m_i=M_i/TDD_i$, M_i is import of good *i*, and TDD_i denotes total domestic demand of good *I*, excluding exports.

From equation (4), the induced impact from domestic final demand to value added and imports are identified.

$$GVA = v^* (I–A^d)^{-1} {}^* Y^d \quad (10)$$
$$M = m^* (I–A^d)^{-1} {}^* Y^d \quad (11)$$

where *v* and *m* are coefficient vectors of value added and imports, *GVA* is the vector of gross value added, and *M* is the vector of imports.

From that, to identify the induced impact coefficient of final demand to output, value added and imports are presented as follows:

$$X^* R_y^{-1} = R_x \quad (12)$$
$$GVA^* R_y^{-1} = R_V \quad (13)$$
$$M^* R_y^{-1} = R_M \quad (14)$$

where R_y is a diagonal matrix with the elements on the diagonal being the sum of final demand factors. The three elements of domestic final demand are: final consumption of domestic products (C^d), saving from domestic products (I^d), and exports (E), so the elements on the diagonal of R can be represented as: $\sum_i C^d_i$; $\sum_i I^d_i$; $\sum_i E_i$.

Finally, to estimate the domestic value-added content of one sector's exports, this chapter assumes that the sector's final output is not differentiated for domestic consumption and exports. This assumption is repeated from the study of Tran Binh Minh et al. (2011). Moreover, all imported inputs for the sector are used up within the year under consideration. Accordingly, the import content of exports can be calculated as the share of gross imported inputs in gross output of the sector; and the domestic value-added content can be calculated.

This method, again, is subject to several limitations. Using the latest input–output table for 2011 to update the calculation of import contents of exports is possible, yet the underlying assumptions can be restrictive. Moreover, the method assumes homogeneity of locally consumed and exported products, which is highly unlikely. In fact, as Viet Nam generally enforces lower quality standards for the domestic market, export products tend to be of better quality. Furthermore, the input–output table is often published rather late, making the data outdated. This issue is even more relevant in the context of rapid structural changes and reforms in Viet Nam. Another issue is the inability to separate re-imports and re-exports from official trade data. The method used in this study is unable to separate different tiers, that is, the second- or even third-tier domestic suppliers for export products. The assumption that all imported inputs are used up within one year is another limitation. Finally, the latest input–output table of Viet Nam may not be consistent with data reported by external partners.

This method can be complemented by an econometric analysis of import dependence in export activities at the firm level. In fact, Viet Nam's 2012 Enterprise Survey included questionnaires to identify the shares of various imported products in firms' activities, but these data have not been released. At any rate, these data would not distinguish the second or third tiers of inputs from other domestic suppliers. Thus, the figures may not be appropriate measures for the local and import contents of manufacturing exports. At the same time, making inferences about the import content of exports requires the assumption that locally consumed and export products of firms are homogeneous. Moreover, the Enterprise Surveys only cover a selected portion of the enterprises in Viet Nam, so the survey-based implications can be limited.

4. Empirical analysis

This section discusses the findings from our investigation into the trade and import content of Viet Nam's exports.

4.1 General trends of exports and imports

Figure 9.2 depicts Viet Nam's exports, imports, and balance of trade of goods during 2000–2013. Exports maintained an upward trend, from US$26.5 billion in 2004 to $62.7 billion in 2008 and $150.2 billion in 2014. The effects of the financial crisis and economic recession resulted in a drop of 8.9 per cent in 2009, before the value of exports bounced back in the following years. During 2007–2014, exports registered an average annual growth rate of 18.0 per cent, which remained slower than in 2002–2006 (21.5 per cent).

The ratio of merchandise exports to gross domestic product (GDP) showed an upward trend, reaching 77 per cent in 2013 and 80.6 per cent in 2014, which was higher than in 2008 (64.3 per cent) and in 1995 (only 26.2 per cent). Exports of goods and services, as a share of GDP, rose from 63.7 per cent in 2005 to 72 per cent in 2010 and 83.1 per cent in 2013.[7] These figures reflect a relatively high openness to exporting in Viet Nam compared with other Asian countries and the world. Viet Nam continued to expand exports relative to GDP after its accession to the WTO, notwithstanding more complicated and rapid impacts from the global economic environment.[8] Thus, even in a volatile economic environment, Viet Nam was able to make the best of its export potential.

Unit: Million US$

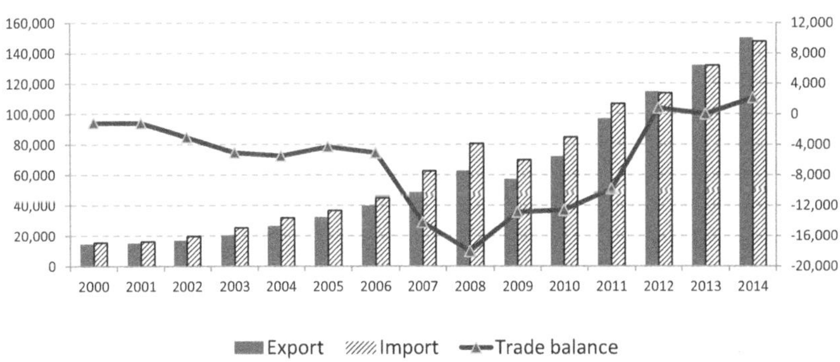

Figure 9.2 Viet Nam's imports, exports, and balance of trade, 2000–2014

Unit: Million US$.

Note: Trade balance is measured on the right-axis. Values of exports and imports are reflected on the left-axis.

Source: General Statistics Office, Viet Nam.

Viet Nam's export achievement can be partly explained by the expansion of world trade growth. Vo and Nguyen (2006) showed that the contribution of world trade growth to the country's export growth jumped from 53 per cent in 2001–2004 to 61 per cent in 2004–2007. In addition, estimated results by CIEM (2013, p. 64) reveal that Viet Nam's exports were significantly induced by world economic growth, with exports increasing by 4.4 per cent for each 1 per cent increase of world GDP. Liberalization and improvement of competitiveness were other drivers of the expansion of Viet Nam's exports to the world. Underlying these factors are also the attempts to engage more deeply with the regional production network.

Similarly, Viet Nam's import values increased almost continuously, except for a decrease in 2009 due to the financial crisis and economic recession. Import growth has been rather rapid until 2011, especially after WTO accession in 2007. After 2012, import growth slowed, mainly due to a domestic economic downturn. Overall, in 2007–2014, imports increased by 3.3 times, registering an average growth rate of 16.1 per cent. Thus, imports grew at slower rate than in 2002–2006 when average annual growth was 22.6 per cent. Imports as a share of GDP dropped from its highest levels of 82.8 per cent in 2008 to 76.9 per cent in 2013, lower than that of exports, but much higher than the rate of 49.6 per cent registered in 2000 and 39.2 per cent in 1995. In 2014, the ratio recovered slightly, to 79.7 per cent. Imports of goods and services as a share of GDP fell to 79.8 per cent in 2013 from 83.5 per cent in 2010 and 80.2 per cent in 2011.[9]

Unlike exports, imports rose sharply only in 2007–2008 and the growth rate slowed after that. One reason is that Viet Nam adapted to the WTO 'games', at both the policy and enterprise levels. The government and policy makers had introduced more measures to restrain imports more effectively without breaching WTO regulations (CIEM 2013). And to some extent, Vietnamese products became more competitive compared with imported ones.

Estimations by Truong et al. (2011) revealed several factors that had influenced imports. On one hand, an increase in the nominal VND/USD exchange rate had made Viet Nam's products more expensive (in real terms), thereby attracting higher imports. Moreover, the disbursement of foreign direct investment (FDI) capital had increased demand for imported goods. A rise of disbursed FDI capital of 1 per cent resulted in an increase in import volume of about 0.3 per cent. High GDP growth is also associated with higher demand for imported goods to serve domestic consumption and production. This effect will be more pronounced as a result of tariff cuts as part of WTO commitments, which will further increase import demand in the absence of technical measures to restrain imports.[10]

Due to the slower growth rate of imports compared with that of exports in recent years, especially since 2009, Viet Nam's trade deficit gradually decreased from $8.0 billion in 2008 to $12.6 billion in 2010 and $9.8 billion in 2011. In 2012, exports *exceeded* imports, resulting in a surplus of $0.8 billion. This trend continued until 2014, with a surplus of more than $2.1 billion.

4.2 Source of value added in exports

Figure 9.2 shows that both exports and imports moved in the same direction in Viet Nam, especially after 2007. This provides an initial hint about the dependence of exports on imported inputs. In fact, notwithstanding the continuous increase in exports, the share of domestic value added in exports appeared to decrease over time. The figure fell continuously from 75.6 per cent in 1995 to 70.4 per cent in 2000 and 60.2 per cent in 2008. This fall was not a material problem, however. The domestic value added in exports continued to rise in absolute terms. The decrease in the share of domestic value added was largely due to an aggressive liberalization of trade and investment, which had led to a gradual substitution of imported inputs with local ones in export-oriented production. The trend was only reversed in 2009 when the share of domestic value added in exports rose to more than 63.0 per cent. This year was marked by the effects of the global financial crisis, and Viet Nam experienced a rapid decrease in the value of imports.

Table 9.1 tabulates structural changes in Viet Nam's economy over 2000–2011. As can be seen from the measurement of pervasiveness, in 2011, both final consumption and investment better induced output and income than in 2000 and 2007. And an increase in overall exports quickly resulted in a drastic increase in output, but its impact on income was less strong. In particular, the ratio of value added induced by exports continued to decrease after 2000. Moreover, the pervasiveness to imports shows that exports quickly stimulated imports. This index rose from 1 in 2000 to 1.45 in 2011. Furthermore, the ratio of value added of exports to total final domestic demand fell sharply, from 0.45 in 2000 to 0.27 in 2011. As an implication, the value-added content in Viet Nam's exports had tended to decline until 2011, which helps explain the country's persistent trade deficit during this period.

By trading partners, the share of ASEAN in Viet Nam's value added from exports was rather stable, if excluding the country itself (Figure 9.3). East Asia's share tended to increase, reflecting the outcomes of integration efforts with East Asian countries, which induced more imports of intermediate products from East Asian countries. Amongst the East Asian countries, China stood out as the main source of intermediate goods imported into Viet Nam. The share of value added from China in Viet Nam's exports increased from 1.5 per cent in 2000 to 6.1 per cent in 2008, then fell to 5.8 per cent in 2009 before recovering to 6.3 per cent in 2011.[11] Consistent with this, Truong et al. (2011) showed that intermediate inputs accounted for an increasing share in Viet Nam's imports from China, with the share rising from 41.3 per cent in 2000 to 59.4 per cent in 2008, before falling to 50.2 per cent in 2009. Over the same period, the role of the EU in Viet Nam's export in terms of value added only increased slightly.

4.3 Domestic and foreign value-added contents of exports of manufactured goods

By source industry, the largest sources of value added in gross exports are mining and quarrying; food products, beverages and tobacco; and wholesale, retail

Table 9.1 Output, gross value added, and import-induced effect by final demand

	Year 2000			*Year 2007*			*Year 2011*		
	C	*I*	*E*	*C*	*I*	*E*	*C*	*I*	*E*
Pervasiveness from final demand to production value	1.27	1.35	1.53	1.09	1.12	1.7	2.04	1.17	2.01
Percentage of changes				–14.10%	–17.10%	11.70%	86.90%	4.50%	18.27%
Pervasiveness from final demand to income	0.6	0.43	0.69	0.48	0.41	0.59	0.64	0.54	0.54
Percentage of changes				–20.40%	–5.60%	–13.30%	33.30%	31.22%	–8.94%
Pervasiveness from domestic demand to import	1.44	1.7	1	1.28	1.63	1.47	1.31	1.5	1.45
Percentage of changes				–12.10%	–3.90%	52.00%	2.27%	–8.13%	–1.29%
Ratio of value added in domestic final demand	0.47	0.32	0.45	0.44	0.37	0.35	0.31	0.46	0.27

Source: Authors' calculations.

Unit: %

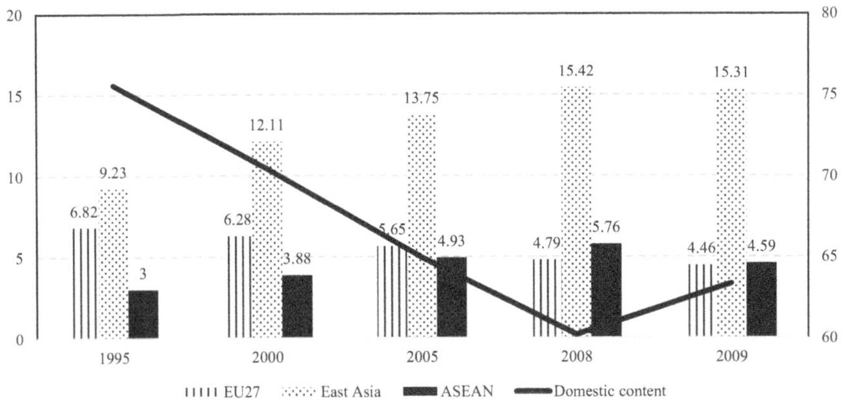

Figure 9.3 Share of value added embodied in gross exports by country, 1995–2011

Note: Values for EU28, East Asia, and ASEAN are on the left-axis. Values for domestic content are on the right-axis.

Source: OECD TiVA database (June 2015).

trade, and repairs. The share of agriculture, hunting, forestry, and fishing only increased slightly, from 5.4 per cent in 1995 to 6.2 per cent in 2011. The share of wholesale, retail trade, and repairs fell from 12.3 per cent to 9.7 per cent over the same period. The share of mining and quarrying in gross value added in exports rose from 15.4 per cent in 1995 to 17.1 per cent in 2000 before contracting continuously to 7.6 per cent in 2011. Viet Nam had aggressively promoted the development of the manufacturing sector, and it generally accounted for around 20 per cent of GDP, but it made only limited contributions to the value added of exports.

When we compared the value-added content in exports generated by Viet Nam in 2007 and 2011 across different sectors, we found that, overall, the domestic value-added content in Viet Nam's exports in 2011 was higher than in 2007. This was largely due to the drastic surge in import flows into Viet Nam in 2007, immediately following its WTO accession.[12] In conjunction with the observed export expansion, this calculation implies that Viet Nam's domestic industries obtained even higher value added from exported products. Note that the overall domestic value-added content in 2011 differs from that reported in the OECD TiVA database, which might be explained by the failure of Viet Nam's input–output table to appropriately account for re-exports and re-imports.

Many manufacturing sectors, however, saw a fall in the domestic value-added content of exports. Notable examples are production of costumes (with domestic value-added content decreasing from 58.9 per cent to 38.8 per cent in 2007–2011), medical equipment (from 84.3 per cent to 43.4 per cent), cars (from

65.2 per cent to 55.5 per cent). But a number of manufacturing sectors saw an increase in the domestic value-added content of exports, including animal feed, leather, paper, electronic devices, and computers and peripherals. This complicated pattern of changes in domestic value-added contents can be largely attributed to structural adjustments in various sectors under the impacts of: (1) integration commitments which induced changes in import structures and resulted in additional inflows of FDI; and (2) domestic policy changes which encouraged purchases of domestically produced industrial products with a view to reducing imports and the trade deficit.[13]

Notwithstanding high export values, Viet Nam's textiles and garments industry is still considered to have significant growth potential.[14] Exports of textiles and garments rose rapidly from $1.9 billion in 2000 to $7.7 billion in 2007 and $20.9 billion in 2014. From the view of producers in the industry, nevertheless, textiles and garment exports still depend on the availability as well as timeliness of inputs, including cloth and accessories for production. Meanwhile, the domestic textiles industry is a long way from being able to supply the needed inputs. Increasing garment exports, therefore, would increase the dependence on imports of textile products, particularly from China. This dependence on import is further evidenced by the low and decreasing domestic value-added content of the garment sector (No.41). Finding alternatives for textile inputs from China would not be easy, at least in the medium term, as the those inputs would need to be of sufficiently high quality and supply would have to be timely (i.e., just-in-time delivery).

Another manufacturing sector that has attracted attention in recent years is electronic devices, computers, and peripherals, which has been targeted for development in Viet Nam since the 2000s. Decision No. 879/QD–TTg of the Prime Minister in 2014 re-emphasised the need for upgrading Viet Nam's electronics industry. When considering the increase in domestic value-added content of exported electronic products, we observe that it was not until 2012 that exports of electronics products saw rapid growth. In 2012–2013, exports of electronic devices, computers, and peripherals grew by 50.8 per cent on average per year. In the same period, export growth of phones and peripherals averaged 82.2 per cent. Underlying these drastic increases were the operations of various big electronic producers such as Samsung, Canon, and Nokia. In 2014, Samsung alone had exports worth $26.3 billion.[15]

Given these multinational companies' networks of suppliers in other countries (mostly in China and the ASEAN region), the growth in their exports was accompanied by a surge of imported components and spare parts from other countries into Viet Nam. The participation of Vietnamese enterprises in the supply chain of electronics products was modest in 2012–2014. There are several reasons for such low participation – (1) lack of mutual trust between major foreign-invested producers (who drive the supply chain) and Vietnamese firms, with the former being doubtful about the capability of the latter, and the latter being reluctant to invest in the absence of guaranteed purchase of output by the former; (2) impediments against participating in the supply chain (e.g., Even tier-1 suppliers could not select the tier-2 suppliers; instead, tier-2 suppliers are nominated by the major driver of the chain).

Another perspective involves the power of dispersion of different sectors on the economy and imports. This measure indicates how expansion in one sector would induce increases in aggregate economic activity or imports. We found that various manufacturing sectors had higher power of dispersion on the economy than on imports. Examples of these sectors include animal feed, processed wood and wooden products, household electrical appliances, and consumer electronics equipmen. For electronic devices, computers, and peripherals, their export increase induced even larger rises in both aggregate economic activity and imports. Conversely, for textile and garment products, a rise in exports did not induce an increase in aggregate economic activity; it was matched by an increase in imports.

5. Conclusions and policy recommendations

This chapter seeks answers to two questions – (1) whether Viet Nam is depending more on imported products to meet the growth of its exports; and (2) whether export growth brings about sufficient benefits for the domestic economy. Statistics show that during the reform period of 1986–2014, Viet Nam experienced an almost continuous expansion of trade. Export growth, in general, was accompanied by parallel import growth. The share of intermediate goods in Viet Nam's trade has risen drastically over the past decade. Efforts to deepen economic integration, particularly after WTO accession in 2007, prepared the way for Viet Nam's growing participation in the regional production network.

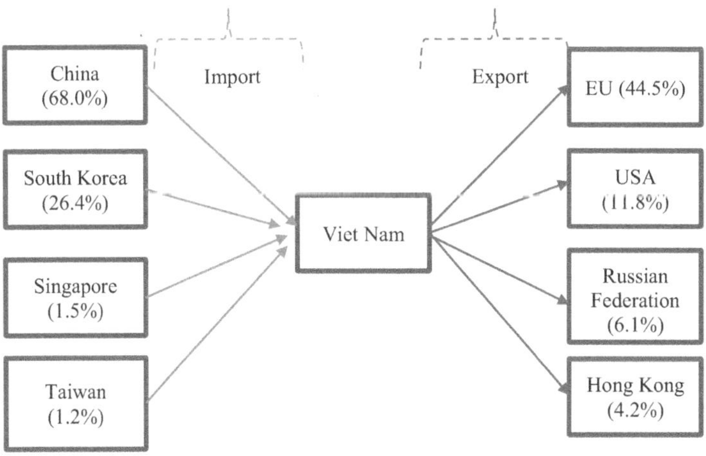

Figure 9.4 Trade flows of phones and accessories in Viet Nam, 2012

Source: General Department of Customs (2013).

Notwithstanding high growth, the inducement impact of exports on the economy and income decreased from 2007 to 2011. Viet Nam benefited in gross terms as the overall domestic value-added content of exports increased in 2007–2011 and was further exacerbated by a drastic increase in exports. During 2012–2014, however, Viet Nam experienced drastic increases for some major products such as electronics and garments, which induced a rapid surge in imports of inputs. These are not reflected in the calculations of this chapter due to the unavailability of a more up-to-date input–output table. To assess if there are long-term, sustainable benefits for the Vietnamese economy, therefore, we should look beyond gross export figures to determine whether various export products, including manufacturing products, induce increases in aggregate economic activities.

Several explanations can be put forward for the fall in value-added content of some export products. First, there were rather loose links between export-oriented sectors and upstream import-substitution sectors. In fact, a range of upstream sectors (e.g., textiles) have been under trade protection, yet they fail to supply inputs of sufficient quantity and/or quality for downstream export-oriented sectors (e.g., garment). Therefore, for certain sectors to increase their exports, input imports need to be increased.[16] Second, past policies to increase local firms' participation in the regional supply chains, with the inducement of large export-oriented multinational enterprises in Viet Nam, were not effective. In fact, those multinational enterprises largely had positive spillover impacts on local firms via: (1) competition; and (2) movement of labours from those enterprises to the local counterparts. Moreover, forward and backward linkages between multinational enterprises and local enterprises were rather limited and improved only slowly, largely due to the local enterprises' inadequate capacity for R&D and meeting technical standards.[17] Hence, multinational enterprises deciding to invest in Viet Nam may need to bring their own suppliers and/or rely on imported inputs.[18] Finally, Viet Nam lacks a properly designated industrial policy that identifies appropriate priorities. Its industrial policy has closely followed the picking-winner approach, but ran into the issue of 'picking too many winners'. At this stage, the list of key industrial sectors with development aspirations is long, covering dozens of sectors. Meanwhile, given its low development base, Viet Nam could only mobilize modest financial resources for industrial projects, resulting in capital dispersion and deficiency in various industrial subsectors. Measures to develop these sectors fail to sufficiently take into account the impacts on upstream and downstream sectors. This resulted in heavily distorted incentives for certain industrial sectors, since their development is conditional on the output and/or prices of upstream and/or downstream sectors.

Improving the domestic value-added content in Viet Nam's exports is no easy task. Intense competition for FDI in the ASEAN region means Viet Nam has very little power to bargain with major FDI enterprises – most of them are drivers of various supply chains and can readily relocate their production base. Requiring FDI enterprises to use locally supplied products and services is a virtually impossible task, as this may be in breach of Viet Nam's integration commitments.

More important, the policy space for upgrading domestic manufacturing capacity became much narrower after Viet Nam's participation in a number of integration arrangements (such as WTO). The available instruments for supporting industrial production are now limited to support for education and training, and small and medium enterprises. The room for using traditional barriers such as tariffs and technical standards has also become more limited. To add to these concerns, the further trade and investment liberalization under high-quality arrangements such as the Trans-Pacific Partnership and the European Union–Viet Nam FTA may further reduce the policy space for supporting industrial development due to regulatory commitments concerning issues such as intellectual property rights and government procurement.

To enhance Viet Nam's benefits from trade, we make the following recommendations:

> First, Viet Nam should not worry about the fall in the domestic value-added content of exports, so long as the gross value added from exports keeps rising. The priority should be ensuring better diffusion of exports to domestic aggregate economic activity and, ultimately, achieving better returns on various production factors.
>
> Second, Viet Nam should attempt to facilitate and deepen linkages between FDI enterprises and local firms. As demonstrated by various studies on FDI in Viet Nam (e.g., Nguyen et al. 2005; Nguyen 2008; Pham 2008), FDI has positive spillover effects on domestic enterprises, including SMEs. Hence, the policies should aim at supporting SMEs in establishing production linkages with other FDI counterparts. Government assistance is also needed to improve the capacity of those SMEs so they can benefit from the diffusion of new technology. The government could also serve as a 'matchmaker' to help and incentivize multinational enterprises to identify appropriate subcontractors in Viet Nam.
>
> Third, industrial policy should be more targeted. The government should refrain from developing too many industrial sectors. Credible assignment of priorities to the targeted industrial sectors will give a clear and transparent signal to resource allocation and, at the same time, induce targeted development of relevant supporting industries. Viet Nam's industrialization strategy since 2013, which selected six sectors for targeted development,[19] may be a good example.
>
> Finally, Viet Nam should preserve the policy space for supporting industrial development. Nguyen Anh Duong et al. (2015) assessed the policy space for promoting the electronics and food processing industries. Further efforts are needed to examine the available policy instruments for other sectors, including export-oriented sectors such as garments. Identification and preservation of policy space should not aim to protect concerned sectors from trade; instead, they only enhance the flexibility for Viet Nam in designating appropriate industrial policy to benefit from trade.

Notes

1 The views and opinions expressed in this chapter are solely those of the authors and may not necessarily reflect those of ERIA, CIEM or GSO.
2 Lead author. Email: votrithanh@mpi.gov.vn; votrithanh98@yahoo.com. Address: 68 Phan Dinh Phung street, Ba Dinh district, Hanoi, Viet Nam.
3 For instance, see Ueki (2011).
4 See CIEM (2010); Truong et al. (2011); CIEM (2013).
5 For details, see CIEM (2013).
6 This figure rose from 39.8 per cent in 2000 to 54.6 per cent in 2009 (see Truong Dinh Tuyen et al. 2011 for further details).
7 http://baodientu.chinhphu.vn/Kinh-te/Do-mo-cua-nen-kinh-te-va-van-de-dat-ra/199039.vgp
8 See CIEM (2013) for details.
9 http://baodientu.chinhphu.vn/Kinh-te/Do-mo-cua-nen-kinh-te-va-van-de-dat-ra/199039.vgp, accessed on 18 August 2014.
10 CIEM (2013, p. 65).
11 From TiVA database (June 2015).
12 See CIEM (2013).
13 An example is the Decision No. 1899/QD–BCT of the Ministry of Industry and Trade, dated 16 April 2010.
14 See CIEM (2015).
15 http://bizlive.vn/doanh-nghiep/nam-2015-xuat-khau-cua-samsung-viet-nam-co-the-dat-gia-tri-32-ty-usd-899250.html [accessed 25 May 2015].
16 This issue is not new. See Vo and Nguyen (2006), CIEM (2013).
17 See Nguyen et al. (2005, 2014).
18 Nokia and Canon are typical examples, as they have to rely on imported inputs and peripherals from China.
19 These sectors are: agricultural machinery, environment and energy saving, shipbuilding, electronics, food processing, and automotive.

References

Bui, T, Pham, LH, and Bui, CG 2008, 'Import multiplier in input–output analysis', *Depocen Working Paper Series* No. 2008/23, Development and Policies Research Center, Hanoi.

Central Institute for Economic Management 2010, 'Assessment of Vietnam's Socio-economic Situation 3 Years after the Accession to the World Trade Organization', report to the Government of Vietnam, Hanoi. In Vietnamese.

Central Institute for Economic Management 2013, *Comprehensive Evaluation of Vietnam's Socio-economic Performance 5 Years after the Accession to the World Trade Organization*, Finance Publishing House, Hanoi.

Central Institute for Economic Management 2015, 'Feasibility study of economic cooperation agreement between Vietnam and Taiwan', draft. April.

Dao, TA and Nguyen, VS 2013, 'Vietnam Agricultural Value Chain in the FTA of Asian Region', mimeo. Submitted as a country paper for the FFTC.NACF International Seminar on Threats and Opportunities of the Free Trade Agreements in the Asian Region, Sept. 29–Oct.3, 2013, Seoul. FFTC NACF = Food and Fertilizer Technology Center for the Asian and Pacific Region. Available at: http://ap.fftc.agnet.org/files/ap_policy/106/106_1.pdf.

General Department of Customs 2013, 'Merchandise Trade in December and the Whole Year of 2012' [Tình hình xuất nhập khẩu hàng hóa của Việt Nam tháng 12 và

12 tháng năm 2012]. Available from: www.customs.gov.vn/lists/tinhoatdong/ViewDetails.aspx?ID=19213&Category=Th%E1%BB%91ng%20k%C3%AA%20H%E1%BA%A3i%20quan [31 May 2015].

Ha, VH 2012, 'Analysis of value chain of textiles and garment export in Vietnam' [Phân tích chuỗi giá trị dệt may ở Việt Nam], *Journal of Economics and Business* (Hanoi National University), vol. 28, pp. 49–59. In Vietnamese.

Hummels, D, Ishii, J, and Yi, KM 2001, 'The nature and growth of vertical specialization in world trade', *Journal of International Economics*, vol. 54, no. 1, pp. 75–96.

Inomata, S 2013, 'Trade in value added: an East Asian perspective', *ADBI Working Paper* No. 451, ADBI, Tokyo.

Nadvi, K and Thoburn, J 2003, 'Vietnam in the Global Garment and Textile Value Chain: Implications for Firms and Workers', report for Department for International Development (DFID), April.

Nguyen, AD, Vo, TT, Dinh, TH, and Tran, BM 2015, 'Impact of Current and Proposed FTAs and BITs on Vietnam's Long-Term Development Goals: A Case Study of Food Processing and Electronics Manufacturing Sector', report prepared for ActionAid International, Vietnam.

Nguyen, PL 2008, *Productivity Spillovers from Foreign Direct Investment: Evidence from Vietnamese Firm Data*, School of Commerce, University of South Australia. Available from: http://papers.ssrn.com/sol3/papers.cfm?abstract_id=1101203 [13 September 2008].

Nguyen, TTA 2014, 'Adjusting Vietnam's FDI Policy', *State-Level Research Project*. In Vietnamese.

Nguyen, TTA, Vu, XNH, Tran, TT, and Nguyen, MH 2005, *The Impacts of Foreign Direct Investment on Economic Growth in Vietnam*, Science and Technics Publishing House, Hanoi.

Pham, XK 2008, *The Impact of Foreign Direct Investment on the Productivity in Host Countries: The Case of Vietnam*, Vietnam Development Forum, Hanoi. Online. Available from: www.vdf.org.vn/workingpapers/vdfwp0814.pdf [27 November 2008].

Rasmussen, PN 1957, *Studies in Inter-sectoral Relations*, North-Holland, Amsterdam.

Tran, BM, Bui, T, Trinh, QL, and Hoang, VC 2011, Estimating the import content of exports to support policies in Vietnam [Ước tính hàm lượng nhập khẩu trong xuất khẩu phục vụ công tác hoạch định chính sách kinh tế ở Việt Nam]. Ministerial-Level Research Project. In Vietnamese.

Truong, DT, Vo, TT, Bui, TG, Phan, VC, Le, TD, Nguyen, AD, and Pham, SA 2011, 'Impacts of market access commitments under WTO and FTAs on production and business in Vietnam', paper prepared for Multilateral Trade Assistance Project. In Vietnamese.

Ueki, Y 2011, 'Intermediate goods trade in East Asia', in *Intermediate Goods Trade in East Asia: Economic Deepening Through FTAs/EPAs*, ed. M Kagami, Bangkok Research Center/IDE–JETRO, Bangkok, pp. 24–67.

Vo, TT and Nguyen, AD 2006, 'Vietnam in the regional production network', paper presented at the International Conference on 'Comparative Study on Industrial Competitiveness between China and Vietnam in the Context of East Asian Trade Liberalization, and the Experiences of Japan', Danang, February.

World Bank 2011, 'Vietnam food security and rice value chain', *Policy Note* No. 2, The World Bank, Washington, DC.

10 'Policies for industrial progress', not 'industry policy'

Lessons from Southeast Asia*

Hal Hill and Archanun Kohpaiboon

1. Introduction

'Industry policy', defined as non-neutral inter-industry (and sometimes inter-firm) incentives, remains a contested field. There is much general agreement in the development economics and political economy literature about the factors that underpin rapid economic development. These include macroeconomic stability; openness to trade, investment, and technology; a stable and business-friendly commercial environment; mechanisms that ensure broad-based, 'inclusive' development; and investment in supply-side capabilities, ranging from infrastructure to human capital. The contestation focuses on whether these general, economy-wide approaches are sufficient, or whether in addition there is a role for sector-specific interventions. The earlier literature focused on the merits of import substitution versus export orientation and the case for outward-looking development strategies. Both the analytical literature, including the famous Bhagwati–Krueger studies and much else, and the empirical investigations demonstrated conclusively that outward-looking economies generally grow faster over the long term.[1] In response, and recognizing the inevitability of global economic integration, the contemporary literature has focused on whether there is a case for 'smart industry policy' in broadly open economies.

Here the arguments typically centre on dynamic externalities, learning by doing, linkages, and various forms of technological acquisition that will not be forthcoming if 'left to the market'. Much of the thinking has drawn on the East Asian development experiences, particularly the early successes of Japan and the Republic of Korea, that achieved very rapid, export-oriented growth, but with a policy mix that did not conform to the more laissez-faire prescriptions embodied in the so-called Washington consensus. The importance of activist industry policies that deliberately 'got prices wrong' was examined by earlier writers, including Chalmers (1982) on Japan and Alice Amsden (1989) on the Republic of Korea. More recently, Ha-Joon Chang (2002) has developed a 'radical heterodox' perspective in his 'kicking away the ladder' proposition. The debate has entered mainstream development economics, through, for example, Dwight Perkins's (2013) narrative on East Asian development and former World Bank chief economist Justin Lin's (2012) attempted synthesis.[2]

The aim of this chapter is to contribute to this literature with an examination of the Southeast Asian experience with industry policy. Our approach is deductive and inferential. That is, we select various instances of dynamic sectoral growth, examining the factors behind this growth and the role – if any – of public policy in these outcomes. The cases are country-specific, but in the concluding section we draw out some broader generalizations. We focus mainly on success stories, and why they occur, for the obvious reason that success is more elusive than failure. We are also interested in cases where the same country has experienced both success and failure in sector promotion efforts, a finding that suggests an institutionally nuanced approach to understanding industry policy success is necessary.

This is a very broad topic. Constrained by length, we have selected four main case studies as analytical prisms through which to investigate the issues. The first is the Thai automotive industry (Section 2), which is often regarded as one of the most important industrial policy successes in Southeast Asia. Then follows case studies of the internationalization of higher education in Malaysia (Section 3), business process outsourcing in the Philippines (Section 4), and the Cambodian garment industry (Section 5). We also review Indonesia's industrial policy experience in Section 6.

There is obviously 'sample selection bias' here, in that the number of case studies is small, and they are all success stories. But apart from the common element of success, this is a diverse set of case studies that, we argue, facilitates broader generalizations. Important also, we do not restrict our analysis to the manufacturing sector. Although much of the industry policy literature does focus on this sector, conceptually the paradigm of sectoral dynamism and its determinants is applicable to all sectors. A final introductory point to note is that we largely eschew discussion of political economy factors, although they are of course essential in the design of industry policy. That is, industry policy informed by arms-length scientific expertise, implemented transparently by an incorruptible bureaucracy, and in an open economy where the costs of failure become more quickly evident is more likely to be successful than the case where these factors are weakly present.

2. The Thai automotive industry

Thailand has established itself as the major automotive production hub in Southeast Asia and beyond. In 2013, vehicle production was approaching 2 million units, ranking ninth in the world. About half of the vehicles produced are exported. The rapidly rising production and export orientation has also been associated with increased production of local parts and components. For some vehicle types, pickup trucks in particular, most auto parts, including engines, are locally manufactured, thereby increasing domestic value added (Kohpaiboon and Jongwanich 2013). The purpose of this section is to examine the factors accounting for the success, and to explore the lessons for latecomers seeking to emulate the Thai record.

Early history: The origins of the industry were a conventional import-substitution approach, commencing in the 1960s with the assembly of imported completely knocked-down (CKD) kits behind high tariff protection. Local content requirement measures (LCRMs) were progressively introduced from the mid-1970s, resulting in the local manufacture of parts, with the intention of creating backward linkages and industrial deepening. These LCRMs were formulated in close consultation with both the vehicle and parts manufacturers. The former's involvement was to ensure the measures were practical and acceptable, while the role of the latter was to inform the government about firms' existing technological capability and to ensure local suppliers would benefit from these measures. As a result, the measures were a good deal less restrictive than the official targets indicated and the policy regimes in neighbouring countries. In fact, 'quasi non-traded' parts accounted for the lion's share of locally produced parts.[3]

Liberalization and export orientation: These policies generated rapid growth, but resulted in a high-cost production model. Inevitably, as the limits to domestic market expansion were reached, growth began to slow. The market was too small to attract major auto manufacturers, unless the high protection was maintained (Orihashi 2010). Reform and rationalization were therefore necessary for the industry to maintain its growth. As it turned out, major structural changes in the global automotive industry hastened the need for such a policy reorientation. By the late 1980s, the principal automobile markets in the 'Triad' regions (North America, Western Europe, and Japan), which accounted for more than 90 per cent of global sales, were growing slowly (Abrenica 1998). By contrast, demand for vehicles in emerging market economies had brighter prospects, so carmakers began to focus on these economies. At the same time, governments in a number of emerging market economies began liberalizing their automotive sector, in some cases through regional preferential agreements such as MERCUSOR for Latin America, NAFTA for North America, and the ASEAN Free Trade Area (AFTA) for Southeast Asia.[4]

These changes forced MNE carmakers to adopt new trade and investment strategies. It was not economically viable for them to set up affiliates in each developing country behind high import protection. As regional liberalization and therefore economic integration proceeded, the MNEs selected one or two developing countries as a production base. A fundamental requirement of the assembly facilities in the selected countries was that they could meet international quality and price standards. The carmakers would decide which models to produce in which locations, at what price and quality standard, and for which markets, either region or global. Under this new strategy, the overall policy environment of countries was the key determinant of production location decisions, and the relevance of the old import substitution model began to wane. In particular, the MNEs sought liberal, competitive, and stable business environments.

The Thai government was quick to respond to these changes. Following the 'pragmatic' LCRMs, the government shifted to promote specific products such as the one-ton diesel engine (see later in this chapter), as well as supporting industries, including petrochemicals, steel, and plastics (Suehiro 2010). Crucially, and unlike neighbouring Malaysia and Indonesia, Thailand did not adopt a 'national car' project. Moreover, unlike Indonesia and the Philippines, its commercial policy settings were relatively stable. Singapore had shown no interest in the automotive industry, and Viet Nam had yet to open up its economy. Therefore, partly by default and partly owing to its more attractive policy settings, in the late 1980s Thailand was the only feasible Southeast Asian location for the industry to employ an export-oriented strategy.[5] Furthermore, the import ban on CBUs was dropped, and replaced with a very high tariff, and by 2000 the LCRMs were abolished as part of WTO commitments, notwithstanding the deep economic crisis of 1997–1998.[6]

In addition to the relatively liberal commercial environment, two other factors enhanced Thailand's attractiveness. The first was massive infrastructure investments during the 1980s. The discovery of natural gas reserves in the Gulf of Siam in the early 1980s led to the government's 'Eastern Seaboard Development Plan'. While the plan was launched to promote the petroleum and chemical industries complex along the eastern coast, the accompanying high-quality infrastructure investment in power, transport, telecommunications, and water resulted in the establishment of flourishing industrial estates. The zones also bypassed many cumbersome bureaucratic procedures, resulting in the country's strong performance in various business climate rankings.

Second, although recently overtaken by Indonesia, for many years Thailand had the largest domestic market for vehicles in ASEAN. This was particularly the case for commercial vehicles, where, for example, it has become the world's second largest production base for one-ton pickup vehicles. This large domestic base has provided a springboard for exports. In the case of one-ton pickup trucks, there are four such assemblers (Isuzu, Toyota, Mitsubishi, and Nissan), and all of them have been able to achieve the necessary economies of scale. The demand for these multi-purpose vehicles has also grown rapidly because road transport dominates over the underdeveloped rail system, and because the tariff on these vehicles has been lower than that on passenger cars.

The one-ton diesel engine: The one-ton diesel engine project is often regarded as a successful case of industrial deepening. Initially introduced in the late 1970s, it was abandoned two years later when officials realized that it was not economically viable. It was then given a new lease on life in 1987 following the decision of several MNE carmakers to use Thailand as an export base for pickup trucks. This would provide the necessary economies of scale, moreover, for an activity in which the technological requirements were not far beyond the capabilities of local suppliers. A key local actor in this initiative was the very large Thai conglomerate Siam Cement Group, whose subsidiary, SNF, was already manufacturing agriculture diesel engines.

As part of the programme, the targeted local content requirement was 80 per cent, to be achieved within seven years. The one-ton diesel engine was to be fitted in all locally assembled pickup trucks. In addition, five compulsory items were stipulated: cylinder blocks, cylinder heads, crankshafts, camshafts, and connecting rods, all of which had to employ domestic production processes, particularly casting, forging, and machining. Export targets for the engine manufacturers were also set, at not less than 120 million baht (around $4.8 million) annually during the first four years, and at least 280 million baht in the subsequent three years (years 5–7). Important, while these requirements might appear onerous, in practice they were by then well within the country's technological competence. Moreover, they were implemented flexibly, with export targets postponed in response to periods of sluggish demand. Cooperation among firms in the manufacture of key parts was also allowed, to cushion the effects of any unexpected shocks.[7] This too was a result of close consultation between BOI staff and the three manufacturers, even though such cooperation was not a common feature of industrial practice among Japanese carmakers. Eventually the cooperation was abandoned in the early 2000s as domestic demand was large enough.

The eco-car project: A second major government intervention aimed at 'picking winners' in the auto industry was the eco-car project. This too was a market-conforming, flexibly implemented initiative. The project was launched in 2007 to promote the production of small, environmentally friendly, and fuel-efficient vehicles. It was also seen as a means of promoting industrial diversification. The main fiscal incentive under this project was a lower excise tax of 17 per cent, compared with 41–48 per cent for normal passenger vehicles. Participating companies were required to meet specified production, fuel efficiency, emission, and safety targets.

The political economy of this policy was quite complex as opinions of the carmakers differed. Those whose production portfolio was dominated by pickup trucks opposed it, fearing that the lower excise tax imposed on eco-cars would give them a major price advantage in the domestic market. Thus the eco-car project proceeded slowly. However, by 2007, when international oil prices were rising, the project attracted renewed attention and it was formally launched. Important, all the eco-car project requirements were in line with the earlier proposals submitted by the carmakers, again underlining the government's pragmatic and market-conforming approach. This included the local engine manufacturing requirement, which anyway was no longer a binding constraint when annual production volumes already exceeded 100,000 units. In the event, five companies were approved – Toyota, Honda, Suzuki, Nissan, and Mitsubishi – and production commenced over the period 2010–2013.

Lessons and future challenges: Three general observations are warranted in assessing the lessons from the Thai auto industry and its future challenges. First, the industry faces a major human capital challenge in the form of a skill mismatch in the labour market (World Bank 2012). Although this is an economy-wide problem, it is particularly pressing in the automotive industry as it becomes ever more export-oriented and the tasks to be undertaken require more sophisticated engineering and design capabilities. The government has responded with various programmes, mainly directed at producing more university-trained engineers. Nonetheless, the labour shortages are even more severe for semi-skilled workers with a vocational training background. Interesting, a handful of large private automotive firms offer on-the-job training in vocational schools and universities.

Under this programme, students work at the auto factories for a certain period of time (e.g., 6–12 months) with a stipend payment, and this work is accredited by the schools and universities. Upon graduation, these students are offered employment by the firms, although they are not obliged to work for the firm. Pending broader educational reform, this seems to be an effective approach to ameliorate the labour shortages. However, it is not generally a feasible option for the small-medium auto parts suppliers.

The second challenge is to enhance the competitiveness of local suppliers in regional production networks. Even though Thailand has become the major regional production hub, few local suppliers have been able to avail themselves of the very large market opportunities. According to a recent presentation by the Automotive Club, the Federation of Thai Industries (FTI), there were 508 firms supplying their products directly to carmakers, that is, as tier-1 suppliers.[8] However, this figure grossly overstates the number of participating firms. In practice only around 10–15 local suppliers are actively involved in production networks.[9] The other suppliers manufacture parts for older vehicle models, especially those that were assembled before the export-oriented strategy was implemented. Most of these firms struggle to participate in the more demanding export-oriented networks, where they are expected to produce their own drawings and to address product defects beyond basic checkpoints, for example, with respect to tensile strength and hardness (Kondo 2012). There are also deficiencies in product design and product inspection.

In fact, contrary to much conventional wisdom, this disappointing performance is in part a legacy of the LCRMs imposed through to 2000. Clearly this policy provided a jump-start for the local firms that entered the auto industry as suppliers, as well as exposing them to newly imported technologies.[10] However, the absence of competitive pressures meant that firms did not have an incentive to use this opportunity to strengthen their technological capabilities significantly (Kohpaiboon 2006; Haraguchi 2010). There have, of course, been efforts to strengthen the competitiveness of these local suppliers, although collective-action institutions among these firms are generally weak. The most important example was the establishment of the Thai Automotive Institute (TAI) in 1998, with the mission to lift the capabilities of local auto parts manufacturers in both management and production technology. In practice, however, it has had a limited impact, owing to frequent policy changes and a very modest budget.

The third issue concerns tariff reform. Tariffs on commercial and passenger CBU vehicles remain very high, at 40 per cent and 80 per cent, respectively, one of the highest in East Asia (Kohpaiboon, forthcoming). The cascading tariff structure results in even higher effective protection for some subsectors. In theory, the presence of high output tariffs could result in an anti-export bias as it alters the relative profitability of domestic and export sales. But the experience to date has been the coexistence of high tariff protection and rapid export growth. The explanation is that CBU vehicle exports have been driven predominately by MNE carmakers, which have set up production plants in Thailand to serve both the domestic and foreign markets. The high CBU tariff allows the carmakers to price discriminate, as the domestic price of CBU vehicles is higher than the exporting price. We do not have comprehensive information on the region's prices and tax regimes, but a simple illustration is sufficient to support this conclusion: in 2014, the retail price of a Toyota Camry 2000 cc. in Australia was cheaper than that in Thailand, in spite of transport costs, as well as the Australian tariffs and value-added tax. Interesting, the estimated fob price of this example in Thailand is slightly higher than the factory price estimate, strongly suggesting a cross-subsidy from the domestic to foreign markets. The industry is now well established, and so it is not plausible to advance any infant industry argument for continued protection, if indeed it could have been advanced in the first place.

3. Malaysia and the internationalization of higher education

The Malaysian experience with industry policy provides important insights from an export-oriented, high-growth economy that is nevertheless struggling to graduate to the ranks of high-income countries. It is also an interesting example of considerable general success and one notable, costly failure. There were in fact two early success stories, in export-oriented electronics manufacturing and tropical cash crops, and two more recent successes, in higher education and Islamic finance. The costly failure was an attempt to develop a national car project, and associated heavy industry. The variety of outcomes serves as a reminder that industrial policy success is institutionally nuanced and that outcomes vary not only between but also within countries. We focus here on one significant success story, that of the internationalization of higher education.[11]

Higher education is one of the most rapidly growing international traded services (Salmi 2009). Developed countries, with their large, well-resourced universities of international standing but fiscally constrained governments, see overseas education markets as a major commercial export opportunity, as well as a means of projecting abroad their 'soft power'. Developing countries, recognizing that the creation of high-quality universities is both a costly and lengthy process,[12] send their students abroad as one of the most effective means of quickly accessing advanced skills, as well as the intellectual networks that leading universities offer.

The Southeast Asian countries are major players in the internationalization of higher education (Welch 2011), and the region's governments are now focusing increasingly on their universities. Although only Singapore can realistically aspire to international excellence in higher education for at least the medium term, another feature of the internationalization of higher education is its 'off-shoring' in the form of overseas campuses. These arrangements have attractions for both parties. For the source university, it opens up new markets, especially for parents who are unable to meet the tuition and living costs of overseas study, or are reluctant to send their children abroad for cultural reasons. The operating costs of these offshore campuses may also be cheaper. For the host nation, such an arrangement widens the study options for its students, and it may offer opportunities for collaborative learning interactions. Important also, the host nation may be able to establish itself as a regional hub for higher education. In what follows, we focus in part on the establishment of comprehensive campuses under the brand name of the foreign university and that offer a wide array of undergraduate courses, together with some graduate programmes. There are in addition many other academic interactions, including specialist programmes, for example, in business education, and a myriad of 'twinning' and exchange programmes of one sort or another.

In Southeast Asia, Singapore was the first to envisage such academic 'off shoring'. However, its initial forays, involving Warwick University and the University of New South Wales, were respectively aborted and extremely short-lived. Into this vacuum, Malaysia entered successfully, and it has now established itself as the major regional hub for international-quality higher education.[13] International tertiary student enrolments have risen quickly. There were practically none around 1990, but from the late 1990s the numbers began to increase rapidly, quadrupling over the period 2002–2010 to almost 87,000 in the latter year. Most of the growth has occurred in the private sector, which in 2010 accounted for almost three-quarters of total foreign enrolments. These institutions have grown very rapidly since this sector was opened up, with a variety of players entering the market. Included among the latter are some well-established, 'top 100' universities such as Monash and Nottingham. Important, these universities have been internationally oriented, seeking to attract not just domestic students, but also those from abroad. As a result, the country has emerged as one of the largest international hosts of foreign students in the developing world.

A priori, there are reasons both to anticipate and to doubt the Malaysian success. The former include the large investments in the country's higher education system over several decades,[14] and its many other attractions. The latter include an open economy and labour market,[15] high standards of liveability (including excellent physical infrastructure, relatively low levels of crime, and competitive living costs), a multicultural environment that enables students and their parents from diverse countries and ethnic backgrounds to feel comfortable in the country,[16] and widespread English-language proficiency. On the other hand, the principal handicap historically was a state-run (albeit generously funded) university system that was highly regulated. Much of this regulation was the result of the country's affirmative action programme introduced in 1971 that specified explicit quotas for students based on their ethnicity, and that also gradually but inexorably resulted in virtually all senior university positions being occupied by Bumiputera staff.[17] As a result, the country has persistently lost talented individuals, particularly from its ethnic Chinese community, to Singapore and elsewhere.

The Malaysian university sector has grown rapidly over the past half century. As recently as 1969, it had just one university, the University of Malaya, which originally had campuses in both Kuala Lumpur and Singapore. Although the state university system began to expand rapidly in the 1970s, demand for university education far exceeded supply. The government was reluctant to relax its regulatory barriers to entry and allow the entry of private universities. But it did begin to liberalize the system cautiously, through twinning arrangements with several lesser-known tertiary institutions in Australia and the United Kingdom (UK). The supply response was strong, and the government saw the opportunities to establish the country as a regional higher education hub. In 1996, it therefore introduced a more radical reform when it passed the Private Higher Educational Act. This enabled private institutions to establish degree-granting programmes, and it also allowed foreign universities to set up branch campuses in the country. Important also, these institutions were permitted to teach in English, a major breakthrough in what was then a requirement that university classes be conducted in Bahasa Malaysia, and thus constituting a reversion to the earlier policy of using English as the medium of instruction.

In addition, the government streamlined accreditation and quality control procedures through the establishment of the Malaysian Qualifications Agency (MQA), which was charged with approving the qualifications of all local programmes. Drawing also on professional bodies in the legal, accounting, and other fields, these accreditation procedures were seen as a necessary step in attracting foreign students. Some of the foreign university campuses are not required to seek MQA approval. Khong, from the private sector, reports that 'while there were grumblings among the private institutions about the additional layer of bureaucracy imposed on them, the agency has worked well, and MQA accreditation is now seen by the public to be an important indicator of educational quality' (Khong 2013, p. 226).

As noted, international student numbers have risen quickly, particularly those from Southeast Asia (with Indonesia the largest source country), China, and the Middle East, including Iran.[18] Moreover, Malaysian students have switched to private institutions in increasing numbers, notwithstanding their higher tuition fees as compared to the state universities. They have been attracted particularly to the local degrees from international institutions that involve outlays of as little as one-third of the cost of an equivalent overseas education. Most of the growth in enrolments has been in the private sector, as these institutions actively sought overseas markets, while foreign enrolments in the state universities were initially capped at 5 per cent of their total student population. However, this requirement was dropped in 2006, resulting in foreign student numbers in the latter more than trebling over the period 2006–2010. By 2011, total enrolments (both domestic and foreign) in the private universities were approaching those in the public institutes of higher education, with 452,000 in the former and 508,000 in the latter. Ironically also, the government indirectly contributed to the growth of the private institutions through the unusually young retirement age in state universities of 55 years, thus providing well-qualified academic staff with an incentive to transfer to the private sector.[19]

There have been further reforms in recent years. In 2010, foreign education providers were permitted to take equity stakes in Malaysian institutions, while it is expected that 100 per cent foreign ownership of universities will soon be possible. Important also, large numbers of Malaysian students continue to go abroad, with a significant number funded by government scholarships.[20] Thus the internationalization of its higher education sector has been a comprehensive two-way process. Large numbers of Malaysian students continue to go abroad, in fact still marginally more than the inflows.[21]

Malaysia appears to have managed quality control issues reasonably effectively. As noted, it has a functioning regulatory system, which even in high-income countries is a complex institution to establish and maintain. To safeguard quality, some foreign universities have a provision for simultaneous examinations where feasible, that is, for both their home and Malaysian campuses. In any case, fee-paying private students in this deregulated system are likely to be more demanding, and so the reforms have forced the state universities to compete.

Malaysia's deregulation does not of course necessarily lift tertiary education quality. It appears to be the case that the leading state universities, with their lower fees and established reputations, still get the best students. The private institutions are more likely to accept students with the minimum entry standards, and therefore they have a student body of more mixed quality, and their teaching staff have to pitch instruction at this middle standard. There is also a lament that almost all the foreign students come from developing countries, except for a small number attracted specifically to Malaysia for research or personal reasons. This is to be expected: students go abroad for a better-quality university experience than they can get at home. Thus the deregulation of the market is no substitute for measures aimed at lifting research-quality indicators of the country.

There have also been additional 'spillover' benefits or positive externalities for Malaysia from its open higher education policies. First, the country is now able to tap into a larger, internationally dispersed alumni network, with both commercial and broader sociopolitical benefits. Second, there is a well-established nexus between international education and tourism, particularly as the families of these international students are likely to visit their children while studying in Malaysia. Third, opening up the private sector has ameliorated the sense of discontent among the non-Bumiputera communities arising out of the discriminatory entry provisions that apply in the state universities. Indeed, the graduates of the stronger private universities, particularly those with international connections, generally have better employment prospects both at home (at least in the private sector) and abroad.

In summary, this case study illustrates the essential ingredients of 'smart' industry policy. Three in particular deserve emphasis. First, the country had many of the attributes to be a successful international higher education hub, including a multicultural, liveable environment, the earlier large government investments in higher education, and an open economy and labour market. Some of these attributes are of course the result of deliberate past policies and public investments. Second, there was a major and decisive deregulation that created the market opportunities for educational providers. In this respect, Malaysia has benefitted from a clear 'early mover' advantage, as few other developing countries have introduced such bold measures. Third, the government not only did not backtrack on this reform, but it also provided the additional regulatory environment that has underpinned its success, particularly with regard to accreditation and quality assurance.

4. The Philippine business process outsourcing services

The Philippines has a mixed record of industrial policy interventions, with its long history of prolonged import substitution generally regarded as one of the key explanations for its indifferent post-independence economic performance (Balisacan and Hill 2003). Trade policy reform, which commenced in the early 1980s, was interrupted by the deep economic and political crisis of that decade, but it resumed again, more decisively, from the late 1980s. The crisis also signalled the end of the government's ill-fated '11 major industrial projects', that were in any case mostly uneconomic.

It has, however, achieved major success in its business process outsourcing (BPO) services. These are internationally traded back-office services that undertake their operations in more cost-effective centres abroad. They have grown rapidly over the past two decades as the ongoing telecommunications revolution has opened up the possibility of an ever-wider array of digital cross-border service provision.

The Philippines has been a major participant in the industry, among developing countries ranking second only to India.[22] Growth has been very rapid. The industry hardly existed prior to the mid-1990s. By 2000, it was employing about 100,000 workers, and this figure has now increased to more than 700,000. The industry now constitutes almost 5 per cent of GDP, with annual export earnings of about US$12 billion.[23] The BPOs range in size from small-medium enterprises of around 100 employees to giants employing about 25,000 workers. The working environment is generally attractive, while wage levels are well above the prescribed minimums. They are therefore creating quality jobs for well-educated employees. The workforce is predominantly youthful; about 80 per cent are college graduates. About half the workforce is women. Entry-level salaries are about P15,000 per month, while experienced analysts earn around P40,000. Managers earn about P80,000. While located mainly in Metro Manila, owing to the strong agglomeration effects, these are in principle spatially footloose operations, and they are now diversifying to major regional centres where costs are lower.

Contact (call) centres are the largest subsector, accounting for almost two-thirds of turnover. But the industry is increasingly diverse, with back office, transcription, and a range of other IT services of increasing importance. As with overseas employment, thus far the industry's growth has been largely recession-proof, with turnover actually increasing during the 2008–2009 global financial crisis. The United States is by far the largest customer base, but the Asian share – particularly Japan – has been rising quickly in recent years. Foreign firms have dominated the industry, accounting for about 90 per cent of total investments in recent years, and rising. US firms have been by far the largest among the foreign-owned group.

This was arguably a case of 'accidental industry policy', in the sense that, at the broadest level of international advertising and promotion, the government did not initially set out to target the industry. There have been three key drivers. First, there was a global technological revolution in information and communication technology commencing in the 1990s that made possible the internationalization of these traded services on a large scale. Although the industry had always existed on a small scale, the arrival of the Internet lowered costs dramatically.

Second, the Philippine government implemented a comprehensive telecommunications liberalization in the 1990s. Historically the country had poor-quality telephone service, with a telephone density of just 1 per cent, run by a private monopoly, and with waiting periods of five or more years to obtain access to what was in any case an expensive and low-quality service (Abrenica and Llanto 2003; Salazar 2007). The telephone liberalization was one of the earliest and most successful reforms of the Ramos administration. Combined with the contemporaneous technological revolution in cellular communications, telephone density rose dramatically, by almost eightfold in the period 1992–2000, while costs fell very sharply.

Third, with these two factors in place, one global, the other domestic, the latent Philippine comparative advantage in semi-skilled labour service provision was revealed. That is, the country possesses a large pool of college-educated workers fluent in English, service-oriented, and culturally assimilated to the needs of the market, particularly that of the United States.[24] The country also has additional cost advantages in office rentals and other operating costs compared with India and China.

Although the government did not target the BPO industry, fiscal incentives offered by the Philippine Export Zone Authority (PEZA) enhanced the country's attractiveness, including tax holidays for up to eight years and import duty drawbacks. BPO operations typically locate in metropolitan (mainly Metro Manila) office blocks, but these are administratively regarded as export zones.[25] The country also enjoyed an early mover advantage, as the new entrants became established, and the country's international reputation was established.

This Philippine case study also illustrates the importance of seeking nuanced explanations for country export success. That is, success in one sector does not necessarily translate into success in other sectors with seemingly very similar characteristics and factor proportions. Specifically, while the BPOs achieved rapid export growth, two similar sectors grew much more slowly. First, in global production (manufacturing) networks, the Philippines achieved early success in the 1990s, mainly owing to the significant trade, investment, and infrastructure reforms of the Ramos presidency, 1992–1998. However, its international share of components trade has slipped over the past decade as other countries have reformed more quickly (Athukorala 2010). In particular, two factors have held back the global manufacturing networks that are not barriers to BPO export growth. One is the country's relatively high minimum wage regulations, which push up labour costs for the lower-skilled manufacturing operations but are not a relevant consideration for the more skilled BPO operations. The other constraint is that the global manufacturing operations have to contend with the country's inefficient port services and its customs complexities, both of which push up costs compared with competitor countries. The BPOs obviously face no such constraint. The second comparator is international tourism where the Philippines has lagged behind neighbours such as Indonesia and Thailand. In spite of government support and the country's many natural and cultural attractions, tourism has been held back by poor transport infrastructure, especially in-country, and security concerns.

5. The Cambodian garments industry

The export-oriented Cambodian garment industry is an interesting case study of success against the odds. This is a country that experienced deeply traumatic war and conflict for several decades, including one of the worst genocides in the second half of the twentieth century. At the time it re-engaged with the global community in the early 1990s, it was one of the poorest countries in the world. It had few functioning institutions, its infrastructure was shattered, and it had lost most of its skilled workforce. How it achieved export success in garments is therefore of broader analytical interest in understanding the dynamics of development in post-conflict societies.[26]

Garments had long existed as a handicraft activity in Cambodia, but its modern form, as an export-oriented factory sector, dates back only to the mid-1990s. The initial stimulus came from the granting of preferential market access to the European Union (EU) and the United States, which in turn induced major garment manufacturers to relocate to the country to avail themselves of these export privileges. Thereafter, export growth has been extremely rapid. Cambodia did not rank among the major garment exporters through to 2000, but its ranking has risen quickly since then. By 2012, it was the world's twelfth largest exporter by value, overtaking established exporters such as Pakistan, Sri Lanka, and Thailand. Garments have also become the country's major merchandise export, and by far the largest manufacturing employer.

As noted, preferential market access, particularly to the United States, was the initial trigger for growth. As recently as 2007, shortly after the abolition of the Multi-Fibre Arrangement, North America absorbed more than three-quarters of the total exports, while the EU took most of the rest. However, there has been rapid market diversification in recent years. The US export share is now just 40 per cent and, while the EU share has been rising, the fastest growth has occurred in the 'RoW' (Rest of World) category, principally East Asia, which has increased from a hitherto insignificant share to about 15 per cent of the total. This increased commodity and market diversification is a healthy indication of Cambodia's improved competitiveness and, therefore, its reduced reliance on temporary (and politically fickle) preferential market access provisions. The labour market effects have also been transformative. In a country with a non-agricultural workforce of about 3.5 million persons, a fraction of whom have formal-sector employment, almost 10 per cent were employed in garments. That figure has grown from less than 20,000 in 1995 to 320,000 by 2010, and in excess of 400,000 currently. Factory sector wages remain low in Cambodia, but they are well above agricultural sector earnings, and garments employment is reasonably secure.

Although this preferential market access was the major explanation for the initial rapid growth, the open, business-friendly economic policies that the Cambodian government has adopted have sustained it. Over time, these preferential facilities have become increasingly unimportant, as the MFA was phased out in 2005 and the margins of preference in major export markets have shrunk. The years after 2005 were seen as a crucial test of whether the industry could survive in the absence of this privileged market access. The export data show that, apart from the global recession of 2008–2009, growth rates have been maintained in this more competitive environment, buoyed also by the continuing relocation of the industry from higher-wage Asian economies.

As the industry has taken root, firm-level productivity is rising. Firms are beginning to move out of simple CMT (cut, make, and trim) production lines. Some are now also soliciting export orders in new markets rather than waiting for them from headquarters. Local personnel are replacing expatriate staff. Firms are moving into more fashion-intensive garments. Associated activities such as thread, embroidery, and accessory firms have begun to appear. New industry-sponsored training centres develop production and design skills.

Crucially, Cambodia was able to use the period of temporary export support to establish a country reputation among international investors and buyers, in garments and other light manufactures. This has resulted not just in export market diversification in garments, but also export diversification into other labour-intensive manufactures. The most important of these has been footwear, the exports of which have increased rapidly, from $88 million in 2008 to $311 million in 2012. Moreover, being open has meant that the country has been able to take advantage of unforeseen opportunities, particularly in the international relocation of labour-intensive manufacturing for export. For example, multinational enterprises with Asian production networks have long adopted a 'China plus one' risk diversification strategy. These firms have also adopted a 'Thailand plus one' strategy in response to the serious floods in 2011 that incapacitated that country's production and logistics networks for several months. Thailand is the major export-oriented hub for the automotive industry in Southeast Asia, and therefore several MNEs, particularly from Japan, sought to establish production facilities in neighbouring countries. Cambodia's open economy and business-friendly environment have made it a preferred destination, particularly as its government established special economic zones along the Thai–Cambodian border.

6. Indonesia's industrial policy experience

These case studies do not include Indonesia, the region's largest economy. This omission may at first sight appear puzzling. After all, the country has achieved rapid economic development since the late 1960s, and its industrial sector has grown rapidly, especially over the period 1968–1996, when for most years there was double-digit growth. Successive Indonesian governments have also adopted wide-ranging sector-specific promotion policies, frequently targeted at the industrial sector.[27] The explanation for Indonesia's omission from these case studies is that it is the exception that proves the rule. That is, its industrialization was driven overwhelmingly by effective economy-wide policies and not by sector-specific interventions. This can be demonstrated by examining the effects of industry-specific policies, that is, government measures that explicitly introduce non-neutral inter-industry incentives. These policies include trade barriers, fiscal incentives, investments by state-owned enterprises, and subsidized finance.

A detailed, quantitative study of these impacts is beyond the scope of this chapter. However, we draw here on earlier research by one of the authors, to demonstrate the methodologies involved.[28] To the best of our knowledge, the conclusions from this earlier work retain their contemporary validity. Certainly we are unaware of subsequent work that empirically challenges their conclusions.

First, a central argument in the literature on the efficacy of industrial policy is that policy makers deliberately nurture infant industries (or even firms) that, while not competitive in the short run, will become internationally competitive after a protracted period of learning and technology acquisition. For this argument to be valid, one needs to examine the subsequent performance of protected industries, allowing for lags that encompass the infant industry phase. There are technical measurement issues in such an exercise that are beyond the scope of this chapter, most importantly involving the measurement of 'success'.[29] Whatever the case, there is no compelling evidence that any of the historically most protected industries in Indonesia, such as the automotive, steel, petrochemical and other heavy industries, subsequently became spearheads of efficient industrial growth. In fact, on the contrary, using export growth as a performance indicator, not surprisingly the most dynamic sectors have been traditional labour-intensive industries. These industries were historically penalized by the protection accorded to the upstream sectors, and they have performed best when liberal trade and investment reforms have been in the ascendancy.

The explanation for such an outcome, and for why this key pillar of industrial policy has not been successful, lies in an understanding of the political economy of import protection in Indonesia. According to the seminal study of the subject, by Basri (2001), the most important determinant of variations in inter-industry protection has been what he termed the 'crony' variable. That is, the industries that received the highest protection were those dominated by Soeharto-linked business cronies. Basri further demonstrated that the causality generally (but not always) flowed from these cronies entering the industries, rather than the reverse. His study was for the period through to the mid-1990s, which also coincided with the rapid industrialization. Whether this analytical framework has the same explanatory force for the democratic period, when political power was rapidly diffused and patterns of corruption much less organized, remains an open question. There is, however, persistence in the inter-industry patterns of protection, with the most protected industries in the latter period similar to those of the earlier period (Marks and Rahardja 2012).

A second instrument of industrial policy has been state-owned enterprises (SOEs). Indonesia has traditionally had a large SOE sector, as a result of the nationalizations during the Sukarno period and the new ventures financed during commodity booms. Apart from the aftermath of the Asian Financial Crisis, few major attempts have been made to privatize them, owing to their political popularity as a counterbalance to the commercial domination of foreign and ethnic Chinese investors in the modern sector. Measuring their performance accurately is hindered by the lack of reliable financial data, and by the many implicit subsidies received and exactions imposed on them. By any indicator, however, these SOEs have generally performed poorly, owing to political interference, multiple (and sometimes conflicting) objectives, and a reluctance to delegate managerial autonomy. Moreover, there is no evidence from the case study material that they have played a catalytic role, directly or indirectly, as technological innovators.

Third, fiscal incentives have been employed intermittently, but with limited impact. They were introduced from 1967 onwards when Indonesia sought to re-establish its international commercial reputation, and they were initially available to most new foreign investors. Domestic investors were also able to take advantage of them. The investment regime became more restrictive from the mid-1970s, and the BKPM (Investment Coordinating Board) attempted to impose conditionality in the granting of incentives. However, it lacked the analytical expertise to implement any sort of formal industry policy. Its effectiveness anyway was diminished by widespread corruption and cumbersome administrative procedures. Fiscal incentives were largely abolished during the major tax reforms of 1984, and they played no role in the successful period of export-oriented industrialization in the decade following the reforms. Some incentives have been reintroduced in recent years, but with little impact.

A fourth tool of industry policy has been the provision of subsidized finance. State-owned banks dominated the financial sector until 1983, and they offered a range of subsidized lending programmes. Command lending to politically well-connected borrowers was widespread, as was bribery to secure this cheaper credit. Studies of these programmes generally concluded that there was no systematic difference in commercial success between recipients of subsidized and non-subsidized loans. This was because the value of the subsidies was diluted by the illegal payments, and because, in the system of credit rationing, loans were extended as much on the basis of political connections as commercial viability.

Finally, from the early 1980s, the Indonesian government embarked on a range of ambitious 'high-tech' projects under the direction of Research and Technology Minister (and later President) Habibie. The most important of these was the IPTN project to develop Indonesia's first domestically produced aeroplanes. There was a range of other projects, including shipbuilding and munitions. These were cases of 'back-to-front' industrialization, in that Indonesia then possessed a very limited industrial base of supplier firms with advanced engineering competence. Therefore the aircraft factory was somehow intended to leapfrog the process of technological development. However, these projects collapsed during the Asian Financial Crisis, when the government was no longer able to bankroll them, and the country subsequently had very little to show for the estimated $3 billion invested in them.

All this is not to deny that there were unquantifiable 'positive externalities' in some of these interventions that may have contributed subsequently to industrial success. For example, the Habibie high-tech projects sent many Indonesian students abroad for advanced scientific training. Some of the SOEs acquired advanced technological know-how through their R&D activities. The fiscal incentives may, at the margin, have attracted FDI embodying positive spillover effects that would not otherwise have come to Indonesia. There was undoubtedly some 'learning by doing' in protected activities such as the automotive industry.

But any assessment of these interventions has to also take account of their costs, and their effectiveness as compared to alternative strategies. For example, Indonesia needs to invest in advanced training for its citizens, but this could have been undertaken – as it now is – independent of the high-tech projects. (Meanwhile, the government's non-politicized research institute, LIPI, the Indonesian Institute of Sciences, has been starved of resources.) There is also little evidence that the SOEs have been technological innovators any more than their private sector counterparts. In the case of foreign investment, the learning and spillovers that occur are equally present in export-oriented ventures as in the protected import-substituting cases. In fact, evidence suggests that export orientation is a spur to greater efficiency and spillovers through competitive pressures and greater opportunity for international demonstration effects. Furthermore, as we have demonstrated, the more outward-looking Thai auto industry has developed stronger technological capabilities and local subcontracting networks than that of the more inward-looking Indonesian industry. Indonesia's limited participation in global production networks, driven by more restrictive trade and investment policies and complex regulatory systems, has also limited the scope for learning in these networks (Soejachmoen 2012).

7. Conclusions and lessons

At least six main conclusions regarding industry policy flow from our analysis. The first is the importance of openness. All cases of success involved export-oriented goods or services. This is a reflection of the openness of these economies, their small size, and their exploitation of comparative advantage. Foreign investors also played a significant role in transmitting technology and international market information. FDI and associated human capital flows were dominant in the Cambodian and Thai cases, and substantial as a catalyst in Malaysia and the Philippines. Cambodia attracted foreign investments in garments in part because its FDI regime was (and is) among the most open in the region. Malaysia went further than most countries in permitting the establishment of full foreign university campuses.

Second, government policies contributed significantly to the successes in one way or another. Often this simply took the form of a major deregulation, and thus it could hardly be described as a policy achievement, other than undoing the harm created by earlier policies. In Malaysia, it took the form of the government opening up the sector to new non-state entrants. It was then able to build on earlier, large investments in the public tertiary education sector, especially generous government scholarships for overseas graduate study, which had created a pool of academics with knowledge of the international industry, while credible regulatory arrangements to facilitate quality standards were also introduced. In the Philippines, the major telecoms deregulation was a key factor. In Thailand, the government switched more quickly than its neighbours to liberalize the auto industry, and actively courted foreign auto companies, particularly those from Japan, which already had deep roots in and familiarity with the country.

Third, there were specific country features in all cases that, combined with open policies, underpinned success. In the Philippines, it was English language proficiency, aided by an outward-looking and cosmopolitan culture. Malaysia's cultural and linguistic openness, to Islam, to Chinese, Indian, and Malay communities, and the widespread use of English all contributed to its educational success.

Fourth, a measure of good luck, including fortunate timing, was present in most of the cases. An obvious example is BPOs in the Philippines, where the government had recently liberalized a chronically repressed telecoms industry just in time to link in to the wave of technologically driven international outsourcing of services. The timing of Thailand's liberalization was also auspicious, in addition to its good fortune that each of its three potential Southeast Asian auto rivals – Indonesia, Malaysia, and the Philippines – had policy regimes that effectively precluded their international participation in the industry. Malaysia opened up a hitherto state-dominated tertiary education system just as the internationalization of higher education was gathering pace. Cambodia was awarded privileged garment export market access at a time when the MFA was still a significant constraint on international competition, and the industry therefore had a decade to establish its competitiveness and commercial reputation.

Fifth, a recurring theme of the studies is that, while openness creates commercial opportunities, further consolidating reforms are generally required. Thailand has emerged as the automotive hub of Southeast Asia, but its national R&D effort remains significantly underfunded, and on current conditions it could hardly aspire to develop its own indigenous auto technology capabilities. Cambodia will need to strengthen its general education system, as well as sector-specific skills, if it is to graduate from the 'CTM' model, and to upgrade to more innovative and fashion-intensive garment products. The internationalization of Malaysian higher education cannot be expected to provide a significant boost to the research intensity and standing of its university system.

Finally, there are considerable within-country variations and therefore the analysis also has to be institutionally (and possibly period) specific. This is most clearly illustrated in the Malaysian case, where four success stories have more or less coexisted with one very costly failure. In the Philippines also there have been marked differences, mostly obviously in the case of the early unsuccessful heavy industry forays and the current BPO success. In fact, even currently, for internationally traded activities, the Philippines has done much better in services (the BPOs) than it has in manufacturing, and also as compared to another major internationally traded service, tourism. Such a conclusion reflects the fact that government and leadership priorities change over time. Governments represent bundles of interests and constituents, while bureaucratic capacities and the specificities of the required policy reforms vary across sectors.

Notes

* We thank Dr Lili Ang Ing for inviting us to join the project, for administrative support, and for extensive comments on earlier drafts. We also thank Professor Fukunari Kimura and other participants at the workshops in Singapore (September 2014) and Tokyo (April 2015) for very useful comments on our earlier drafts. The usual caveat applies.

1 Bhagwati (2004) provides an excellent retrospective on the debates from a largely neoclassical perspective.

2 The literature on the subject is very broad. For recent and comprehensive surveys examining various aspects of industrialization and industrial policy, see Szirmai, et al. (2013) and Felipe (2015). One of the clearest expositions of the widely invoked infant industry argument is still that of Corden (1974).

3 Further details of this early system and its implementation are provided by Nawadhinsukh (1983), Doner (1991), Veloso et al. (1998), Doner et al. (2004), and Kohpaiboon (2006).

4 See, for example, Humphrey and Oeter (2000) and Takayasu and Mori (2004).

5 As the President of Toyota Motor Thailand observed of this period, 'Thailand is the best candidate for hub status because it has no 'national-car' policy and offers a level of playing field.' (*Bangkok Post* Economic Review, 1999)

6 See Doner (1991), Kohpaiboon (2006), and Suehiro (2010).

7 For example, forged crankshaft and connecting rods were manufactured by Isuzu and sold to Toyota and Nissan; Toyota and Nissan made casting cylinder blocks and cylinder heads, respectively, for the others.

8 Of this total, 122 firms supply both car and motorcycle makers.

9 An earlier study by Brimble and Sherman (1999) reached a similar conclusion.

10 The first generation of parts suppliers was typically Sino-Thai auto parts traders around Bangkok's Chinatown area. Most of them were persuaded by carmakers to become auto parts suppliers in the 1970s, producing mostly bulky and non-traded products.

11 There is a large literature on the early successes and failures. In the case of tropical cash crops, see Barlow (2010) and IDE (2015). For examinations of the early success in export-oriented electronics, see Ismail (1999) and Hobday (2001). See Athukorala (2014) on the costly failure in the automotive industry, and Rasiah (2012) for the country's struggle to develop an effective national innovation policy.

12 Estimates of the cost of creating de novo a university of international quality range up to $1 billion (Salmi 2009).

13 The following draws heavily on two informative accounts of the internationalization of higher education in Malaysia, by Khong (2013) and Tham (2013). Both authors are senior academics in the country, Khong in one of the more successful private universities and Tham in a leading state university.

14 Indeed, some researchers maintain that the government has over-invested in higher education at the expense of broad-based primary and secondary education. See Lee and Nagaraj (2012).

15 An open labour market is an essential prerequisite for the easy movement of foreign students and faculty. Malaysia has one of the most open labour markets in the world, with foreign workers constituting about 30 per cent of its workforce.

16 These include importantly those of Islamic faith and those seeking a Mandarin-speaking environment. The Islamic factor was increasingly crucial in the wake of the September 2001 US terrorist attacks, which resulted in more restrictive visa arrangements in the United States (US), the United Kingdom (UK), and elsewhere, which in practice particularly affected nationals from Islamic countries.

17 That is, the entry score requirements are lower for Bumiputera students than they are for students from the ethnic Chinese and Indian communities. There has also been a perception that Bumiputera academics enjoyed more rapid promotion opportunities.

18 In 2011, the major student sources were Iran, Indonesia, Yemen, and Iraq in state universities, and China, Indonesia, Nigeria, Iran, and Yemen in private universities.

19 Subsequently, the retirement age has been progressively increased to 60 years.

20 Khong (2013, p. 223) reported that almost 90,000 Malaysian students were enrolled abroad, 27,000 with scholarships from the government or other local organizations.

21 The interactive dimensions of these flows are further illustrated by the fact that Australia is now the largest destination country for Malaysian students abroad and the largest source of joint venture university projects in the country.

22 This section draws on di Gropello et al. (2010) and Yi (2012).

23 For comparative orders of magnitude, remittances from overseas workers, where some of the growth drivers are similar, are about double this figure.

24 The large pool of Philippine workers with overseas employment experience has also been a contributing factor. Currently, about 10 million Philippine workers work abroad out of a population of 100 million.

25 One would hesitate to conclude that the fiscal incentives were the crucial inducement. A more likely explanation is that they operated to negate some of the less attractive features of the country's business environment, including high utility costs and regulatory complexities.

26 This section draws on joint ongoing work one of the authors has with Jayant Menon. See, for example, Hill and Menon (2013).

27 We put aside here the issue of slower industrial growth since the Asian financial crisis, since the slowdown is unrelated to any change in industrial policies.

28 See Hill (1996) and Aswicahyono et al. (2010).

29 For example, whether the indicator is output growth, TFP growth, export performance, some broader 'technology capacity' variable; how long the infant industry phase should be; and what other policies and global factors may have impacted industry performance.

References

Abrenica, JV 1998, 'The Asian automotive industry: assessing the roles of state and market in the age of global competition', *Asian-Pacific Economic Literature*, vol. 12, no. 1, pp. 12–26.

Abrenica, J and Llanto, G (2003), 'Services', *in The Philippine Economy: Development, Policies, and Challenges*, eds A Balisacan and H Hill, Ateneo de Manila University Press, Quezon City, pp. 254–280.

Amsden, A 1989, *Asia's Next Giant: South Korea and Late Industrialization*, Oxford University Press, New York, NY.

Aswicahyono, H, Hill, H, and Narjoko, D 2010, 'Industrialization after a deep economic crisis: Indonesia', *Journal of Development Studies*, vol. 46, no. 6, pp. 1084–1108.

Athukorala, PC (ed.) 2010, *The Rise of Asia: Trade and Investment in Global Perspective*, Routledge, London.

Athukorala, PC 2014, 'Industrialisation through state–MNC partnership: lessons from Malaysia's national car project', *Malaysian Journal of Economic Studies*, vol. 51, pp. 113–127.

Balisacan, AM and Hill, H (eds.) 2003, *The Philippine Economy: Development, Policies and Challenges*, Oxford University Press, New York, NY.

Barlow, C 2010, *Malaysian Agriculture in Transition*, The James Jackson Memorial Lecture, Asian Studies Association of Australia, Flinders University.

Basri, MC 2001, 'The political economy of manufacturing protection in Indonesia, 1975–1995', unpublished doctoral dissertation, Australian National University.

Bhagwati, J 2004, *In Defense of Globalization*, Oxford University Press, New York, NY.

Brimble, P and Sherman, J 1999, 'Mergers and acquisitions in Thailand: the changing face of foreign direct investment', UN Conference on Trade and Development, Bangkok.

Chang, HJ 2002, *Kicking Away the Ladder: Development Strategy in Historical Perspective*, Anthem Press, London.

Corden, WM 1974, *Trade Policy and Economic Welfare*, Clarendon Press, Oxford, UK.

di Gropello, E, Tan, H, and Tandon, P 2010, *Skills for the Labor Market in the Philippines*, The World Bank, Washington, DC.

Doner, RF 1991, *Driving a Bargain: Automobile Industrialization and Japanese Firms in Southeast Asia*, University of California Press, Berkeley, CA.

Doner, RF, Noble, GW, and Ravenhill, J 2004, 'Production Networks in East Asia's Auto Parts Industry', in *Global Production Networking and Technological Change in East Asia*, eds. S Yusuf, MA Altak, and K Nabeshima, Oxford University Press, Washington, DC, pp. 159–208.

Felipe, J (ed.) 2015, *Development and Modern Industrial Policy in Practice: Issues and Country Experiences*, Edward Elgar, Cheltenham, UK.

Haraguchi, N 2010, 'Impact of the Global Economic Crisis on the Thai Automotive Industry: From the Perspective of the Interplay between Shocks and the Industrial Structure', *UNIDO Research and Statistics Branch Working Paper* 07/2009, UNIDO, Vienna.

Hill, H 1996, 'Indonesia's industrial policy and performance: "orthodoxy" vindicated', *Economic Development and Cultural Change*, vol. 45, no. 1, pp. 147–174.

Hill, H and Menon, J 2013, 'Cambodia: rapid growth with weak institutions', *Asian Economic Policy Review*, vol. 8, pp. 46–65.

Hobday, M 2001, 'The electronics industries of Pacific Asia', *Asia-Pacific Economic Literature*, vol. 15, no. 1, pp. 13–29.

Humphrey, J and Oeter, A 2000, 'Motor Industry Policies in Emerging Markets: Globalisation and the Promotion of Domestic Industry', in *Global Strategies and Local Realities: The Auto Industry in Emerging Markets*, eds. J Humphrey, Y Lecler, and MS Salerno, St. Martin's Press, New York, NY, pp. 1–15.

IDE 2014, *Catch-Up Industrialization in the 21st Century*, Institute of Developing Economies, Tokyo.

Ismail, MN 1999, 'Foreign Firms and National Technology Upgrading', in *Technology, Competitiveness, and the State: Malaysia's Industrial Technology Policies*, eds. KS Jomo and G Felker, Routledge, London, pp. 21–37.

Johnson, C 1982, *MITI and the Japanese Miracle*, Stanford University Press, Stanford, CA.
Khong KH 2013, 'The Transformation and Internationalization of Higher Education in Malaysia', in *Education in Indonesia*, eds. D Suryadarma and GW Jones, Institute of Southeast Asian Studies, Singapore, pp. 216–235.
Kohpaiboon, A 2006, *Multinational Enterprises and Industrial Transformation: Evidence from Thailand*, Edward Elgar, Cheltenham, UK.
Kohpaiboon, A and Jongwanich, J 2013, 'International production network, clusters and industrial upgrading: evidence from automotive and hard disk drive industries', *Review of Policy Research*, vol. 30, no. 2, pp. 211–239.
Kondo, H 2012, *Auto Industry in Thailand and Automotive Human Resource Development*, JETRO, Bangkok.
Lee, KH and Nagaraj, S 2012, 'The Crisis in Education', in *Malaysian Development Challenges: Graduating from the Middle*, eds. H Hill, SY Tham, and HMZ Ragayah, Routledge, London, pp. 213–232.
Lin, J 2012, *New Structural Economics: A Framework for Rethinking Development and Policy*, The World Bank, Washington, DC.
Marks, SV and Rahardja, S 2012, 'Effective rates of protection revisited for Indonesia', *Bulletin of Indonesian Economic Studies*, vol.48, no.1, pp.57–84.
Nawadhinsukh, S 1983, 'Ancillary Firm Development in the Thai Automobile Industry', in *The Motor Vehicle Industry in Asia: A Study of Ancillary Firm Development*, ed. K Odaka, Singapore University Press, Singapore, p. 405.
Orihashi, S 2010, 'Manufacturing and Management Systems of Japanese Manufacturers in Southeast Asia: The Case of Automobile Industry in Thailand', in *Sustainability of Thailand's Competitiveness: The Policy Challenges*, eds. P Intarakumnerd and Y Lecler, Institute of Southeast Asian Studies, Singapore, pp. 209–234.
Perkins, D 2013, *East Asian Development: Foundations and Strategies*, Harvard University Press, Cambridge, MA.
Rasiah, R 2012, 'Is Malaysia's Electronics Industry Moving Up the Value Chain?', in *Malaysia's Development Challenges: Graduating from the Middle*, eds. H Hill, SW Tham, and HMZ Ragayah, Routledge, London, pp. 194–212.
Salazar, L 2007, *Getting a Dial Tone: Telecommunications Liberalisation in Malaysia and the Philippines*, Institute of Southeast Asian Studies, Singapore.
Salmi, J 2009, *The Challenge of Establishing World-Class Universities*, The World Bank, Washington, DC.
Soejachmoen, M 2012, 'Why is Indonesia left behind in global production networks?', PhD dissertation, Australian National University, Canberra.
Suehiro, A 2010, 'Industrial Restructuring Policies in Thailand: Japanese or American Approach', in *Sustainability of Thailand's Competitiveness: The Policy Challenges*, eds. P Intarakumnerd and Y Lecler, Institute of Southeast Asian Studies, Singapore, pp. 129–173.
Szirmai, A, Naude, W, and Alcorta, L (eds.) 2013, *Pathways to Industrialization in the Twenty First Century: New Challenges and Emerging Paradigms*, Oxford University Press, Oxford, UK.
Takayasu, K and Mori, M 2004, 'The Global Strategies of Japanese Vehicle Assemblers and the Implications of the Thai Automobile Industry', in *Global Production Networking and Technological Change in East Asia*, eds. S Yusuf, MA Altaf, and K Nabeshima, Oxford University Press, Washington, DC, pp. 209–253.

Tham, SY (ed.) 2013, *Internationalizing Higher Education in Malaysia: Understanding, Practices and Challenges*, Institute of Southeast Asian Studies, Singapore.

Veloso, F, Romero, JS, and Amsden, A 1998, *A Comparative Assessment of the Development of the Auto Parts Industry in Taiwan and Mexico: Policy Implications for Thailand*, MIT Press, Cambridge, MA.

Welch, A 2011, *Higher Education in Southeast Asia: Blurring Borders, Changing Balance*, Routledge, London.

World Bank 2012, *Leading with Ideas: Skills for Growth and Equity in Thailand*, The World Bank, Bangkok.

Yi, S 2012, 'Reaching the World through Private Sector Initiative: Service Exports from the Philippines', in *Exporting Services: A Developing Country Perspective*, eds. AG Goswani, A Mattoo, and S Saez, The World Bank, Washington, DC, pp. 121–159.

11 The employment effects of GVCs on Asian countries and the phenomenon of value-added erosion

Xiao Jiang[1] *and Jose Caraballo*[2]

1. Introduction

Global value chains (GVCs) and international production sharing are not new phenomena. According to Athukorala and Menon (2010), researchers have been documenting such phenomena since the 1920s. However, what is special about GVCs today is their rapid expansion in terms of both quantity and locations. Trade in intermediates, as one of the key indications for the degree of GVC participation, started to become prominent in the 1990s with China's entry into the global production system. The process was accelerated by the setting up of low-cost offshore facilities by electronics companies after the burst of the dotcom bubble in the early 2000s (Milberg and Winkler 2010). World exports from developing countries continued to grow throughout this period, but their composition also started to change as imports of intermediates increased steadily in the 1990s and accelerated in the 2000s. By the end of the 2000s, trade in intermediates accounted for about 56 per cent of total world trade in the case of goods and 73 per cent in the case of services (Miroudot et al. 2009). In terms of the total trade volume, world trade in parts and components increased from US$502 billion in 1992–1993 to US$11,762 billion by 2005–2006 (Wim et al. 2013).

With the prevalence of GVCs, the relationship between trade and employment becomes more complicated. The labour content associated with a country's foreign trade is no longer simply of two kinds: (1) domestic labour contained in exports; and (2) foreign labour contained in imports. With the addition of GVC trade, there are three more categories of employment to consider: (3) foreign labour contained in exports; (4) domestic labour contained in imports; and (5) third-countries' labour contained in a country's imports. The recent publication of the World Input–Output Database (WIOD)[3] allows us to calculate all five categories of employment generated by trade over the period 1995–2011 for a panel of countries that covers 85 per cent of world gross domestic product (GDP).

In this chapter, we focus on six Asian economies – China, Indonesia, India, Japan, the Republic of Korea (henceforth, Korea), and Taiwan in the WIOD. These economies are chosen partly because of data constraints. Several amongst them are facing the challenge of shifting towards production of higher value-added goods and services, bringing them into more direct competition with advanced economies, while at the same time their competitiveness in lower value added goods and services is being eroded by the increasing presence of lower-labour cost countries in global markets. In addition, all of these countries, despite their development heterogeneity, failed to increase their share of value added in the period 1995–2009.

This chapter is made up of five sections. Section 2 introduces the WIOD data set, as well as the methodology of global input–output analysis that is applied to this data set to estimate the employment effects of GVCs (or trade in intermediate inputs). Section 3 presents and analyses the estimated employment generated by GVCs in these Asian economies. Section 4 uses some of the results obtained from Section 3 to address a specific research question: what causes the phenomenon of 'value-added erosion'[4] for some countries? We explore this question by conducting a panel data co-integration econometric analysis. Section 5 discusses some policy implications of our research.

2. GVCs, employment and input–output model

Studying the employment effects of GVCs requires a clear definition of GVCs and a method for their measurement. While this is an area subject to numerous disagreements in the literature, in this chapter we use trade flow of intermediate inputs as the indicator for GVCs. Thus, employment (both domestic and foreign) generated by a country's trade in intermediates is treated as the employment generated by this country's participation in GVCs.

Since the empirical method for studying the effects of trade in intermediates is grounded in the input–output framework[5] and the WIOD is a database of input–output tables, the method we adopt in this chapter is grounded in input–output analysis by necessity. An input–output table is a representation of national accounts showing the flows of all economic transactions for a country and is a quantitative economic technique that is used to analyse the impact of a change in final demand (including exports) on the levels of production. Assuming fixed labour, capital, and inter-industry input requirements for production, this model enables us to calculate the economy-wide labour content of a country's export demand.

An input–output model can also be extended to take into account not only inter-industry linkages, but also inter-regional or international linkages. This is known as the multiregional input–output model. With a harmonised international input–output data set and sufficient information on international sectoral linkages, this model can be applied to extract all kinds of labour content embodied in each country's foreign trade. Later we briefly introduce the model we use in this chapter to estimate the employment effect of GVCs, but further details can be found in Miller and Blair (2009).

We assume there are three countries (1, 2, and 3) in the world, trading in both intermediates and final outputs with each other. Let A^{ij} be the imported intermediate coefficient matrix, intermediates are exported from country i and used by country j. When $i = j$, it becomes the domestic input–output coefficient matrix. Υ^i is country i's total output. Following the system of three equations the final demands of the three countries can be expressed in the following way:

$$\begin{aligned} \Upsilon^1 - A^{11}\Upsilon^1 - A^{12}\Upsilon^2 - A^{13}\Upsilon^3 &= f^1 \\ -A^{21}\Upsilon^1 + \Upsilon^2 - A^{22}\Upsilon^2 - A^{23}\Upsilon^3 &= f^2 \\ -A^{31}\Upsilon^1 - A^{32}\Upsilon^2 + \Upsilon^3 - A^{33}\Upsilon^3 &= f^3 \end{aligned} \tag{1}$$

System (1) states that, for each country, final demand is the residual of total outputs minus domestic and exported intermediates.[6] If we collect the terms and rearrange the system into a system of partitioned matrix multiplication, we yield Equation (2):

$$\begin{bmatrix} I - A^{11} & -A^{12} & -A^{13} \\ -A^{21} & I - A^{22} & -A^{23} \\ -A^{31} & -A^{32} & I - A^{33} \end{bmatrix} . \begin{bmatrix} \Upsilon^1 \\ \Upsilon^2 \\ \Upsilon^3 \end{bmatrix} = \begin{bmatrix} f^1 \\ f^2 \\ f^3 \end{bmatrix} \tag{2}$$

At this point, one can easily see that the multi-countries input–output system is almost identical to the single-country system except that scalars are now replaced with matrices. If we factor out the identity matrices from the left of Equation (2) and move them to the right hand side, then we obtain a global Leontief inverse matrix:

$$L^G = \left[\begin{bmatrix} I & 0 & 0 \\ 0 & I & 0 \\ 0 & 0 & I \end{bmatrix} - \begin{bmatrix} A^{11} & A^{12} & A^{13} \\ A^{21} & A^{22} & A^{23} \\ A^{31} & A^{32} & A^{33} \end{bmatrix}^{-1} \right] = \begin{bmatrix} A^{11} & A^{12} & A^{13} \\ A^{21} & A^{22} & A^{23} \\ A^{31} & A^{32} & A^{33} \end{bmatrix} \tag{3}$$

This global Leontief inverse matrix L^G is of great importance. This is a $n \times n$ matrix with $n = 3 \times m$, and m is the number of sectors. At this stage, we partition this matrix into a 3-by-3 partitioned matrix for future use. As a partitioned matrix, it contains nine small matrices each of dimension $m \times m$. This matrix multiplies a partitioned vector (where the partitioning is done in the same order as the three countries) of final demands, which will give us the total outputs for each country that are required to produce that vector of final demands directly and indirectly.

Let T_i ($3 \times m$) be the trade vector from the perspective of country 2. At position 2, it will be the vector of country 2's exports, and at positions 1 and 3, there will be the vector of country 1 and 3's exports, respectively. Multiplying the global Leontief inverse matrix by the trade vector T_i will give us ζ_i, the total value generated by country 2's foreign trade globally.

$$\zeta_i = L^g . T_i \tag{4}$$

As mentioned in the introduction, multi-countries trade with the existence of intermediate trades that contain five components – *exports* and *imports* of final goods, *import content of exports*, *export content of imports*, and *third-country*

intermediate contents in a home country's imports. Equation (4)cannot help us to decompose the trade into these five components because matrix multiplication would aggregate all the rows. So we construct matrix Θ, using the partitioned global Leontief inverse matrix (3) and we obtain:

$$\Theta_2 = \begin{bmatrix} l^{11}t_1 & l^{12}t_2 & l^{13}t_3 \\ l^{21}t_1 & l^{22}t_2 & l^{23}t_3 \\ l^{31}t_1 & l^{32}t_2 & l^{33}t_3 \end{bmatrix} \tag{5}$$

Essentially, Θ_2 is constructed by multiplying l^{ij} $(m \times m)$ by $t_j(m \times 1)$ at each position. In this way, Θ_2 is of dimension $m \times 3 \times 3$, with each sub-matrix of dimension $m \times 1$. Matrix Θ_2 is of great importance because it contains the five components of country 2's trade.

The diagonal items are total value generated by country 2's final goods exported (position {2, 2}), and countries 1 and 3's final goods exported to country 2 (Positions{1,1} and {3,3}). The vertical elements up and down from the 'home position' {2, 2} are foreign values generated from country 2's intermediate use of countries 1 and 3's exports, and the horizontal elements left and right from the home position are total value generated in the home country due to country's 1 and 3's exports, channelled through foreign demand of the home country's exports as intermediates. The rest of the elements in the Θ_2 matrix (e.g., {1, 3} and {3, 1}) are values generated in country 1 due to country 3's production of exports for country 2, and vice versa. At this stage, the five components of total values generated by a country's trade are identified and explained.

Finally, we could add the employment dimension to this model. Let $\widehat{E}_i$ be the diagonal matrix of labour coefficients for country *i*.We multiply each partitioned row $(m \times 1)$ of the Θ_2 matrix by the corresponding $\widehat{E}_i$ $(m \times m)$matrix, the Θ_2 matrix is transformed into an employment matrix Λ_2, as shown in (6):

$$\Lambda_2 = \begin{bmatrix} \widehat{E}_1 l^{11}t_1 & \widehat{E}_1 l^{12}t_2 & \widehat{E}_1 l^{13}t_3 \\ \widehat{E}_2 l^{21}t_1 & \widehat{E}_2 l^{22}t_2 & \widehat{E}_2 l^{23}t_3 \\ \widehat{E}_3 l^{31}t_1 & \widehat{E}_3 l^{32}t_2 & \widehat{E}_3 l^{33}t_3 \end{bmatrix} \tag{6}$$

Λ_2 is country 2's matrix of trade-induced employments, and the kind of employment generated by each of the five components of trades is shown in the same order as in (5).

3. Trade and employment in GVCs

The World Input–Output Database has several unique features. First, it provides input–output tables and bilateral trade data for 40 economies, which covers 85 per cent of world GDP. Second, all the data are harmonised into 35 input–output sectors. Thus, cross-countries comparisons are possible. Third, the bilateral trade data are split into intermediate and final goods traded, and traded intermediates are reported as intermediate trade basic flow matrix for each country. Therefore, the imported intermediate coefficient matrix can be constructed for each country. Fourth, in a separate account called the Social Economic Accounts (SEA), employment data in terms of number of persons engaged, total hours worked, and total hours worked by skill types for every country on sectoral level are provided. With these accounts, labour coefficient matrices $\widehat{E}_i$ can be constructed. Finally, most of the data from this database are provided as a 17-year time-series (1995–2011).[7]

Table 11.1 presents the 17-year average of the total employment generated in each of the five components of trade for the most recent year, in this case 2011, for the panel of Asian economies – China, India, Indonesia, Japan, Korea, and Taiwan. Information at the sectoral level has been aggregated to a single employment figure for each country.

We can also view the total domestic labour demand for each country as the sum of labour demand by domestic exports and domestic content of imports. The sum of the rest is counted as the total foreign labour demand resulting from each country's trade, on average over 17 years. These results are presented in Table 11.2. It comes as no surprise to see that China, India, and Indonesia's foreign trade generated more domestic employment than foreign employment

Table 11.1 Employment generated by five components of foreign trade, 17-year average (in 1000s)

	Exports	*Imports*	*Import content of exports*	*Export content of imports*	*Import content of imports*
China	110,576	9,780	2,276	1,604	1,362
India	36,957	4,369	655	65	329
Indonesia	12,155	3143	447	24	266
Japan	4305	25,008	1,539	72	1,846
Korea	3,675	11,442	2,445	30	787
Taiwan	2,888	4,285	1,633	18	626

Source: Authors' own calculations.

Table 11.2 Domestic and foreign employment generated, 17-year average (in 1000s)

	Domestic labour	*Foreign labour*	*Difference*
China	112,180	13,419	98,761
Indonesia	12,179	3,876	8,303
India	37,023	5,353	31,670
Japan	4,377	28,394	–24,017
Korea	3,705	14,674	–10,969
Taiwan	2,906	6,545	–3,639

Source: Authors' own calculations.

given their export-oriented development strategies. The three more developed economies – Japan, Korea, and Taiwan – on the other hand resulted in more foreign employment than domestic employment via trade.

The social economic account of the WIOD contains sectoral employment information based on three skill types: high, medium, and low. Skill types are defined based on education attainment level of the average worker of the sector, where high, medium, and low skill refers to post-secondary education attained, secondary education attained, and secondary education unattained, respectively (Timmer 2012). Using this information, in addition to estimating employment generated by each country's five components of foreign trade, we can study the skill composition of our employment results. Table 11.3 reports the skill composition of each economy's employment generated by each of the five components of trade, for 2009.[8]

Let us first look at the results for final goods exported. It is clear that the emerging economies such as China, India, and Indonesia tend to export final goods with very limited high-skilled labour content – all less than 10 per cent – whereas more developed economies such as Japan, Korea, and Taiwan tend to export final goods with far more high- and medium-skilled labour content. This result seems to conform to the Grossman and Rossi–Hansberg argument that developed countries endowed with more high-skilled labour tend to export goods with higher high-skill labour content, and vice versa for developing countries (Grossman and Rossi-Hansberg 2006), an extension of the standard Heckscher–Ohlin theory. However, the Grossman and Rossi–Hansberg/Heckscher–Ohlin pattern is less evident when we look into the skill composition of employment generated by final exports. All economies in our panel tend to import final goods and service containing mostly medium- and low-skilled labour content (above 80 per cent). China is the economy in which final imports contain the highest share of high-skilled labour content, at 17.7 per cent. What we observe here, amongst many other possible explanations, could be evidence of intra-industry trade for these Asian economies.

Let us now turn to the import content of export (ICE), which is an important indicator for the degree of a country's participation in GVCs. For all countries in our panel, their ICE contains mostly medium- and low-skilled labour. China once again stands out as the country with ICE containing about 20 per cent of foreign high-skilled labour. What determines the skill composition of a country's ICE? The broader question is: what determines the pattern of global production sharing? This is a theoretical area of GVC analysis that urgently needs more contributions. However, it is clear from our results that global production sharing is determined by mechanisms other than relative factor/skills endowments. The Grossman and Rossi–Hansberg/Heckscher–Ohlin pattern appears again in export content of imports (ECI) with developing economies' ECI containing more low-skilled labour content and developed economies' ECI containing more high-skilled labour content. However, for most countries, the employment effect of ECI is very small compared with other components of trade. To develop

Table 11.3 Skill composition of employment generated in 2009 (in 1000s)

		Exports	*Imports*	*Import content of exports*	*Export content of imports*	*Import content of imports*
China	High skilled	5.80%	17.71%	19.36%	7.13%	19.22%
	Medium skilled	34.63%	44.67%	45.82%	39.80%	48.58%
	Low skilled	59.57%	37.61%	34.81%	53.06%	32.19%
Indonesia	High skilled	5.87%	7.69%	6.86%	6.13%	13.16%
	Medium skilled	21.31%	33.90%	32.46%	22.80%	43.38%
	Low skilled	72.83%	58.41%	60.67%	71.07%	43.45%
India	High skilled	8.83%	5.77%	5.56%	9.55%	14.20%
	Medium skilled	34.10%	33.39%	33.38%	34.84%	45.57%
	Low skilled	57.04%	60.84%	61.05%	55.58%	40.22%
Japan	High skilled	24.52%	7.49%	10.78%	24.62%	12.60%
	Medium skilled	66.35%	33.17%	40.38%	66.38%	41.29%
	Low skilled	9.13%	59.34%	48.83%	9.00%	46.10%
Korea	High skilled	41.06%	9.25%	10.15%	41.51%	12.71%
	Medium skilled	49.14%	42.14%	43.03%	49.18%	43.50%
	Low skilled	9.83%	48.61%	46.81%	9.34%	43.78%
Taiwan	High skilled	27.60%	10.50%	11.53%	28.12%	12.80%
	Medium skilled	39.37%	41.13%	43.34%	40.06%	43.49%
	Low skilled	33.01%	48.36%	45.13%	31.81%	43.70%

Source: Authors' own calculations.

some intuitive feeling for how the relationship between GVC participation and employment changes over time for these Asian economies, employment generated by import content of exports for each country is plotted on a time-scale in Figure 11.1.

Import content of exports is often viewed as an important indicator of the degree of a country's integration to the GVCs (Hummels et al. 2001). Many have observed and documented a pattern of steady increases of the import content of exports over the past 20 years (Miroudot et al. 2009; Meng et al. 2011; Wim et al. 2013). Figure 11.1, therefore, is the employment aspect of import content of exports for the WIOD Asian economies over time. It is evident that almost all Asian economies have been increasing their demand for foreign labour via their import content of exports over the period 1995–2011, as expected. Indonesia is an exception. However, it does not imply that Indonesia's import content of export did not increase over time. The relatively flat number of jobs generated by its import content of export might be the result of Indonesia importing foreign intermediates that are increasingly less labour intensive. What is the consequence of increasing foreign labour embodied in a country's exports in the form of import content of exports? The answer to this question depends on the skill type of the foreign labour embodied, and we will explore this question in the next section.

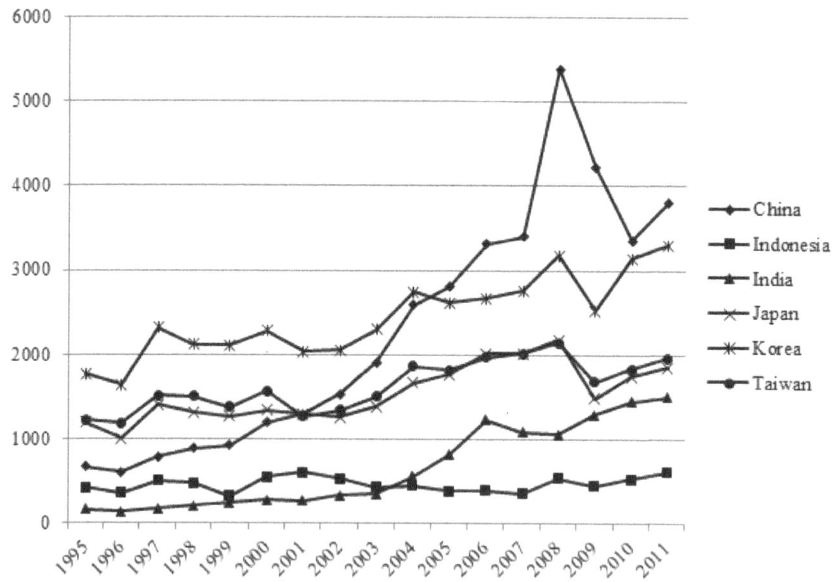

Figure 11.1 Employment generated by import content of exports, 1995–2011

Source: Authors' own calculations.

4. High-skilled labour in import content of exports and value-added erosion

In the previous sections, we introduced the methodology of calculating employment generated by the five components of trade for each of the Asian economies, and analysed the general results of our calculations. In this section, we will use some of these results to explain a phenomenon – the reduction of domestic value-added share of value generated by exports due to the increase of foreign intermediates; we call this phenomenon 'value-added erosion'.

4.1 Value added vs. value-added share

Before considering value-added erosion, which is defined by the decline of a country's total domestic value-added share in exports, we engage in a brief discussion on the usefulness of the concept of value-added share. The low domestic value-added share for countries participating in GVCs was brought to our attention by various studies in the late 2000s, the most widely cited being Koopman et al. (2010), Xing and Detert (2010), and Kraemer et al. (2011). Such findings generated some anxiety amongst policymakers and triggered policy discussions about whether increasing domestic value-added share should be a new policy objective. At the same time, some voices argued that increasing value-added share might be a misguided policy measure, and that the emphasis should be on the total *volume* of domestic value added rather than the *share* of domestic value added, as it is the former that generates absolute income for a country. This line of argument is further backed up by the endogenous growth theory arguing that foreign intermediates facilitate technological diffusion in the home country. Therefore, it would presumably stimulate exports and output growth. In other words, they argue for *complementarity* between domestic value added and foreign intermediates, because the later supposedly fuels the growth of the former. The argument seems to be that, as long as foreign intermediates are generating high volumes of domestic value added, the decline of domestic value added share should not be a concern.

We acknowledge this argument has some merits, but we would like to provide an alternative perspective, summed up in four points. First, the complementarity between domestic value added and foreign intermediates is extremely difficult to verify empirically as these variables are highly endogenous. Any trade expansion would likely lead to a simultaneous increase of both domestic value added and foreign intermediates in exports. Hence, to empirically establish a causal relationship between these two variables is very difficult.

Second, regarding the endogenous growth story, although some foreign intermediates in the form of high-tech components might stimulate domestic growth and might increase domestic productivity via technological and information diffusion, it is difficult to see how foreign intermediates in the form of financing, marketing, and advertising would help to spread advanced foreign technology in a country. There is growing evidence that foreign intermediates are increasingly taking on the form of financing, marketing and advertising (Heinz 2006; Milberg and Winkler 2010).

Third, if one cannot provide a solid empirical justification for the complementarity of foreign intermediates and domestic value added in exports, the validity of the argument emphasising volume over share has to rest on the assumption that global effective demand is always expanding, and growth of the total size of the pie will be greater than the decrease in the share of the pie. There is no reason to believe in an ever-expanding global effective demand in the future. In fact, the 2008 Great Recession showed the high vulnerability of global demand to shocks that emanated from some developed countries, particularly the United States (Baldwin 2009). Furthermore, from a Keynesian perspective, a very uneven global distribution of income may slow growth of global effective demand.[9] If global effective demand slows at some point, capturing a larger share of the gains from globalisation will be a major concern.

Lastly, the debate over share and volume has its roots in economic ideology. This debate is essentially a mirror image of the heated debate on wage rate versus wage share between neoclassical and post-Keynesian economists. If one believes that the labour market is perfect and that wage rate is determined by the marginal product of labour, the decline of wage share should not be a major concern. Similarly, if one believes the market structure in GVCs is more or less perfect, the decline of domestic value-added share should not be a worrisome phenomenon either. However, numerous authors point out the asymmetries of market and bargaining power in GVCs and have raised concerns of the distributional conflicts in GVCs.[10] Therefore, if we believe the distribution of the gains from global production sharing are, to a large extent, the result of asymmetric power relations, the decline of domestic value-added share should be a policy concern. We made the four points earlier hoping to simply show that we might not want to totally disregard concerns about the decline of the domestic value-added share.

4.2 Value-added erosion and its measurement

The steady increase of imported intermediate content in exports for most of the countries surveyed has been well documented in the literature, but its effect on domestic value added is not very clear. If we decompose the total value generated by a unit of exports into three shares – domestic intermediate share, foreign intermediate share, and value added share – we will see that an increase of the foreign intermediates share can result in three possible outcomes. First, we may observe a reduction of the domestic intermediate share compensating for the increase of the foreign intermediate share. We call this effect 'domestic–foreign-substitution'. Second, the increase of the foreign intermediate share maybe compensated for by a reduction of the value-added share. This case is in fact similar to Xi and Detert (2010)'s iPod example. We call this effect 'value-added erosion'. The third outcome is, of course, a combination of the previous two effects. From this point of view, the effect of an increase of the foreign intermediate share on the domestic value-added share should vary from country to country.

With the national input–output data we are able to compute the three components of a country's exports over the period of 15 years.$[I - A]^{-1} X$ is the vector of total values generated by a country's exports. This vector contains three components in a globalised economy: Domestic Intermediate (DI), Foreign Intermediate (FI), and Value Added Generated by Exports (VAE). These three components can be empirically calculated by the following equations.

$$DI = B^{DT} [I - A]^{-1} X \tag{7}$$

$$FI = B^{MT} [I - A]^{-1} X \tag{8}$$

$$VAE = \widehat{va} [I - A]^{-1} X \tag{9}$$

In these equations, *B* is called the direct output coefficient matrix. Unlike the input coefficient matrix *A*, matrix *B* is acquired by dividing each row of the basic flow matrix *Z*, by gross outputs. The superscripts *D* and *M* denote domestic and imported, respectively. $\widehat{va}$ in (9) is the diagonal matrix of value added coefficients. The reason we use *B* here in our decomposition comes from the following input–output identity:

$$Y = B^T Y + VA \tag{10}$$

where *VA* is the vector of value added. Since we are interested in how much value added is generated in a country's exports, using *B*matrix instead of *A* is necessary. An important balancing condition underlies equations (7) to (9) – the sum of the shares has to equal to one.

$$\frac{\sum DI + \sum FI + \sum VAE}{\sum [I - A]^{-1} X} = 1A \tag{11}$$

This balance condition restates the earlier discussion – that an increase in *FI* (which should be expected from the observed pattern of GVC expansion) has to be compensated for by a reduction of *DI*, or *VAE*, or some combination of both. Figure 11.2 takes China and India as examples to illustrate the phenomenon of value-added erosion and domestic–foreign substitution in aggregate.

It is clear from the upper panel that, in the case of China, the increase of the foreign intermediate share (FI) over time is compensated for by the decline of the domestic value-added share (VAE). Therefore, this is called value -added erosion. In the case of India in the lower panel, the increase of the foreign intermediates share over time has been compensated for by the decline of domestic intermediates. Therefore, this is an example of domestic–foreign substitution.

Now, the question becomes, when facing the steady increase of foreign intermediate share of countries' exports, what determines a country's response? Or more precisely: why would some countries experience value-added erosion as they participate in GVCs by importing more foreign intermediates to produce

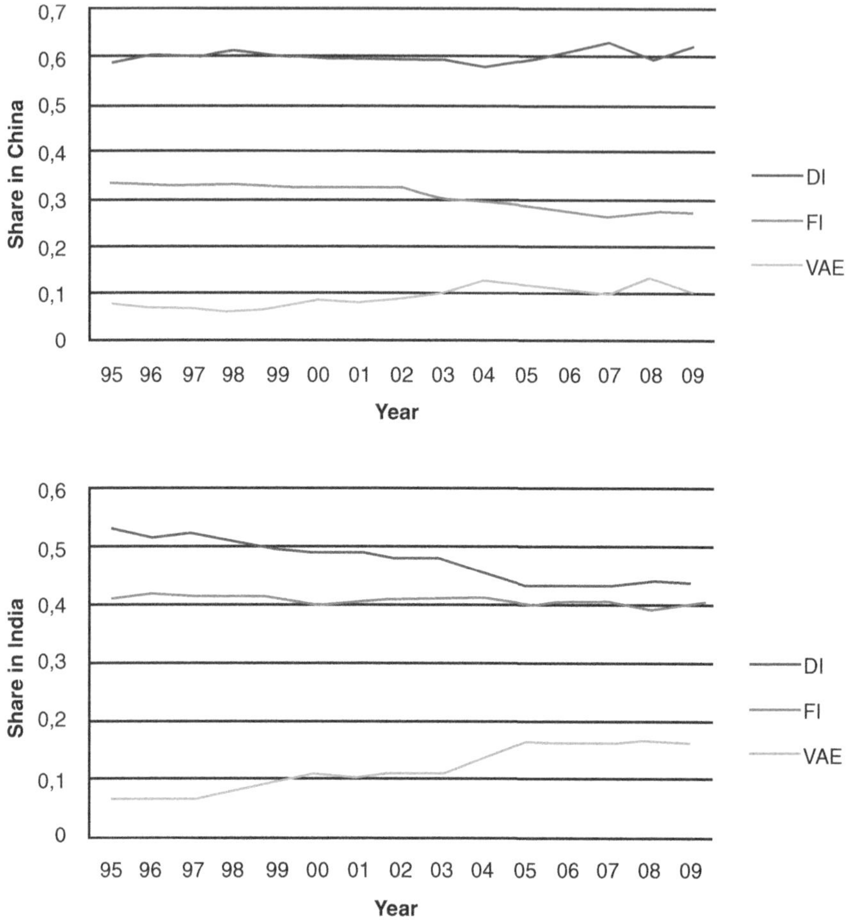

Figure 11.2 Composition of value generated by exports, China and India, 1995–2009

Source: Authors' own calculations.

their export goods? Heinz (2006) hypothesised a dynamic process of foreign high value-adding activities eroding the domestic value-added portion of exports. However, this hypothesis has never been validated empirically at the macro-level (to our knowledge). We should note that the notion of 'value-added erosion' is already apparent in his theory. Heinz's distinction between high value-adding and low value-adding activities is also of great importance to us. If we follow the international trade theorists who believe international trade is 'trading tasks',[11] we could approximate high value-adding activities by hours worked by high-skilled workers, and low value-adding activities by hours worked by low skilled workers. This approximation is consistent with Heinz's argument, since the types of activities that leading firms are likely to perform, according to Heinz, tend to require high-skilled labour. If this is a good approximation, we should be able to explain domestic value-added erosion by the injection of high-skilled foreign labour in a country's import content of exports.

Figure 11.3 illustrates the evolution of foreign high-skilled labour share in China and India's aggregate import content of exports. The aggregate results for China and India seem to verify our hypothesis based on Heinz (2006): China is a country that experienced value-added erosion and is also a country that experienced a surge of foreign high-skilled labour share embodied in its import content of exports. However, this did not occur in India, where the country experienced domestic-foreign substitution.

While the aggregate results for China and India seem to indicate a positive relationship between the surge of foreign high-skilled labour embodied in a country's import content of exports and the decline of domestic value-added share over time, we still need to find out whether this relationship holds for all the countries in our panel across sectors. We also need to understand the other determinants of domestic value-added share. We address these questions systematically using econometrics in the next subsection.

4.3 The model

To answer the question of what causes value-added erosion, we apply panel data regressions to the two variables in question. The first is the by-sector share of foreign high-skilled labour embodied in a country's import content of exports (FHS), which is estimated using (1)–(6). The second is the by-sector share of

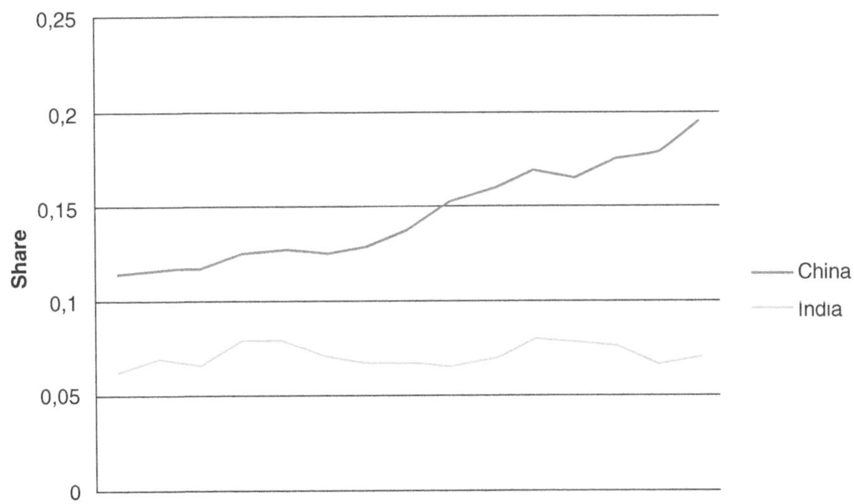

Figure 11.3 Share of foreign high-skilled labour in import content of exports, China and India

Source: Authors' own calculations.

domestic value added generated by a country's exports, which is estimated using (7)–(11). To avoid over- or underestimating the effect of FHS, we added a set of control variables from the WIOD. Using 33 out of the 35 sectors[12] from the WIOD, we applied the following panel data regressions,

$$V_{it} = C + \theta F_{it}\, \varphi X_{it} + \rho P_{it} + \delta K_{it} + \vartheta T_{it} + \mathrm{u}_{it} \tag{12}$$

where V is the domestic share of domestic value added generated by exports in sector i, F is FHS, X is labour productivity, P is the price level, K is real capital stock, T is total real output in the previous period, and u is deviations. Through the Variance Inflation Factor, no problems of collinearity were found.

P serves as the control variable for the effect output prices (at the sector level) may have on domestic value-added share by impacting final demand, K controls for capital-intensive sectors (also sectors where large investment has taken place), T controls for the output sold that may affect the current share of domestic value added, and finally, X indicates the role of labour productivity in stimulating or discouraging the relative importance of certain sectors. Labour productivity is fundamental to many trade theories, including classical trade theory, since it determines a country's comparative advantage and the sectors that will predominate in total exports.

The series in (12) are I(1). According to the Panel Augmented Dickey–Fuller (ADF) test, the Panel Phillip–Perron (PP) test, the Group ADF test, and the PP test, the series in (12) are co-integrated. To rectify this process, we applied a dynamic ordinary least squares (DOLS) estimator to correct for potential endogeneity and serial correlation if there is any (Kao et al. 1999). The augmented co-integrating equation associated with (12) is:

$$\begin{aligned} V_{it} = {} & C + \rho_f F_{it} + \rho_x X_{it} + \rho_p P_{it} + \rho_k K_{it} + \rho_t T_{it} + \gamma_{f,0}\Delta F_{it} \\ & + \gamma_{f,-j}\Delta F_{i,t+j} + \gamma_{f,j}\Delta F_{i,t-j} + \gamma_{x,0}\Delta X_{it} + \gamma_{x,-j}\Delta X_{i,t+j} \\ & + \gamma_{X,j}\Delta X_{i,t-j} + \gamma_{p,0}\Delta P_{it} + \gamma_{p,-j}\Delta P_{i,t+j} + \gamma_{p,j}\Delta P_{i,t-j} + \gamma_{k,0}\Delta K_{i} \\ & + \gamma_{k,-j}\Delta K_{i,t+j} + \gamma_{t,j}\Delta T_{i,t-j} + \gamma_{t,0}\Delta T_{it} + \gamma_{t,-j}\Delta T_{i,t+j} + \end{aligned} \tag{13}$$

where j are the leads and lags chosen by following different information criteria, such as the Akaike Information Criterion (AIC), to remove long-run correlations amongst the innovations. The long-run variance of the error term is estimated by:

$$\hat{\lambda}_v^2 = \frac{\hat{\sigma}_e^2}{\left(1 - \hat{\varphi}_1 - \ldots - \hat{\varphi}_q\right)^2},\ \hat{\sigma}_e^2 = \frac{1}{T-q}\sum_{t=q+1}^{T} k(t,b)\,\hat{e}_t^2$$

where $\hat{\varphi}_j\,(j = 1,2,\ldots,q)$ are estimated from an AR(q) process fitted to V_{it}, whose estimated residuals are given by $\hat{e}_t^2$. k(t,b) is a weight where k is a symmetrical Bartlett kernel function and b > 0 is a Newey–West fixed bandwidth. The q lag length was selected from AIC. Attempting to capture some idiosyncratic elements at the industry level, we also apply the weighing method à la Mark and Sul (1999) to allow heterogeneity in the long-run variances across industries.

4.4 The results

Amongst the six economies considered here, China had the greatest losses in the share of domestic value added generated by exports. From 1995 to 2009, the share of value added in Chinese exports decreased by 11 percentage points, while it decreased by 9 percentage points in Korea, 3 in Taiwan, 4 in Japan, 2.4 in Indonesia, and 1 in India. Again, the question to be addressed is: what factors drive this erosion in value-added share that has occurred in each of the countries we analysed?

We conducted two estimations for Equation (13) – one for all sectors included in the WIOD, and the other for the 16 manufacturing sectors given their importance in global production sharing, especially for most of the Asian economies. As suspected, an increase of the Foreign High-skilled Labor Share (FHS) had a negative impact on the domestic share of value added in the six Asian countries under study. After controlling for the impacts of productivity, capital intensiveness, output dependence, and inflation, we found that the lower domestic value-added share is generally associated with higher FHS in each industry. These results are illustrated in Table 11.4.

Table 11.4 Estimation of Equation (13), all sectors, by country, 1995–2009

Dependent variable: share of value added in value generated by exports						
Determinants	*India*	*Japan*	*Korea*	*China*	*Indonesia*	*Taiwan*
FHS	-.21*** (.03)	-.17*** (.04)	-.33*** (.06)	-.11*** (.03)	-.16** (.08)	-.25*** (0.04)
Labour productivity	-.0005 (.0004)	.02** (.008)	.003* (.002)	.006*** (.002)	-.0003 (.001)	.12 (.08)
Capital stock	-.43*** (.05)	.03*** (.01)	-.06 (.07)	-.15*** (.03)	0.02 (.05)	.21*** (.06)
Output in t-1	.18*** (.05)	-.18*** (.07)	-.26*** (.08)	.09*** (.02)	-.22*** (.07)	-.51*** (.13)
Output prices	.14*** (.02)	-.39*** (.04)	-.51*** (.08)	.06*** (.02)	-.05*** (.02)	.39*** (.04)
N	416	429	429	416	416	416
S.E. of regression	0.34	0.34	0.57	0.26	0.62	0.44

Notes: S.E. indicates the standard error of the regression. All variables were standardised. Standard errors are in parentheses. A linear term was added as deterministic variables. Standard errors are in parentheses. The *** indicates statistical significance at 99% confidence interval, ** at 95% and * 90%. Sample differences only depend on the availability of data.

Source: WIOD (2014)

Since each variable was standardised, we can compare the magnitude of each regressor. We observed that FHS was the factor with the second largest influence in China, India, and Korea. In the cases of Japan, Korea, and Taiwan, the major determinant for the movements in the share of value added was the change in output prices. For Indonesia and Taiwan, the main factor was the output produced in the previous period, which had a negative impact on the current share of value added. In India, the greatest influence came from capital stock: a sector's lower domestic value-added share is associated with higher capital intensiveness. Overall, labour productivity does not appear to be a primary determinant for the share of value added, either for the economy as a whole, or for the manufacturing sectors represented in Table 11.5.

According to Table 11.5, FHS appears to play a central role in India, but not by decreasing the domestic share of value added. This result is consistent with Goldberg et al. (2010) finding that trade reform has benefited Indian exporting firms in a way that enables them to substitute domestic intermediates with foreign intermediates. In other words, India's participation in GVCs has resulted in domestic–foreign inputs substitution rather than value-added erosion. However, in economies where manufacturing is more intense, such as Indonesia and China, FHS was once again the factor with the second largest influence in reducing the domestic share of value-added exports. The main driver for the

Table 11.5 Regressions for manufacturing sectors only, by country, 1995–2009

Dependent variable: share of value added in value generated by exports						
Countries	*India*	*Japan*	*Korea*	*China*	*Indonesia*	*Taiwan*
FHS	.22**	−.23	.07	−.23***	−.27***	−.08
	(.09)	(.15)	(.13)	(.02)	(.09)	(0.19)
Labour Productivity	−.07***	.01	.003	−.007***	−.0006	.21**
	(.03)	(.009)	(.002)	(.002)	(.003)	(.09)
Capital Stock	−.02	−.85***	−1.13***	−.41***	−.43***	−.39
	(.11)	(.14)	(.31)	(.03)	(.11)	(.32)
Output in t-1	−.12	.29	.41**	.02*	−.19***	.23
	(.12)	(.19)	(.22)	(.01)	(.08)	(.37)
Output Prices	.12***	−.32***	−.51***	−.008	−.02	.52***
	(.03)	(.04)	(.08)	(.02)	(.01)	(.07)
N	169	169	169	169	169	169
S.E. of Regression	0.35	0.31	0.44	0.13	0.30	0.36

Notes: S.E. indicates the standard error of the regression. All variables were standardised. Standard errors are in parentheses. A linear term was added as deterministic variables. Standard errors are in parentheses. The ***indicates statistical significance at 99% confidence interval, **at 95% and *90%.

Source: WIOD (2014).

decline in the domestic share of value added for manufacturing sectors was capital stock, especially for China, Indonesia, Japan, and Korea – capital-intensive sectors are more likely to experience a decline of the domestic share of value added in exports. This implies that capital-intensive manufacturing industries tend to experience more difficulties in capturing the gains from GVCs. A high reliance on foreign intermediates rather than the domestic economy might be one of the reasons.

Total production in the previous periods does not appear to have a primary role in the decline of domestic value added, while output prices appear to be a relevant determinant for Japan, Korea, and Taiwan. In Taiwan, output prices had a positive correlation with the domestic share of value added, and in Japan and Korea, output prices had an inverse relationship.

5. Concluding remarks and policy implications

In a world where GVCs are prevalent (as measured by trade in intermediates), value created by a country's foreign trade is made up of five components – direct exports, direct imports, import content of exports, export content of imports, and import content of imports. Regarding employment, each of the five elements would generate its own jobs domestically or internationally. Using the World Input–Output Database (WIOD), we are able to empirically decompose the trade of each of the six large Asian economies (China, India, Indonesia, Japan, Korea, and Taiwan) into those five components, and further compute the labour content in each of them. We found that, on average, developing economies tend to generate more domestic than foreign employment through their foreign trade in both intermediate and final goods. We also decomposed labour content embodied in each of those five components of trade by skill types and found that, according to the most recent 2009 figures, the skill composition of final exports tends to conform to the standard Heckscher–Ohlin prediction that developing economies export more goods and services with low-skilled labour content, whereas developed economies export more goods and services with high-skilled labour content. However, the Heckscher–Ohlin pattern is not found in the skill composition of import content of exports. It appears that labour embodied in import content of exports for all the economies in our panel is mostly low- and medium-skilled labour. This finding is a crucial step in a better understanding of trade and employment in relation to GVCs.

It is evident that the relation between trade and employment is much more complicated in a world where GVCs are prevalent. However, it is worth emphasising that this complication also potentially provides more policy instruments in using trade policies to influence employment outcomes. Back to the matrix Λ in (6), traditionally, trade policies only focus on the diagonal items – exports and imports. But in a world with trade in intermediate inputs, trade policies can also focus on the off-diagonal elements. In principle, jobs can be increased by policies reducing the import content of exports and/or expanding the export content of imports, whereas exports and imports on the diagonal of Λ are unchanged, *ceteris paribus*. However, policies that influence the employment outcomes of trade in intermediates (the off-diagonal elements in the Λ matrix) would have to focus on GVC analysis and industrial upgrading.

In this chapter, we also take advantage of some of the results obtained from the Λ matrix, specifically foreign high-skill labour embodied in a country's import content of exports to empirically examine the phenomenon of value-added erosion, which is characterised by a decline of the domestic value-added share as a country participates in GVCs. A decline of the domestic value-added share is sometimes a misguided concern of policy makers, given the complementarity between foreign intermediates and the volume of domestic value added. We made several points to suggest that concerns about a decline of the domestic value-added share could be justified, especially given the asymmetric market structure in many GVCs. Our econometric results consistently suggest that the higher the FHS in each industry, the lower the domestic value-added share. This relationship holds for each of the countries in our sample and is robust to the addition of covariates.

The relation between value-added erosion and the increase of the high-skilled labour share in the import content of exports has several policy implications. First, from a GVC perspective, if a policy were to be designed for a developing country to prevent value-added erosion, this policy has to encourage domestic producers to engage in more value-adding activities. In other words, it has to promote so-called *Functional Upgrading*(Humphrey 2004; Xing and Detert 2010).Second, capital intensiveness of manufacturing firms also affects negatively the share of domestic value added generated by exports. Authorities seeking to attract foreign investment should be cautious of these types of industries, as well as of firms that operate in high disconnection with the domestic economy. Incentives should be provided to domestic and foreign firms with great concatenation with the domestic economy. One measure that authorities could implement is to require that a certain percentage of high-skilled intermediates be bought from domestic producers, which is a regulatory tool China has been using in its Special Economic Zones since the start of its economic reform. This measure can also be agreed regionally or globally to foster global development by distributing the gains from globalisation more fairly. Third, not all foreign intermediates are the same and, hence, different types of foreign intermediates should be treated differently by the local governance body. Foreign intermediates injected into the domestic production structure via very asymmetric market structures and those (e.g., in the form of financing, marketing, and advertising) that do not help in the diffusion of advanced technology should be carefully regulated, whereas others should be fostered and encouraged. After all, as Rodrik puts it nicely: 'a complicated world requires fox-like policies' (2011).

Notes

1 Assistant Professor of Economics, Denison University, Granville, OH, United States. .jiangx@denison.edu. The author is thankful for the comments and

suggestions from the participants at ERIA's 1st and 2nd Workshop of Trade in Value Added in Singapore, September 2014, and Tokyo, April 2015. Financial support from the ERIA is gratefully acknowledged.

2 Assistant Professor of Economics, University of Puerto Rico at Cayey, Puerto Rico. jose.caraballo8@upr.edu.

3 See Timmer (2012) for details on the WIOD.

4 The reduction of value-added share in the value generated by a country's exports due to the increase of foreign content of exports.

5 See, Hummels et al. (2001); Escaith et al. (2010); Koopman et al. (2010); WTO (2013) and Meng et al. (2011).

6 For example, $A^{11}\Upsilon^{1}$ is country 1's domestic intermediates, and $A^{12}\Upsilon^{2}$ is the intermediate use of country 1's exports by country 2.

7 With the most up-to-date version of the database, most of the data have been updated to 2011, except for a few accounts in its Social Economic Accounts.

8 The WIOD has most of its data updated to 2011, but the skill type-based employment data in its social economic account is not. At the time of writing, the most up-to-date data are still 2009.

9 See Von Arnim et al. (2012) and La Marca (2013).

10 See Kaplinsky (2000); Milberg (2004); Nolan and Zhang (2010); Nicita et al. (2012); and Gereffi (2012).

11 See Grossman and Rossi–Hansberg (2006).

12 In our analysis, we took out 'Sale, Maintenance and Repair of Motor Vehicles and Motorcycles', and 'Private Households with Employed Persons' sectors from WIOD due to too many missing values for these sectors.

References

Athukorala, P and Menon, J 2010, 'Global production sharing, trade patterns, and determinants of trade flows in East Asia', *ADB Working Paper Series on Regional Economic Integration* No. 41, ADB, Manila.

Baldwin, R (ed.) 2009, *The Great Trade Collapse: Causes, Consequences and Prospects*, Centre for Economic Policy Research, London.

Escaith, H, Lindenberg, N, and Miroudot, S 2010, 'International supply chain and trade elasticity in times of global crisis', *WTO Staff Working Paper* 8/2010, WTO, Geneva, Switzerland.

Gereffi, G 2012, *Global Value Chains in a Post-Washington Consensus World: Shifting Governance Structures, Trade Patterns and Development Prospects*, Duke University, Durham, NC.

Goldberg, K, Khandelwal, A, Pavcnik N, and Topalova, N 2010, 'Imported intermediate inputs and domestic product growth: evidence from India', *The Quarterly Journal of Economics*, vol. 125, no. 4, pp. 1727–1767. Source: www.nber.org/papers/w14416.

Grossman, G and Rossi-Hansberg, E 2006, 'The rise of offshoring: it's not wine for cloth anymore', Proceedings of *Economic Policy Symposium*, Federal Reserve Bank of Kansas City, pp. 59–102.

Heintz, J 2006, 'Low-wage manufacturing and global commodity chains: a model in the unequal exchange tradition', *Cambridge Journal of Economics*, vol. 30, pp. 507–520.

Humphrey, J 2004, 'Upgrading in global value chains', *ILO Policy Integration Department Working Paper Series* No. 28, ILO, Geneva, Switzerland.

Hummels, D, Ishii, D, and Yi, K 2001, 'The nature and growth of vertical specialization in world trade', *Journal of International Economics*, vol. 54, pp. 75–96.

Kao, C, Chiang, M, and Chen, B 1999, 'International R&D spillovers: an application of estimation and inference in panel cointegration', *Oxford Bulletin of Economics and Statistics*, vol. 61, pp. 691–709.

Kaplinsky, R 2000, 'Globalization and unequalization: what can be learned from value chain analysis?', *Journal of Development Studies*, vol. 37, no. 2, pp. 117–146.

Koopman, R, Powers, W, Wang, Z, and Wei, S 2010, 'Give credit where credit is due: tracing value adding global production chains', *NBER Working Paper Series* No. 16426, NBER, Cambridge, MA.

Kraemer, KL, Linden, G, and Dedrick, J (2011), 'Capturing Value in Global Networks: Apple's iPad and iPhone', research supported by grants from the Alfred P. Sload Foundation and the US. Lin, T, Zhuang, J, Yarcia D, and Lin, F (2008), 'Income inequality in the People's Republic of Chian and its decomposition: 1990–2004', *Asian Development Review*, 25(1 and 2), pp. 119–136.

La Marca, M and Lee, S 2013, 'Wage and Growth in Open Economies: A Policy Dilemma?', in *Beyond Macroeconomic Stability: Structure Transformation and Inclusive Development*, eds. I Iyanatul and D Kucera, International Labour Organization, Geneva, Switzerland, pp. 263–287.

Mark, N and Sul, D 1999, 'A computationally simple cointegration vector estimator for panel data', Ohio State University manuscript.

Meng, B, Yamano, N, and Webb, C 2011, 'Application of factor decomposition techniques to vertical specialization measurements', *IDE Discussion Paper* No. 276, Institute of Developing Economies, Japan External Trade Organization (JETRO), Tokyo.

Milberg, W 2004, 'The changing structure of trade linked to global production systems: what are the policy implications?', *International Labour Review*, vol. 43, no. 1, pp. 45–90.

Milberg, W and Winkler, D 2010. 'Trade, Crisis, and Recovery: Restructuring Global Value Chains,' in O. Cattaneo, G. Gereffi and C. Staritz (eds): *Global Value Chains in a Post Crisis World: A Development Perspective.* Washington, DC: World Bank, pp.23–72.

Miller, R and Blair, P 2009, *Input–Output Analysis* (2nd ed.), Cambridge University Press, New York, NY.

Miroudot, S, Lanz, R, and Ragoussis, A 2009, 'Trade in intermediate goods and services', *OECD Trade Policy Working Paper* No. 93, OECD, Paris.

Nicita, A, Ognivetsev, V, and Shirotori, M 2012, 'Global supply chains: trade and economic policies in developing countries', UNCTAD Policy Issues in International Trade and Commodities Study Series.

Nolan, P and Zhang, J 2010, 'Global competition after the financial crisis', *New Left Review*, vol. 64, pp. 97–108.

Rodrik, D 2011, *The Globalization Paradox: Democracy and the Future of the World Economy*, W.W. Norton & Company, New York, NY.

Timmer, M 2012, *The World Input–Output Database (WIOD): Contents, Sources and Methods*, OECD, Paris.

Von Arnim, R, Tavani, D, and Barbosa, L 2012, 'Globalization as coordination failure: a Keynesian perspective', *The New School Working Paper* No. 2, The New School, New York, NY.

Wim, N, Szirmai, A, and Lavopa, A 2013, 'Industrialization lessons from BRICS: a comparative analysis', *IZA Discussion Paper Series* No. 7543, Institute for the Study of Labor, Bonn, Germany.

Xing, Y and Detert, N 2010, 'How the iPhone widens the US trade deficit with the PRC', *Asian Development Bank Institute Working Paper* No. 257, ADBI, Tokyo.

12 How labour market policies affect innovation and trade competitiveness

Siwage Dharma Negara[1]

1. Introduction

Endogenous growth theory postulates that innovation is one of the key drivers of technological progress and productivity growth of a country (Romer 1990; Aghion and Howitt 2005). Technological improvement stemming from a firm's innovative activities can lead to more efficient production processes and generate better-quality outputs. In an increasingly competitive global economy, innovation capacity is crucial for firms to improve productivity and hence sustain their export competitiveness (Aw et al. 2009; Cassiman et al. 2010).

Innovation requires an environment conducive to supporting firms to conduct innovative activities. Studies show that public policies, including labour market policies, can influence the operating conditions and institutional structures of firms to foster innovation that leads to productivity gains. Organisation for Economic Co-operation and Development (OECD) studies in this area suggest that different regulatory environments may explain performance differentials between countries, or between industries with regard to innovation (OECD 2002, 2009).

Certain regulatory policies may affect firms' incentives to innovate. For example, regulations that restrict competition (e.g., creating entry barriers or operational restrictions), or limit the ability of firms to adjust their workforce (e.g., stricter hiring and firing rules), can adversely affect incentives to innovate. This in turn could hinder the process of technological transfer and deny the economy potential productivity improvements.

Labour market policies can influence the allocation of resources across firms and sectors in the economy. They affect firms' choices over labour inputs, investments, technology, and outputs (Boeri et al. 2008). Different labour market policies induce greater or lesser adjustment costs for firms to reallocate workers, thus affecting overall economic efficiency. An efficient labour market is defined as one that ensures the flexibility to shift workers from one production activity to another at low cost (World Economic Forum 2014). Efficiency in the labour market is critical in ensuring that labour is assigned to its most effective use in the production process. An efficient labour market provides opportunities for firms to foster innovation activities, which may involve labour replacement or job reallocation, without excessive social disruption. In this chapter, the phrase 'efficient and flexible labour market' is used interchangeably.

The literature on the impact of labour market policies on firms' incentives to innovate provides mixed empirical evidence. Some studies found that less-stringent labour market policies are associated with stronger innovation competitiveness (OECD 2002; Saint-Paul 2002; Barbosa and Faria 2011; Murphy et al. 2013; Griffith and Macartney 2014). Conversely, other studies also found that stricter labour market policies can foster innovation and lead to greater country-level economic growth. It was argued that stringent labour market policies that limit firms' ability to dismiss workers act as an 'insurance device' for firms not to punish employees when certain innovative activities fail. In turn, this increases workers' incentives to engage in innovative activity (Acharya et al. 2013; Tang 2012).

Countries at different development stages have different labour market requirements, which influence their respective labour market policies. In this chapter, an attempt is made to construct a set of balanced panels consisting of OECD and non-OECD countries to investigate the relationship between labour market policies and countries' innovation capacity. It also looks at how the latter relates to countries' trade competitiveness. We argue that a more efficient labour market characterized by less-rigid labour market policies stimulates innovation. In turn, higher innovation positively impacts countries' trade competitiveness.

This chapter is organized as follows. Section 2 reviews selected studies examining the relationship between labour market policies, and innovation intensity and performance. Section 3 proposes simple empirical models to test the relationship between labour market policies and countries' innovation competitiveness. Section 4 discusses the data used for the estimation and presents stylized facts. Section 5 discusses the empirical findings, and Section 6 concludes.

2. Literature review

There are several studies assessing the relationship between labour market policies and innovation. This literature survey was not meant to be exhaustive, but seeks to provide an assessment of the current state of knowledge and highlight the existing knowledge gaps where future research might be focused. Note that most, if not all, of the studies assessing the relationship between labour market policies and innovation primarily focus on OECD member countries. Availability of good-quality firm- and industrial-level data in OECD countries may explain this trend. However, for non-OECD countries, especially for developing countries in Southeast Asia, there is a lack of good-quality microdata on innovation and productivity (Lee and Narjoko 2015).

Labour market policies vary from country to country. In general, these policies aim to achieve social (rather than economic) objectives, such as providing employment protection measures for workers. Employment protection policies are primarily introduced to protect workers from adverse labour market risks, such as lay-offs or low earnings. Empirically, these policies vary widely across and within countries (see Botero et al. 2004 for detailed discussion of each policy). To protect the earnings of the lowest-level workers, some governments set minimum wages.[2] Some governments also set regulations obliging employers to provide extra benefits to their workers, including training, health care, paid vacations, and maternity leave. Some governments provide unemployment insurance for those who lose their jobs.[3] And some governments prefer to set regulations that restrict firms' ability to dismiss employees at will.

The existing literature suggests that the relationship between labour market policies and innovation capacity can be both positive and negative. On one hand, a stricter labour market policy provides job security; hence this should increase incentives for workers to invest in firm-specific human capital and to be more engaged in innovation activities (see Acharya et al. 2013). On the other hand, a stricter labour market policy, which causes higher hiring and firing costs, will increase the cost for firms that need to adjust their workforce in relation to innovation activities. This in turn will discourage firms from conducting significant innovative activities (see Bassinini and Ernst 2002; Saint-Paul 2002; Barbosa and Faria 2011).

Murphy et al. (2013) identified two channels through which labour market policies (employment protection regulations) may affect innovation. The first channel is linked through human capital investment. In this channel, labour market policy is likely to increase the probability of workers engaging in firm-specific or industry-specific skills. It also increases the probability of workers engaging in innovation activities (Murphy et al. 2013). Furthermore, a stricter labour market policy is likely to increase workers' bargaining power and their incentives to invest in firm-specific skills. Acharya et al. (2013) found that stronger employment protection regulations had a positive impact on innovation at the industry level. It also led to relatively more innovation in innovation-intensive industries than in traditional industries.

Acharya et al. (2013) found that stronger employment protection laws not only foster innovation, but also lead to greater country-level economic growth. They argued that regulations governing the dismissal of workers are the only dimension of labour laws that enhance firm-level innovation and country-level economic growth. They explained that employment protection laws impose limits on firms' ability to dismiss workers. This acts as an 'insurance device' encouraging firms not to punish their workers as a result of an unsuccessful innovative project. Consequently, this incentivizes workers to increase their investment in innovative projects relative to their investment in routine projects. Therefore, stringent employment protection laws encourage firms to find innovative projects that are more value enhancing than routine projects.

Tang (2012) showed that cross-country differences in labour market policies shape the pattern of international trade, with a focus on workers' skill acquisition. Similar to Acharya et al. (2013), Tang found that countries with more protective labour laws export more in firm-specific skill-intensive sectors through both the intensive and extensive margins of trade.[4]

Contrary to the findings of Acharya et al. (2013) and Tang (2012), Barbosa and Faria (2011) found that stricter labour market policy leads to less innovation intensity at the industry level in European countries. They found that in European countries that have rigid labour market policy, the innovation intensity decreases by 1.89 for a unit increase in the indicator of employment rigidity holding other variables constant (see Barbosa and Faria 2011). These findings are in line with Bassanini and Ernst (2002), who investigated the impact of product and labour market regulations on innovation by using a set of OECD indicators of the regulatory framework on a cross-section of 18 OECD countries and 18 manufacturing industries. Their results showed a positive association between more flexible labour market policies and research and development (R&D) intensity, which is used as proxy for innovation.

The second channel through which labour market policy may affect innovation is through firms' adjustment costs when they need to adjust against idiosyncratic shocks (Murphy et al. 2013).[5] In this case, a more rigid labour market policy inflates hiring and firing costs for firms. As a result, this is likely to discourage innovative activities that require adjustments or reallocations of labour. The adjustment cost will be higher in technologically advanced industries (Saint-Paul 2002). Stricter employment protection regulations discourage firms from experimenting with new technologies with higher returns, but also with higher adjustment costs. Empirically, employment protection restrictions are more costly in industries with rapid technological change, such as information and telecommunications technology (ICT). Therefore, countries with stricter employment protection regulations are likely to specialize in industries with a lower rate of technical change (Murphy et al. 2013).

Pierre and Scarpetta (2006) found that innovative firms are the most negatively affected by stricter labour market policies. Griffith and Macartney (2014) found that multinational firms are likely to locate more innovation activities in countries with stricter employment protection policies. However, the same firms locate more technologically advanced innovation activities in countries with less-stringent employment protection policies. Furthermore, Murphy et al. (2013) found that the effects of labour market policies on innovation vary by industry and country depending on factors such as lay-off propensity, technological intensity, skills intensity, competition pressures, openness, and other labour market institutions, such as wage-setting institutions. They found that a more rigid labour market policy leads to significantly lower innovation intensity in industries with higher job reallocation rates or higher probability of dismissal. In addition, Murphy and colleaues found that innovation intensity was higher in industries with higher import competition and in industries with less rigid product market regulations. Furthermore, in countries with abundant human capital, innovation intensity was higher in human capital-intensive industries. And in countries with abundant physical capital, innovation intensity was higher in physical capital-intensive industries. These results are in line with the findings of Saint-Paul (2002) and Griffith and Macartney (2014). In summary, the empirical literature on the relationship between labour market policies and innovation gives a mixed and complicated picture. The overall evidence leans slightly towards a positive association between more flexible labour market policies and innovation capacity.

With regard to the relationship between innovation and trade competitiveness, studies have found a positive association between innovation capacity and firms' exports. For examples, Wagner (2007), Aw et al. (2009) and Cassiman et al. (2010) found evidence of the link between product innovation, productivity, and exports. In this case, innovation policy targeted at enhancing productivity is likely to be important for improving firms' exporting competitiveness. However, using the Slovenian microdata, Damijan et al. (2010) found a reverse causal direction in which exporting induces process innovation and, in turn, affects the productivity growth relationship. This finding suggests that firms with past exporting experience have a higher probability of being more productive due to process innovation. Therefore, the link between innovation and exporting seems fairly robust, and existing studies show that the direction of causality can run both ways. Along these lines, Lee and Narjoko (2015) reviewed some microdata studies on the relationship between innovation and trade competitiveness measured by exporting capability for five developing countries from Southeast Asia – Indonesia, Malaysia, the Philippines, Thailand, and Viet Nam. They concluded that most existing studies in the region were not able to pinpoint the direction of causality between innovation and exporting capability, most likely due to data constraints. However, there is strong evidence that the incidence of innovation is positively correlated with firms' exporting performance. They also found that foreign competition in domestic markets is important for innovation.

3. Data, empirical model, and variables

3.1 Data

To investigate whether cross-country variations in labour market policies can explain differences in innovation intensity, we constructed a set of panel data of 32 countries covering the period 2009–2013. The data set includes seven Association of Southeast Asian Nations (ASEAN) member states, all members of the BRICS (Brazil, Russia, India, China, and South Africa), 16 OECD countries, plus Argentina, Hong Kong, Mexico, and Pakistan (Table 12.1). The selection of countries tries to capture all different stages of development, as discussed in the Global Competitiveness Report (WEF 2014).

As can be seen from Table 12.1, ASEAN member states' levels of development vary from factor-driven economies such as Cambodia to innovation-driven economies such as Singapore. As mentioned previously, countries with different development levels have different labour market requirements. This influences their respective labour market policies. Countries with different development levels also have different innovation intensities and capacities. Combining such a wide range of development levels within a panel increases the variation in our data. Due to the lack of data, three ASEAN member states – Brunei Darussalam, Lao PDR, and Myanmar – cannot be included in the panel estimation.

As a proxy for innovation intensity, in this chapter we use scores data for innovation competitiveness published by the World Economic Forum (WEF) since 2005. The WEF's Global Competitiveness Index (GCI) is designed to measure both microeconomic and macroeconomic foundations of national

Table 12.1 Countries covered in the panel data set

Stage 1: **factor-driven economies**	*Transition from stage 1 to stage 2*	*Stage 2: efficiency-driven economies*	*Transition from stage 2 to stage 3*	*Stage 3: innovation-driven economies*
Cambodia (CMB)	Philippines (PHP)	China (CHN)	Argentina (ARG)	Australia (AUS)
India (IND)		Indonesia (IDN)	Brazil (BRZ)	Canada (CAN)
Pakistan (PAK)		South Africa (SAF)	Malaysia (MLY)	Finland (FIN)
Viet Nam (VNM)		Thailand (THL)	Mexico (MEX)	France (FRA)
			Russia (RUS)	Germany (GER)
			Turkey (TUR)	Greece (GRE)
				Hong Kong (HKG)
				Italy (ITA)
				Japan (JPN)
				Rep. of Korea (KOR)
				New Zealand (NZ)
				Portugal (POR)
				Singapore (SIN)
				Spain (SPN)
				Sweden (SWE)
				United Kingdom (UK)
				United States (US)

Note: The classification of countries follows the development stages in the Global Competitiveness Report 2014–2015 published by the World Economic Forum. Country codes are shown in parentheses.

competitiveness. It is a composite index consisting of 12 pillars of competitiveness – institutions, infrastructure, macroeconomic environment, health and primary education, higher education and training, goods market efficiency, labour market efficiency, financial market development, technological readiness, market size, business sophistication, and innovation. Each pillar consists of several indicators. The indicators are derived using a standardized survey that targeted more than 14,000 business executives in 144 countries. Similar to other perception surveys, there is a potential subjectivity bias in the data. Nevertheless, the GCI is a valuable data set that can easily be used in a quantitative way, especially when other alternative standardized measures on innovation and labour market policy are not readily available, as in most developing countries. Moreover, the opinions of business executives are invaluable in understanding the factors that determine business competitiveness.

In addition to innovation scores data, in this chapter we use alternative proxies for innovation, including the number of patents residents filed at the national office, the number of venture capital deals, and the number of people employed in knowledge-intensive services from the Global Innovation Index published by Cornell University, L'Institut européen d'administration des affaires (INSEAD), and the World Intellectual Property Organization (WIPO).

Note that comparable cross-country firm-level data on innovation are relatively limited in the non-OECD countries, including in the Southeast Asian region. So far, there is no standardized industry- or firm-level survey on innovation activity and intensity within ASEAN member states. In view of this, it is critical for countries in ASEAN to improve the availability and quality of their microdata on innovation to be on a par with OECD countries.

Lee and Narjoko (2015) explained that there are two types of innovation measures – innovation input and innovation output. The most widely used measure of innovation input is research and development (R&D) expenditure. As for innovation outputs, the patents count data have been widely used to measure innovation intensity. However, there are drawbacks to using either R&D expenditure or the number of patents to measure innovation intensity. R&D expenditure cannot capture the output side of the innovation process. Moreover, R&D expenditure is only one of several inputs for innovation and patents only measure invention rather than innovation and do not capture many non-patented inventions and innovations.[6] Furthermore, the use of patents to protect inventions varies across industries. Nevertheless, the advantage of using innovation indicators from GCI is that they capture both input and output measures of innovation – the capacity for innovation, the quality of scientific research institutions, company spending on R&D, university–industry R&D collaboration, government procurement of advanced technology products, availability of scientists and engineers, patent applications, and intellectual property protection.

As a proxy for labour market policies, this chapter uses scores data for labour market efficiency published by the GCI and cost of redundancy dismissal from the World Bank's annual *Doing Business* report. Specifically, the labour market policies index covers two components – the flexibility and the efficiency of the labour market. The flexibility of a labour market is measured by indicators of: cooperation in labour–employer relations; flexibility of wage determination; hiring and firing practices; costs of redundancy dismissal; and the effect of taxation on incentives to work. The higher the score of labour market efficiency, the more efficient the labour market is, implying a less rigid labour market policy. The cost of redundancy dismissals measures the strictness of regulations in terms of dismissal for a regular and a temporary contract (Murphy et al. 2013). Table 12.2 describes each variable used in the empirical model and gives data sources and the expected impact on innovation.

Note that the innovation scores from the GCI focus on technological innovation capacity. Technological innovation is not related to skills, know-how, or organizational conditions. The importance of technological innovation in improving standards of living is well documented. Technological breakthroughs have been the basis of many of the productivity gains that modern economies currently enjoy. For example, technological innovation in ICT has led the digital

Table 12.2 Variables acronyms, description and expected impact on innovation

Variable	*Description*	*Expected impact*
Innov	Dependent variable, measured by the innovation scores (weighted average of capacity for innovation; quality of scientific research institutions; company spending on R&D; university–industry collaboration in R&D; government procurement of advanced technology products; availability of scientists and engineers; PCT patent applications and intellectual property protection), values vary between 1 and 7 from least to most innovative environment. Source: Global Competitiveness Index, World Economic Forum	
Patent	Dependent variable, measured by weighted scores of number of patent filed by resident at national office (per billion GDP, 2005 PPP$), values vary between 0 and 100 from least to most patent application. Source: World Intellectual Property Organization (WIPO), Global Innovation Index.	
Empkw	Dependent variable, measured by weighted scores of employment in knowledge-intensive services (% of workforce), values vary between 0 and 100 from least to most knowledge employment. Source: ILO, Global Innovation Index.	
Vencap	Dependent variable, measured by weighted scores of venture capital per investment location: number of deals (per trillion GDP, 2005 PPP$), values vary between 0 and 100 from least to most venture deals. Source: Thomson Reuters, Global Innovation Index.	
Lxport	Dependent variable, measured by log of country exports. Source: World Bank, World Integrated Trade Solution (WITS) database.	
Lbreffi	Independent variable (alternative for lmefi), measured by weighted scores of cost of redundancy dismissal (sum of notice period and severance pay for redundancy dismissal in salary weeks), values vary between 0 and 100 from most to least costly. Source: World Bank, Global Innovation Index.	+/-
Lmefi	Independent variable, measured by weighted scores of labour market efficiency (cooperation in labour–employer relations; flexibility of wage determination; hiring and firing practices; redundancy costs; effect of taxation on incentives to work; pay and productivity; reliance on professional management; country capacity to retain and attract talent, and female participation in labour force), values vary between 1 and 7 from least to most efficient and flexible labour market. Source: Global Competitiveness Index, World Economic Forum.	+/-

(*Continued*)

Table 12.2 (Continued)

Variable	*Description*	*Expected impact*
Gmefi	Independent variable, measured by weighted scores of goods market efficiency (domestic and foreign competition), values vary between 1 and 7 from least to most healthy market competition. Source: Global Competitiveness Index, World Economic Forum.	+
Institut	Independent variable (alternative for gmefi), measured by weighted scores of institutional environment (weighted average of property rights, ethics and corruption, undue influence, government efficiency, security, corporate ethics, and accountability), values vary between 1 and 7 from least to most conducive institutional environment. Source: Global Competitiveness Index, World Economic Forum.	+
Highed	Independent variable, measured by weighted scores of higher education and training quality, values vary between 1 and 7 from low to high quality of higher education and training environment. Source: Global Competitiveness Index, World Economic Forum.	+
Macro	Independent variable, measured by weighted scores of stability of macroeconomic environment, values vary between 1 and 7 from least to most stable macroeconomic environment. Source: Global Competitiveness Index, World Economic Forum.	+
Mktsize	Independent variable, measured by weighted scores of market size (domestic and foreign market size), values vary between 1 and 7 from smallest to largest market size. Source: Global Competitiveness Index, World Economic Forum.	+/-
Techred	Independent variable (alternative for highed), measured by weighted scores of technological readiness (technological adoption and ICT use), values vary between 1 and 7 from least to most agile economy to adopt existing technologies to enhance the productivity of its industries. Source: Global Competitiveness Index, World Economic Forum.	+
Infras	Independent variable (alternative for techred), measured by weighted scores of infrastructure quality (weighted average of transport, electricity and telephony infrastructure), values vary between 1 and 7 from least to most extensive and efficient infrastructure. Source: Global Competitiveness Index, World Economic Forum.	+
Openness	Independent variable, measured by weighted scores of applied tariff rate (weighted mean, all products [%]), values vary between 0 and 100 from least to most open economy. Source: World Bank, Global Innovation Index.	+

revolution that, in turn, has benefitted modern economies in terms of increasing productivity and efficiency. The digital revolution would have never happened without technological innovation, which has significantly transformed the ways in which things are done, and opened a wider range of new opportunities in terms of products and services development.

Technological innovation is particularly important for economies approaching the frontiers of knowledge. At this stage of development, the possibility of generating more value by merely integrating and adapting exogenous technologies tends to disappear. For firms in those countries that have reached this innovation stage of development, they must design and develop cutting-edge products and processes to maintain a competitive edge and move towards even higher value-added activities. This advancement requires an environment conducive to innovative activities and supported by both the public and the private sectors. It needs sufficient investment in R&D, especially by the private sector; the presence of high-quality scientific research institutions that can generate the basic knowledge needed to build the new technologies; extensive collaboration in research and technological developments between universities and industry; and the protection of intellectual property (WEF 2014).

Table 12.3 reports some descriptive statistics of the dependent and independent variables used in the empirical analysis. Part I of the table reports the summary statistics and Part II reports the summary of correlation coefficients amongst selected variables.

The correlation matrix shows that a high positive correlation between the institutional environment (*institute*) and goods market efficiency (*gmefi*). The institutional environment is determined by the legal and administrative framework within which firms and government interact to generate wealth. Goods market efficiency is determined by healthy market competition and demand conditions (customer orientation and buyer sophistication).

There is also a high positive correlation between infrastructure quality (*infra*) and higher education quality (*highed*); between infrastructure quality and technological readiness (*techred*); and between higher education quality and technological readiness. Infrastructure quality is measured by how extensive and efficient a country's infrastructure is. A well-developed transport and communications infrastructure network, quality of roads, railroads, ports, electricity supplies, and telecommunications network – all are critical for effective functioning of an economy. Considering the correlation between variables, our choice of explanatory variable in the estimation avoids putting together highly correlated variables. All in all, the correlation matrix shows that none of these variables is perfectly collinear.

3.2 Some stylized facts

Figure 12.1 depicts the trend scores of labour market efficiency based on the GCI. Overall, the global trend shows that employment protection has been increasing in some countries and declining in others. In the data, Argentina, Italy,

Table 12.3 Descriptive statistics of dependent and independent variables

Part I: Summary statistics

Variable	*Obs*	*Mean*	*Std. dev.*	*Min.*	*Max*
Patent	160	31.2105	33.99498	0	100
empkw	155	51.66613	26.38669	.33	100
vencap	160	42.50606	30.90898	0	100
tariff	160	85.90469	12.23839	47.47	100
1brefi	160	76.63431	18.04935	40	100
innov	160	4.065812	.8956397	2.73	5.76
institut	160	4.422312	.9454906	2.79	6.13
infra	160	4.913875	1.086386	2.66	6.77
macro	160	4.866938	.860057	2.42	6.44
highed	160	4.794125	.8149996	2.76	6.27
1mefi	160	4.446938	.6627093	2.97	5.92
gmefi	160	4.5485	.5541214	3.06	5.65
techred	160	4.592937	1.047706	2.75	6.29
mktsize	160	5.181375	.8032916	3.01	6.94

Part II: Correlation coefficient amongst selected variable

	innov	*lmefi*	*gmefi*	*highed*	*infra*	*institut*	*macro*	*mktsize*
innov	1.0000							
lmefi	0.6600	1.0000						
gmefi	0.7598	0.8024	1.0000					
highed	0.8251	0.5502	0.6317	1.0000				
infra	0.7848	0.5873	0.7161	0.8717	1.0000			
institut	0.8279	0.7819	0.9052	0.7565	0.7405	1.0000		
macro	0.3226	0.4062	0.4225	0.3591	0.3145	0.3882	1.0000	
mktsize	0.2561	0.0308	−0.0464	0.1656	0.2490	−0.0680	0.0535	1.0000
techred	0.8234	0.5824	0.7092	0.9062	0.9204	0.7863	0.2552	0.0892

Source: Author.

Pakistan, and Turkey are classified as the strictest countries in terms of labour market policies (measured by lower scores for labour market efficiency), whereas Hong Kong, Singapore, the United Kingdom (UK), and the United States (US) are classified as the least strict countries for labour market policies (higher scores for labour market efficiency).

Figure 12.2 shows the trend scores of innovation capacity based on the GCI. Overall, the global trend shows that innovation intensity has been increasing in some countries and declining in others. Finland, Germany, Japan, and the United States lead in global innovation competitiveness, whereas Argentina, Cambodia, and Pakistan seem to be laggards in innovation.

Figure 12.3 depicts a positive association between labour market efficiency (*lmefi*) and innovation capacity. The figure shows that countries such as Singapore, the United Kingdom, and the United States are relatively advanced in terms of innovation performance and relatively flexible in terms of labour market policies.

This pattern remains unchanged when we use a different proxy for labour market efficiency – the cost of redundancy dismissal (*lbrefi*).

Figures 12.4 to 12.6 show that labour market efficiency has a positive association with the number of patents filed in the national office, the number of venture capital deals, and the number of people employed in knowledge-intensive services.

Figure 12.7 shows that the relationship between labour market efficiency (*lmefi*) and trade openness also seems positive. A more efficient labour market is associated with greater openness to trade. Similarly, Figure 12.8 shows that the relationship between innovation and trade competitiveness (measured by the natural logarithm of export value) is also positive. A more innovative country is associated with greater export competitiveness.

3.3 Empirical model

We are interested in examining how much variation in labour market policies relates to variation in innovation capacity. For this purpose, we propose a simplified version of the fixed effect model developed by Acharya et al. (2013), as follows:

$$Innov_{it} = \alpha_{it} + \beta * Lmefi_{it} + \gamma * X_{it} + \mu_i + \varepsilon_{it} \qquad (1)$$

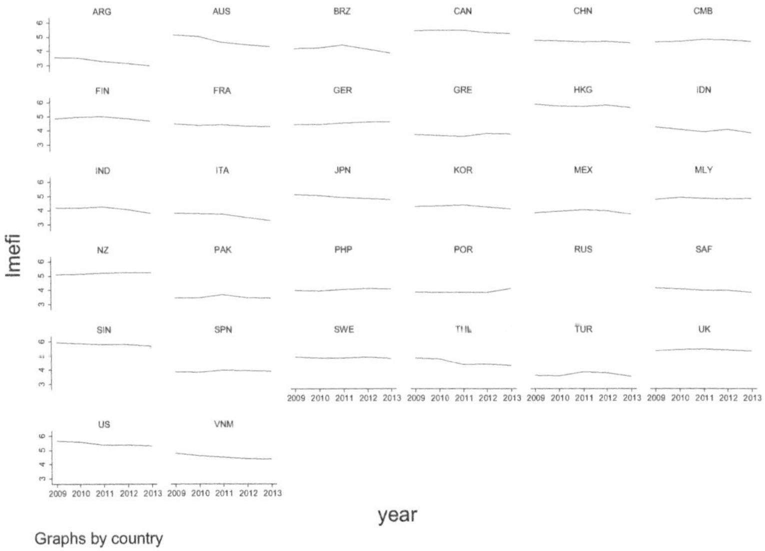

Figure 12.1 Cross-country labour market efficiency trend, 2009–2013

Source: Author's calculation based on Global Competitiveness Index 2009–2013.

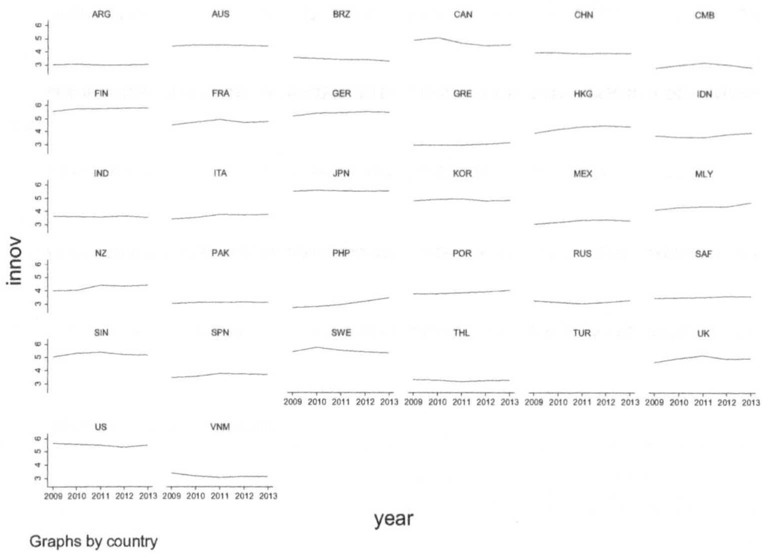

Figure 12.2 Cross-country innovation scores trend, 2009–2013

Source: Author's calculation based on Global Competitiveness Index 2009–2013.

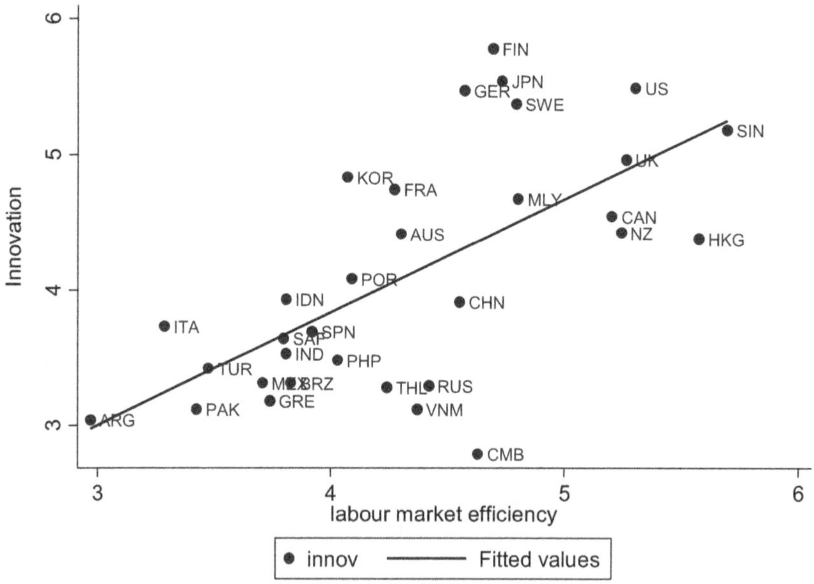

Figure 12.3 Labour market efficiency and innovation capacity, 2013

Source: Author's calculation based on Global Competitiveness Index.

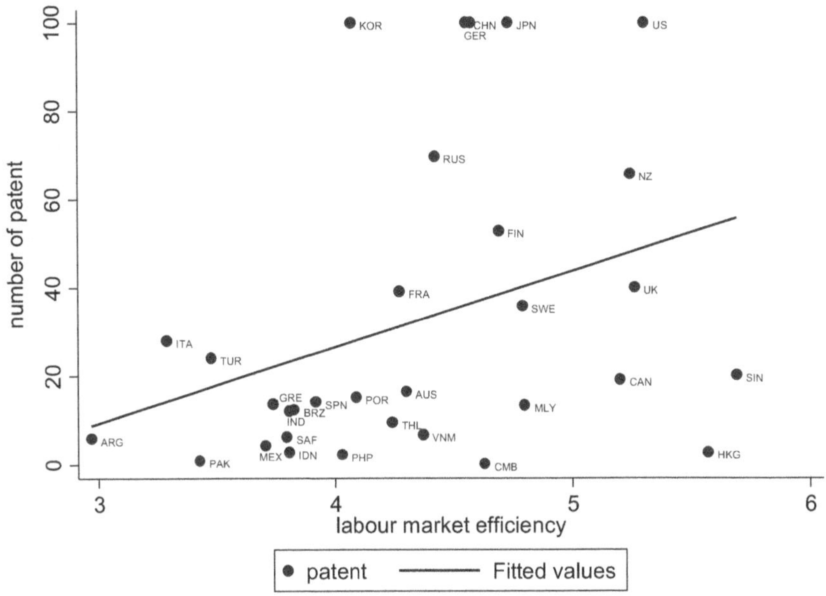

Figure 12.4 Labour market efficiency and number of patents filed, 2013

Source: Author's calculation based on Global Innovation Index and Global Competitiveness Index.

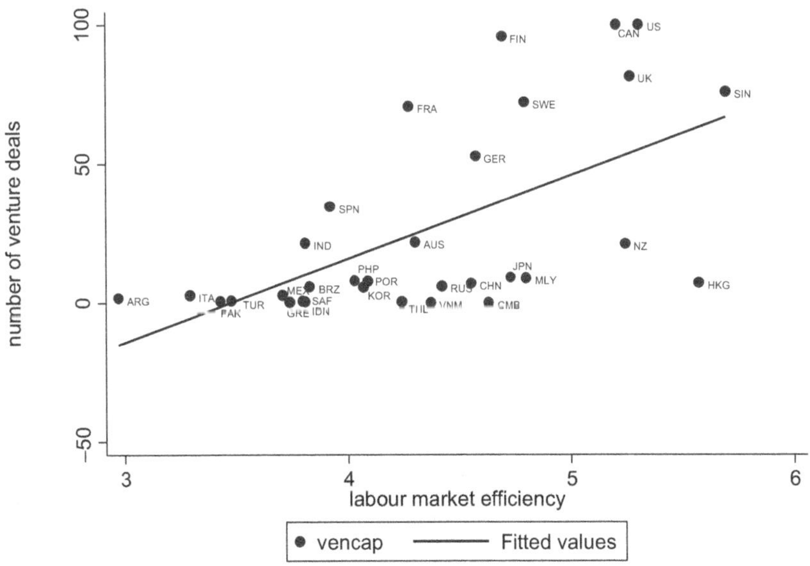

Figure 12.5 Labour market efficiency and number of venture deals, 2013

Source: Author's calculation based on Global Innovation Index and Global Competitiveness Index.

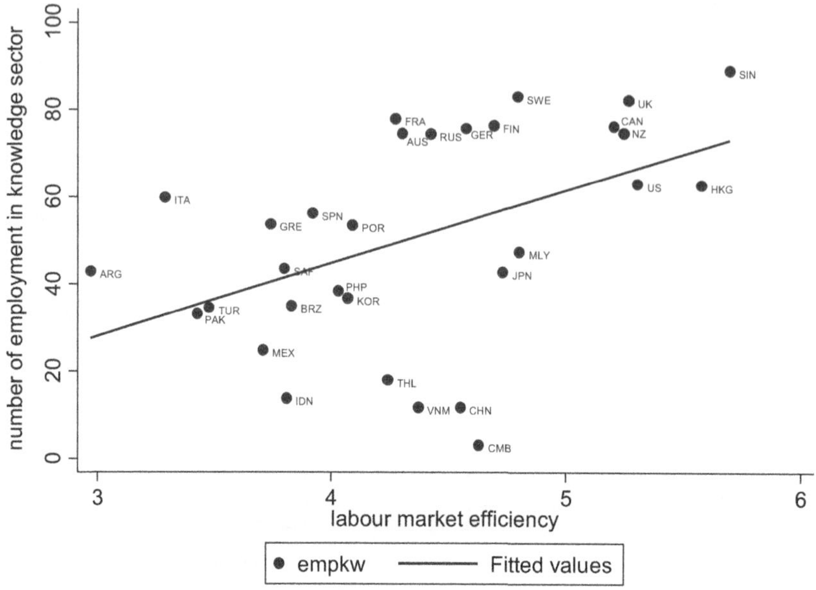

Figure 12.6 Labour market efficiency and employment in knowledge-intensive services, 2013

Source: Author's calculation based on Global Innovation Index and Global Competitiveness Index.

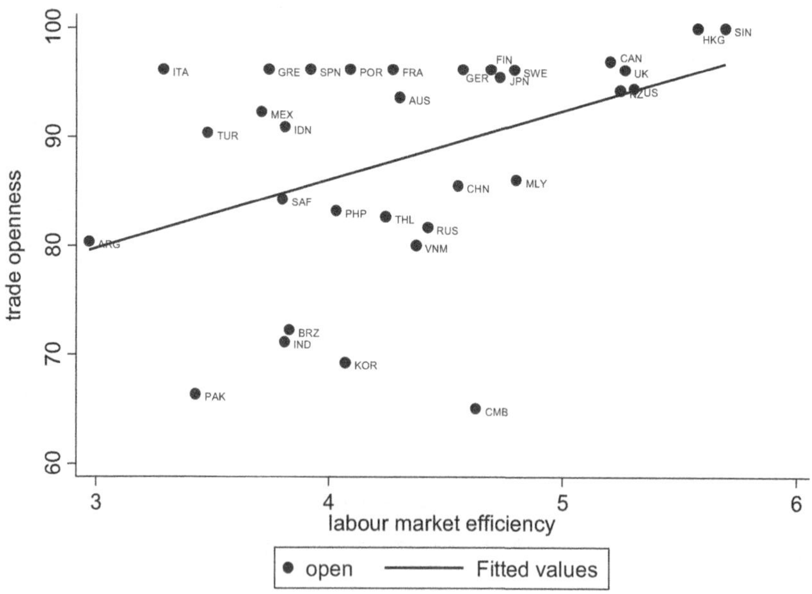

Figure 12.7 Trade openness and labour market efficiency, 2013

Source: Author's calculation based on Global Innovation Index and Global Competitiveness Index.

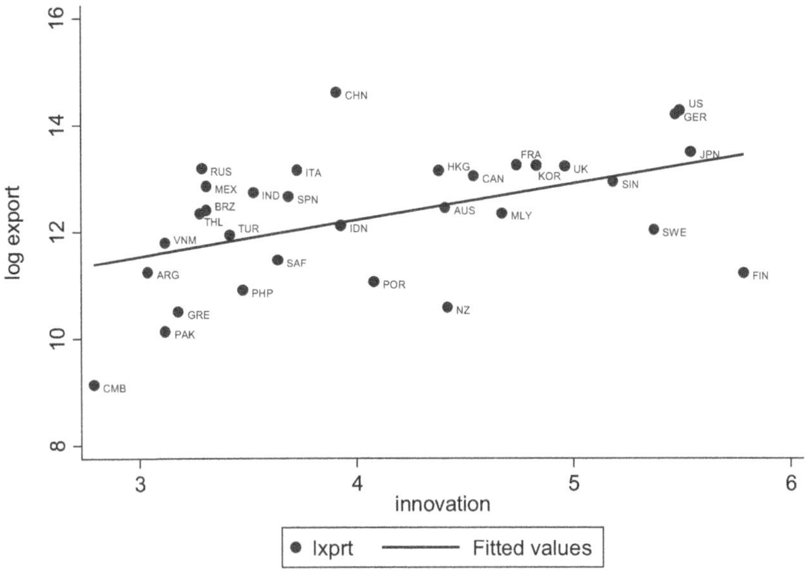

Figure 12.8 Trade competitiveness and innovation, 2013

Source: Authors' calculation based on Global Innovation Index and Global Competitiveness Index.

The dependent variable, $Innov_{it}$, represents innovation competitiveness in country i at time t. $Lmefi_{it}$ represents labour market efficiency in country i at time t. Higher values of this variable correspond to more efficient labour markets, implying more flexible labour market policies. X_{it} is a set of control variables including institutional environment, infrastructure quality, macroeconomic condition, higher education quality, goods market efficiency, and market size of country i at time t. μ_i controls for unobserved country-specific characteristics, and ε_{it} is a disturbance term capturing unobservable variables affecting innovation. The country fixed-effects control for time-invariant unobserved factors at the country level.[7]

The parameters of interest are β coefficients on the labour market efficiency indicators. Model 1 assumes that labour market policies affect the way resources are allocated, and hence the efficiency and flexibility of the overall labour market. This in turn leads to either higher or lower country innovation capacity.

Table 12.4a shows the estimation results for Equation (1) using the innovation scores measure as the dependent variables. The β coefficient for the labour market efficiency (*lmefi*) is positive and significant. This result indicates that as the score of labour market efficiency increases (less rigid labour market policies), a country's innovation score increases, holding other things constant. The estimate for β coefficient remains positive and significant when we control for additional variables, such as higher education quality, the institutional environment, infrastructure

Table 12.4a Fixed effect regressions results: innovation capacity as dependent variable

	Dependent variable: innovation capacity scores				
	1	*2*	*3*	*4*	*5*
Lmefi	0.232**	0.225***	0.296***	1.718***	1.643***
	(0.099)	(0.079)	(0.082)	(0.401)	(0.409)
Gmefi	0.362**	0.290**	0.108	0.150	0.311**
	(0.136)	(0.109)	(0.122)	(0.117)	(0.144)
Highed	0.371***		0.288**	0.284**	0.312***
	(0.112)		(0.112)	(0.112)	(0.109)
Infra		0.451***	0.388 ***	0.378***	
		(0.089)	(0.098)	(0.091)	
Mktsize			0.158		
			(0.286)		
Openness				0.019***	0.019***
				(0.006)	(0.006)
Open*Lmefi				−0.323***	−0.308***
				(0.089)	(0.090)
Techred					0.114**
					(0.066)
Constant	−0.395	−0.469	−1.847	−2.744***	−2.170**
	(0.786)	(0.626)	(1.514)	(0.877)	(0.984)
N	160	160	160	160	160
Rho	0.946	0.952	0.950	0.953	0.945
Prob > F	0.000	0.000	0.000	0.000	0.000

Note: ***, **, *mean coefficients statistically significant at the 1%, 5%, and 10% levels, respectively. See Table 12.2 for the detailed description of the variables.

Source: Author's calculation.

Table 12.4b Fixed effect regressions results: venture capital deals and employment in knowledge services as dependent variable

	Dependent variable: venture capital deals		*Dependent variable: employment in knowledge services*	
	1	*2*	*3*	*4*
Lmefi	2.490***	2.136***	0.357**	0.341**
	(0.471)	(0.472)	(0.150)	(0.156)
Gmefi	−3.247***	−2.059***	−0.532***	−0.535**
	(0.656)	(0.774)	(0.197)	(0.247)
Mktsize	−5.038***	−4.032**	2.708***	2.677***
	(1.804)	(1.781)	(0.482)	(0.486)
Highed		−1.552***		−0.196
		(0.546)		(0.177)
Infra		−0.379		0.199
		(0.531)		(0.173)
Constant	33.818***	34.175***	9.417***	−9.205***
	(9.482)	(9.185)	(2.536)	(2.554)
N	145	145	155	155
Rho	0.973	0.974	0.989	0.989
Prob > F	0.000	0.000	0.000	0.000

Notes: Robust standard errors are in parentheses. Based on them ***, **, *mean coefficients statistically significant at the 1%, 5%, and 10% levels, respectively. See Table 12.2 for the detailed description of the variables.

Source: Author's calculation.

Table 12.5 Arellano-Bond dynamic panel-data estimation

	Dependent variable: innovation capacity scores				
	1	*2*	*3*	*4*	*5*
Lmefi	0.222***	0.243***	0.210**	0.234**	0.201**
	(0.074)	(0.085)	(0.092)	(0.095)	(0.085)
Gmefi	0.549***	0.537***	0.237*	0.230*	0.243*
	(0.125)	(0.134)	(0.134)	(0.139)	(0.133)
Mktsize		0.227	0.168	0.294	
		(0.346)	(0.306)	(0.317)	
Highed	0.181*	0.190*	0.105	0.100	0.101
	(0.097)	(0.103)	(0.090)	(0.094)	(0.088)
Infra			0.436***	0.468***	0.430***
			(0.117)	(0.119)	(0.114)
Institut			0.195*	0.192*	0.199*
			(0.105)	(0.109)	(0.103)
Macro			−0.220	−0.056	
			(0.182)	(0.043)	
Openness					0.001
					(0.002)
Innov (t−1)	−0.181	−0.253		−0.315*	−0.195
	(0.154)	(0.197)		(0.186)	(0.154)
Constant	0.457	−0.505	−1.423	−1.617	−0.700
	(0.711)	(1.625)	(1.472)	(1.517)	(0.611)
N	96	96	96	96	96
Prob>Chi2	0.000	0.000	0.000	0.000	0.000

Notes: ***, **, *mean coefficients statistically significant at the 1%, 5%, and 10% levels, respectively. See Table 12.2 for the detailed description of the variables.

Source: Author's calculation.

quality, market size, openness, and interaction term (*openness*lmefi*). The latter tries to capture the possibility of non-linearity in the model specification. As columns 4 and 5 show, the effect of labour market efficiency on innovation is higher when we include the openness and interaction terms. In this case, the degree of openness matters for innovation.

The estimated coefficient for higher education is positive and significant for different model specification. This means higher education quality is positively associated with innovation. In view of this, policy that supports the quality of higher education is crucial for improving a country's innovation competitiveness. Another significant determinant for innovation is infrastructure quality. Columns 2–4 show that better infrastructure quality is associated with higher innovation competitiveness. Meanwhile, columns 1, 2, and 5 show that estimated coefficients on goods market efficiency are positive and significant, meaning that more efficient goods market policies (more domestic and foreign competition) are associated with higher innovation capacity. This result is in line with other previous findings (see Aghion et al. 2005; Barbosa and Faria 2011).

The Hausman test is conducted to choose between the fixed effect and random effect model. The Hausman test rejects the null hypothesis, that is, that the unique error (u_i) is not correlated with the regressors.[8] Therefore, the fixed effect is selected for Equation (1).

Different proxies for innovation are also tested in the estimation of Equation (1). Table 12.4b shows the results of fixed effect regression on Equation (1) using different proxies for the dependent variable, namely, the number of venture deals (*vencap*) and the number of people employed in knowledge-intensive services (*empkw*) as the dependent variables. Similar to previous results, the β coefficient for the labour market efficiency (*lmefi*) remains positive and significant. However, we cannot find a significant association between labour market policies and the number of patents filed in the national office. This is probably due to weaknesses of the patent data.[9]

Note that our fixed-effect model is likely to suffer from an endogeneity problem. An important concern stems from the fact that changes in a country's labour market policies are likely to be correlated with changes in other unobserved factors. To address this endogeneity problem, we run a dynamic panel-data using the Arellano–Bond method by including the lag of the dependent variable to control for other unobserved factors.

Table 12.5 shows the results of the Arellano–Bond dynamic panel-data estimation. The results corroborate the findings from previous models (compared with Table 12.4a). Our estimation results show that a more efficient labour market is associated positively with innovation capacity. The estimate for β coefficient remains positive and significant when we control for additional variables, such as higher education quality, the institutional environment, infrastructure quality, institutional quality, and openness (columns 3–5). We find that improved goods market efficiency (more competition), better infrastructure quality, and improved institutional quality are all positively correlated with a country's innovation performance. Note that a higher education variable becomes non-significant after including more control variables – infrastructure and institutional quality. This is probably due to the strong positive correlation between higher education and these two variables. The insignificant sign of the openness variable and the lag of innovation are rather puzzling.

In addition, we also test whether labour market policies and innovation will have an effect on a country's trade competitiveness.

$$Lxport_{it} = \alpha_{it} + \beta * Innov_{it-1} + \gamma * Lmefi_{it} + \delta * X_{it} + \mu_i + \varepsilon_{it} \quad (2)$$

The dependent variable, $Lxport_{it}$, represents trade competitiveness measured by the natural logarithm of country *i*'s exports at time *t*. $Innov_{it-1}$ represents innovation intensity in country *i* at time $t - 1$. A potential caveat, due to data constraints, is that we can only use a one-year lag. This one-year lag is probably too short for innovation to show any meaningful impact on a country's trade competitiveness. X_{it} is a set of control variables of country *i* at time *t*. μ_i

Table 12.6 Exploring the relationship between innovation and trade competitiveness

	Dependent variable: log exports			
	1 *(F.E)*	*2* *(R.E)*	*3* *(F.E)*	*4* *(R.E)*
Lmefi	−0.212**	−0.186**	−0.237***	−0.216***
	(0.082)	(0.080)	(0.080)	(0.079)
Gmefi			0.319**	0.303**
			(0.127)	(0.125)
Techred	0.178***	0.194***	0.103	0.121*
	(0.066)	(0.064)	(0.071)	(0.069)
Innov (t − 1)	0.195**	0.237***	0.165**	0.204***
	(0.083)	(0.080)	(0.081)	(0.079)
Constant	11.547***	11.190***	10.676***	10.407***
	(0.542)	(0.536)	(0.630)	(0.620)
N	128	128	128	128
Rho	0.991	0.991	0.992	0.991
Prob > F or Prob>Chi2	0.000	0.000	0.000	0.000
Hausman Test	Prob >Chi2 = 0.173 Cannot reject Ho. Use random effect		Prob >Chi2 = 0.000 Reject Ho. Use fixed effect	

Notes: Standard errors are in parentheses. Based on them ***, **, *mean coefficients statistically significant at the 1%, 5%, and 10% levels, respectively.

Source: Author's calculation.

controls for unobserved country-specific characteristics and ε_{it} is an idiosyncratic error term.

Our hypothesis is that a more efficient labour market is associated with greater innovation. And greater innovation is, in turn, associated with a country's trade competitiveness measured by its exports. Again, this is a rather bold simplification of the reality in which innovation may indirectly affect productivity and export competitiveness. Nevertheless, we are interested in finding preliminary evidence that overall innovation scores could be higher or lower in countries with more efficient labour markets.

To capture the dynamic nature of innovation relative to export competitiveness, we then estimate equation (2) with fixed and random effect models and perform the Hausman test. As stated earlier, trade competitiveness is measured by the natural logarithm of a country's exports value (see Table 12.2 description). Table 12.6 shows that a different model specification leads to different choices between the fixed or random effect. The Hausman test selects the random effect for the model without goods market policy variable (*gmefi*) (see columns 1 and 2), while it selects the fixed effect for the model with goods market policy variable (see columns 3 and 4). Overall, the estimation results show that the lag variable of innovation is positively associated with export competitiveness. The estimated coefficient for innovation variable remains consistent, positive, and significant across different model specifications. In addition, technological readiness and goods market competition policy are associated with higher exports. Meanwhile, a more flexible labour market policy is associated with lower exports. This result contradicts our hypothesis about the impact of labour market policy on trade competitiveness.

4. Conclusions

This chapter investigates the relationship between labour market policies and innovation. Most of the studies in this area have been undertaken for OECD member countries. This study attempts to fill the knowledge gap by expanding the analysis to include both OECD and non-OECD countries, including some of the Southeast Asian countries. The data used in the estimation come from the Global Competitiveness Report and the Global Innovation Index. A set of balanced panel data is constructed including 32 countries covering the period 2009–2013. Data availability limits our empirical model in two ways. First, it is related to the difficulty of finding reliable and standardized innovation data at the firm or industry level for developing countries. Second, it is related to the difficulty of finding good proxies for labour market policy, as there is substantial heterogeneity in labour market policies across countries. Due to institutional complexity, there is no perfect proxy for labour market policy. The use of labour market efficiency indicators from the Global Competivenesss Index (GCI) should therefore be treated with caution.

Despite these challenges, GCI data provide valuable information when better alternative standardized measures on innovation and labour market policy are not easily available in most countries. In this case, opinions from business executives are instrumental in understanding the factors behind business competitiveness. This study finds preliminary evidence that a more efficient labour market (more flexible labour market policy) is positively correlated with innovation. Innovation competitiveness in turn leads to a country's trade competitiveness. Our results are robust even after controlling for other explanatory variables. The quality of higher education is positively associated with a country's innovation competitiveness. Therefore, policies that support improving higher education quality are crucial for economies that want to develop their innovation competitiveness. Other important determinants for innovation are infrastructure quality and goods market policies (more domestic and foreign competition). The two are positively correlated with innovation. The country-level analysis in this chapter demonstrates the average correlation between labour market efficiency and innovation. However, it does not take into account structural differences across industries/sectors. Subject to data availability, future research should be directed towards understanding variances of labour market efficiency and innovation between different industries and sectors across countries.

On trade aspects, we find preliminary evidence that past innovation is positively associated with trade competitiveness. This is in line with previous studies that find a positive link between innovation and exporting (Wagner 2007; Aw et al. 2009; Cassiman et al. 2010). We cannot find a significant positive association between labour market policies and trade competitiveness. However, we postulate that there may be an indirect link between labour market policies and trade competitiveness – one that our simple model fails to capture.

Research on labour market policies and innovation is still limited in the region. We hope that future research will fully explore the relationship between labour market policies and innovation in developing countries in the Southeast Asian region using good-quality, firm-level data. As ASEAN has started the implementation of ASEAN Economic Community, characterized by the free flow of skilled labour amongst others, there is a greater need to assess how labour market policies in one member state may affect other countries, and also how they affect innovation in the region. In addition, it is important to collect better quality microdata to test the relationship between innovation, productivity, and firms' exporting competitiveness. Innovation policy targeted at enhancing productivity, especially in export-oriented sectors, is likely to be important, but empirical evidence in this area is still lacking in ASEAN.

Appendix Hausman test results

	Dependent variable: Innovation capacity			
	1 (*F.E*)	*2* (*R.E*)	*3* (*F.E*)	*4* (*R.E*)
Lmefi	0.296***	0.275***	0.297***	0.276***
	(0.071)	(0.060)	(0.072)	(0.060)
Gmefi	0.108	0.198**	0.109	0.204**
	(0.114)	(0.093)	(0.115)	(0.099)
Mktsize	0.158	0.135	0.165	0.139
	(0.219)	(0.086)	(0.232)	(0.086)
Highed	0.288***	0.307***	0.288 ***	0.307***
	(0.082)	(0.073)	(0.082)	(0.074)
Infra	0.388**	0.297***	0.387***	0.294***
	(0.080)	(0.067)	(0.081)	(0.067)
Macro			-0.003	-0.009
			(0.034)	(0.032)
Constant	-1.847	-1.685***	-1.868	-1.681***
	(1.160)	(0.565)	(1.189)	(0.565)
N	160	160	160	160
Rho	0.950	0.935	0.950	0.934
Prob > F or Prob>Chi2	0.000	0.000	0.000	0.000
Hausman Test	Prob >Chi2 = 0.000 Reject Ho. Use fixed effect		Prob >Chi2 = 0.000 Reject Ho. Use fixed effect	

Notes: Standard errors are in parentheses. Based on them ***, **, *mean coefficients statistically significant at the 1%, 5%, and 10% levels, respectively.

(*Continued*)

Appendix (Continued)

	Dependent variable: Number of venture deals			
	1	*2*	*3*	*4*
	(F.E)	*(R.E)*	*(F.E)*	*(R.E)*
Lmefi	2.490***	2.880***	2.136***	2.826***
Gmefi	(0.471)	(0.332)	(0.472)	(0.073)
Mktsize	−3.247***	−2.486***	−2.059***	−2.753***
Highed	(0.656)	(0.415)	(0. 774)	(0.515)
Infra	−5.038***	−0.107	−4.032**	−0.239
	(1.804)	(0.290)	(1.781)	(0.293)
			−1.552***	−0.601
			(0.546)	(0.388)
			−0.379	0.679**
			(0.531)	(0.332)
Constant	33.818***	2.357***	34.175***	4.043*
	(9.482)	(2.250)	(9.185)	(2.369)
N	145	145	145	145
Rho	0.973	0.701	0.974	0.693
Prob > F or Prob>Chi2	0.000	0.000	0.000	0.000
Hausman Test	Prob >Chi2 = 0.000 Reject Ho. Use fixed effect		Prob >Chi2 = 0.000 Reject Ho. Use Fixed effect	

Notes: Standard errors are in parentheses. Based on them ***, **,* mean coefficients statistically significant at the 1%, 5%, and 10% levels, respectively.

	Dependent variable: Employment in knowledge services			
	1	*2*	*3*	*4*
	(F.E)	*(R.E)*	*(F.E)*	*(R.E)*
Lmefi	0.357**	0.157	0.341**	0.135
	(0.150)	(0.137)	(0.156)	(0.134)
Gmefi	−0.532***	0.002	−0.535**	−0.369*
	(0.197)	(0.172)	(0.247)	(0.215)
Mktsize	2.708***	0.648***	2.677***	0.316*
	(0.482)	(0.194)	(0.486)	(0.162)
Highed			−0.196	0.147
			(0.177)	(0.161)
Infra			0.199	0.383***
			(0.173)	(0.147)
Constant	−9.417***	−0.351	−9.205***	0.534
	(2.536)	(1.273)	(2.554)	(1.108)
Observations	155	155	155	155
Rho	0.989	0.935	0.989	0.884
Prob > F or Prob>Chi2	0.000	0.006	0.000	0.000
Hausman Test	Prob >Chi2 = 0.000 (Reject Ho) Use fixed effect		Prob >Chi2 = 0.000 (Reject Ho) Use fixed effect	

Notes: Standard errors are in parentheses. Based on them ***, **,* mean coefficients statistically significant at the 1%, 5%, and 10% levels, respectively.

Notes

1 Researcher at the Economic Research Center of the Indonesian Institute of Sciences (P2E–LIPI) and visiting fellow at the Institute of Southeast Asian Studies (ISEAS). Email: siwage@yahoo.com.
2 For example, several countries, such as Australia, the United States (US), France, Taiwan, Japan, the Republic of Korea (henceforth, Korea), China, the United Kingdom (UK), Thailand, Malaysia, and Indonesia, implement minimum wage systems. The arguments for minimum wage policy are to ensure low-wage workers have a minimum living standard; to prevent employers from exploiting low-skilled workers; to increase purchasing power of low-wage workers; and to compel employers to raise efficiency and productivity, amongst others.
3 The United States, Taiwan, Japan, Korea, and China also operate unemployment insurance benefits, which are not linked to the minimum wages. Australia operates an unemployment assistance system, in which benefit rates are set below the earnings of minimum wage workers. France operates both systems.
4 The intensive margin refers to the export volume per firm and the extensive margin refers to the number of exporting firms.
5 Idiosyncratic shocks that affect individual firms can be natural, social, economic, political, or environmental.
6 See Kleinknecht et al. (2002) for detailed discussion about the weakness of each innovation indicator.
7 Due to lack of standardized industrial data, the model cannot control for industrial heterogeneity.
8 See the Appendix for the Hausman test results.
9 Firms in different industries and countries have different propensities to patent, and the value of a patent is heterogeneous across countries (see Griffith and Macartney 2014, p. 141).

References

Acharya, V, Baghai-Wadji, R, and Subramarian, K 2013, 'Labour laws and innovation', *Journal of Law and Economics*, vol. 56, pp. 997–1037.

Aghion, P and Howitt, P 2005, 'Growth with Quality-Improving Innovations: An Integrated Framework', in *Handbook of Economic Growth* Vol. 1A, eds. P Aghion and SN Durlauf, Elsevier, Amsterdam, pp. 67–110.

Aw, BY, Roberts, MJ, and Xu, DY 2009, 'R&D investment, exporting, and productivity dynamics', *NBER Working Paper Series* No. 14670, National Bureau of Economic Research, Cambridge, MA.

Barbosa, N and Faria, AP 2011, 'Innovation across Europe: how important are institutional differences?', *Research Policy*, vol. 40, pp. 1157–1169.

Bassanini, A and Ernst, E 2002, 'Labour market institutions, product market regulation, and innovation: cross-country evidence', *OECD Department of Economics Working Paper* No. 316. OECD, Paris.

Boeri, T, Helppie, B, and Macis, M 2008, 'Labor regulations in developing countries: a review of the evidence and directions for future research', *World Bank SP Discussion Paper* No. 0833, The World Bank, Washington, DC.

Botero, J, Djankov, S, La Porta, R, Lopez-De-Silanes, F, and Shleifer, A 2004, 'The regulation of labor', *Quarterly Journal of Economics*, vol. 119, no. 4, pp. 1339–1382.

Cassiman, B, Golovko, E, and Martinez-Ros, E 2010 'Innovation, exports and productivity', *International Journal of Industrial Organization*, vol. 28, pp. 372–376.

Damijan, JP, Kostevc, C, and Polanec, S 2010, 'From innovation to exporting or vice versa?', *The World Economy*, vol. 33, no. 3, pp. 374–398.

Griffith, R and Macartney, G 2014, 'Employment protection legislation, multinational firms, and innovation', *Review of Economics and Statistics*, vol. 96, no. 1, pp. 135–150.

Kleinknecht, A, Van Montfort, K, and Brouwer, E 2002, 'The non-trivial choice between innovation indicators', *Economics of Innovation and New Technology*, vol. 11, no. 2, pp. 109–121.

Lee, C and Narjoko, D 2015, 'Escaping the middle-income trap in Southeast Asia: micro evidence on innovation, productivity, and globalization', *Asian Economic Policy Review*, vol. 10, pp. 124–147.

Murphy, G, Siedschlag, I, and McQuinn, J 2013, 'Employment protection and innovation intensity', *Neujobs Working Paper* No. D6.4. Neujobs.

OECD 2002, *Economic Outlook* (June), vol. 1, no. 71, OECD, Paris.

OECD 2009, *Innovation in Firms: A Microeconomic Perspective*, OECD, Paris.

Pierre, G and Scarpetta, S 2006, 'Employment protection: do firms' perceptions match with legislation?', *Economics Letters*, vol. 90, pp. 328–334.

Romer, PM 1990, 'Endogenous technological change', *Journal of Political Economy*, vol. 98, no. 5, pp. S71–S102.

Saint-Paul, G 2002, 'Employment protection, international specialization and innovation', *European Economic Review*, vol. 46, pp. 375–395.

Tang, H 2012, 'Labor market institutions, firm-specific skills and trade patterns', *Journal of International Economics*, vol. 87, pp. 337–351.

Wagner, J 2007, 'Exports and productivity: a survey of the evidence from firm-level data', *The World Economy*, vol. 30, no. 1, pp. 60–82.

World Economic Forum (WEF) 2014, *The Global Competitiveness Index 2014–2015*. Available from: www3.weforum.org/docs/WEF_GlobalCompetitivenessReport_2014–15.pdf.

13 Developing domestic and export markets and levelling up trade in value added

Lessons learnt

Ben Shepherd[1]

1. Introduction

Value chain trade has become an increasingly important feature of the world economy in recent decades.[2] The Asia–Pacific region, including Southeast Asia, is globally recognized as a leader in many respects, particularly in sectors such as electrical equipment, including consumer electronics. Production in these sectors relies on complex networks of trade and investment that result in increasingly tight links between countries.

Although the development of global and regional value chains has been particularly striking in the Asia–Pacific, important changes have also taken place elsewhere in the global economy. Other regions have experienced growth in value chain trade as well, although its intensity varies considerably from one part of the world to another. Against that background, the purpose of this chapter is to examine the spread and performance of value-added trade in the Association of Southeast Asian Nations (ASEAN) region, and to make comparisons with other regions, particularly Latin America and the Caribbean (LAC), and Central and Eastern Europe (CEE). The emphasis in this chapter is on examining the current reality in ASEAN compared with two other important regions, and to examine any policy implications that might be applicable to ASEAN from the experiences of these comparator groups.

Analyzing value chains quantitatively is still a relatively recent undertaking. Most research on value chains as such is qualitative, focusing on case studies of their development and spread, and analysis of the way in which value added is divided up amongst the various actors in the chain. Although some of the qualitative literature is primarily positive, a strong strand in that literature focuses on the implications of value-chain trade from a more normative standpoint. This chapter discusses some possible policy implications of its findings, but it is, by contrast, primarily quantitative and positive in approach.

The study of production networks in East Asia and elsewhere (e.g., Ando and Kimura 2013b, and previous works by those authors) can be seen as laying the groundwork for the quantitative work undertaken here. That literature focused on trade in intermediate inputs is one of the key processes underlying the spread of value chains. However, identifying trade in intermediate inputs using traditional trade data is difficult. The approach most commonly adopted was to designate certain products in standard international trade classifications as intermediate inputs, to relate them to other products designated as final, and to conduct quantitative work exploiting the distinction (e.g., Saslavsky and Shepherd 2014).

This chapter takes a different empirical approach. It exploits newly released Organisation for Economic Co-operation and Development (OECD)–World Trade Organization (WTO) data on trade-in-value-added (TiVA). Unlike traditional trade data, TiVA data fully account for the use of intermediate goods and services in production – including imports and exports – and make it possible to isolate domestic value added that is then exported. The TiVA data provide an ideal framework for examining trade in global and regional value chains. This is because the business models behind these production platforms are increasingly seen in terms of exchanges of value added within a complex network rather than simple shipments of goods from one point to another.

The TiVA data offer the basis for a rich exploration of value-chain trade, including in its cross-regional comparative aspects. However, analytical tools are necessary to make the data really talk. This chapter adopts four summary measures from the recent literature to provide an entry point into the analysis of value-added trade in ASEAN, and to make comparisons possible with LAC and CEE. The four summary measures constitute different ways of looking at the extent of a country's integration into global and regional value chains. The analysis is undertaken for all countries for which data are available in ASEAN, LAC, and CEE. In sectoral terms, the focus is on three areas of special interest from the point of view of value-chain analysis: electrical equipment, transport equipment, and textiles and clothing.

Interpreting the results from the quantitative exercise and deriving policy implications is challenging, because the measures are summary in nature, and many factors could potentially explain any observed differences. With that in mind, this chapter also includes a selective review of the literature on value chains in the three regions. The review focuses on recent contributions examining the growth and development of value chains, in particular in the sectors under consideration, and the factors that have given rise to the observed results. Combining the results from the quantitative exercise with those from the literature review makes it possible to discuss ASEAN's relative position with greater clarity, and to assess the implications for Southeast Asia of the different ways in which value chains have developed elsewhere.

Against this background, this chapter proceeds as follows. Section 2 presents the TiVA data set, and discusses the indicators to be used in the quantitative part of the chapter. Section 3 contains the comparative review, focusing first on the literature (Section 3.1), and then on an analysis of the TiVA data (Section 3.2). The following section, Section 4, draws out policy implications of the findings in Section 3, and Section 5 concludes.

2. Trade-in-value-added data

2.1 Overview of the TiVA data set

Traditional trade data, such as those found in cross-country databases such as UN Comtrade, are recorded in gross value terms. For example, an iPhone shipped from its final assembly point in China to the United States (US) is recorded as an export from China to the United States equal to the full landed price of the phone in the importing market. Although historically appropriate as a first approximation, the advent of value chains has made traditional trade data less fit for purpose in terms of understanding and analyzing global flows of goods and services. The iPhone is manufactured using a complex network of value-chain interactions and in fact, relatively little value is added in China compared with the high value added of research and design services located in the United States itself, notionally the importing market.

Taking this example further, it is clear that gross value-trade data tend to overstate the true level of trade between countries. The reason is that intermediate inputs are double counted. For example, the full landed price of the solid-state hard drive used in the iPhone is counted as an export from a source country, such as Thailand, to the assembly point, China. The value of the hard drive is then incorporated again in the price of the final product sent from China to the United States, so the same figure is included in export data for two countries through incorporation in a final product.

Because of the double counting of intermediate inputs, traditional trade data are not compatible with national accounts data. Gross domestic product (GDP) is defined as the sum of all value added in the economy, that is, production less intermediate inputs. Exports and imports in the national accounts are, in principle, smaller than Comtrade trade data because they net out intermediate input use. This disconnect is the reason why some countries have exports to GDP ratios that approach, or even occasionally exceed, 100 per cent, which is not possible in terms of true national accounts data.

To deal with these problems, recent research has focused on the idea of measuring trade in value-added terms, rather than gross value terms. The idea is to construct export and import data that net out trade in intermediate inputs – including services – and are therefore fully compatible with national accounts. A number of researchers have moved forward in this area, such as Johnson and Noguera (2012) and Koopman et al. (2014). Building on these efforts, OECD and WTO created a joint global database of trade-in-value-added (TiVA). The raw TiVA data are available for 57 economies and 11 goods sectors for the years 1995, 2000, 2005, 2008, and 2009. They are constructed using national accounts data, trade data, and input–output matrices. The data set has figures on various dimensions of value-added trade, and is set out bilaterally, that is, as a matrix of all relationships between exporters and importers.

The OECD–WTO TiVA data set is much better suited to the study of value chains than are traditional gross value-trade data. For example, the iPhone referred to earlier does not double count intermediate input use in the relevant export values, but instead shows export flows from countries like Thailand (for components such as the hard drive) to the United States, and only records the value added located in China – primarily assembly – as an export flow from China to the United States. As a result of changes like these, data on openness and bilateral trade balances can be substantially different using TiVA data and traditional trade data.

The TiVA data make it possible to examine aspects of value-chain trade that have proved elusive in previous research. Initially, value chains were studied by focusing on trade in intermediate inputs (e.g., Ando and Kimura 2013b). Intermediates were identified intuitively, using standard international trade product classifications. The TiVA approach provides a much finer and more accurate level of analysis on intermediate input use, and importantly includes services as well as goods. Services trade is a crucial component of value chain activity. In many ways, services such as design, transport, and distribution are the glue that holds value chains together. The TiVA data therefore offer the possibility of rich insights into the operation of value chains, and the possibility of cross-regional and through-time comparisons.

To remain concise, this analysis of the TiVA data will be limited in time and sectoral scope. Previous work suggests that the structure of value-chain trade remains remarkably stable over time (Shepherd and Archanskaia 2014). Because of that, the long term will be the focus of the analysis. The two end-points of the TiVA data, 1995 and 2009, will be compared to give a general picture of dynamic evolutions.[3] Of course, many significant economic events have taken place during that period, including substantial liberalization in some countries, as well as macroeconomic events such as the Asian Financial Crisis. In terms of sectoral scope, the analysis will consider three sectors in which value-chain trade is particularly important, but contrasts can be expected amongst ASEAN, CEE, and LAC countries: electrical equipment, transport equipment, and textiles and clothing. From an intra-ASEAN perspective, electrical equipment is perhaps the most crucial sector, but experience differs somewhat from country to country. In the other regions. the relative importance of the three value chains varies.

Comparing performance across regions and value chains can be informative from a policy perspective. However, it is important to keep in mind that each value chain represents a distinct set of business models. For instance, electronics production is typically more integrated than other industries, so value chains can be expected to behave differently, and be tracked differently through quantitative indicators. This chapter is therefore an attempt to map out some of the cross-regional and cross-sectoral differences in value-chain behaviour. However, further detailed research is needed on the ways in which particular value chains operate in the special circumstances of individual countries and sectors.

2.2 Indicators based on the TiVA data

The TiVA data set is a new source that offers real potential for new insights into value-chain trade. However, it contains a large quantity of information. Analytical tools are therefore needed to examine value-added trade in particular countries and regions. The TiVA bilateral trade matrix and the tables used to produce it are raw materials that can be manipulated and transformed to produce indicators that can usefully be compared across regions and time periods. That is the approach adopted in this chapter, building on data work done at OECD (De Backer and Miroudot 2013).

The first indicator to be examined is the value added to gross exports (VAX) ratio (Johnson and Noguera 2012). It provides an overall indication of the intensity of production sharing that takes place through value chains. It is the proportion of total exports measured in gross value terms that is accounted for by domestic value added. Intuitively, a lower VAX ratio means that the level of domestic content in exports is lower, and the foreign content is therefore higher. It clearly shows the effect of netting out intermediate input use – particularly imported intermediate inputs – in value-added trade data. Although comparisons of VAX ratios across time periods and countries should not be over-interpreted, given that many factors are at play, the exercise is useful in providing a first pass indication of the degree of production sharing that is under way in particular countries and sectors.

The second indicator to be used in this chapter is an index of the number of international production stages (Antras et al. 2012). This indicator measures the length of international value chains in a particular country and sector. The index is equal to one if there is a single production stage involved, and takes progressively higher values as inputs from the same or other sectors are used, according to a weighted average formula (De Backer and Miroudot 2013):

$$(1)\ N = u.\,(\mathrm{I} - A)^{-1}$$

where N is a column vector containing the indices according to country (1) and sector (k), u is a unit vector, I is an appropriately dimensioned identity matrix, and A is the matrix of technical coefficients from the international input-output matrix that lies behind the TiVA data.

The last two indicators that will be analyzed are the backward and forward participation indices (Koopman et al. 2010). The first measures the value of imported intermediate inputs in a country's exports (backward participation). The second measures the share of exported goods used as intermediate inputs to produce other countries' exports (forward participation). These two indicators can be summed to give an overall picture of the level of involvement of a country in value chain trade. A higher score is consistent with a greater degree of internationalization.

The participation index is based on the following equations:

$$P_{ik} = \frac{VS_{ik}}{E_i} + \frac{VS1_{ik}}{E_i} \qquad (1)$$

$$VBE = V(1 - A)^{-1} E \qquad (2)$$

where P is the participation index, E is gross exports, VS is an element of the vector obtained by summing the columns of the VBE matrix (without domestic industries), and VS1 is an element of the vector obtained by summing the rows of the VBE matrix (again without domestic industries). For the VBE matrix, A is defined as earlier, E is gross exports, and V is the diagonal of a matrix with value-added shares by country and industry.

Each of the four indicators analyzed in this chapter presents a different way of looking at a country's degree of involvement in global and regional value chains. As emphasized at the outset, many reasons lie behind cross-country and through-time variations, so results need to be interpreted cautiously. Nonetheless, the indicators are useful in providing a first picture of the degree to which internationalization of the value chain has taken place in the three sectors under consideration in ASEAN, LAC, and CEE. Section 3.2 analyzes the data in terms of broad trends that are evident. The main interpretation provided in Section 4 combines insights from the data analysis and the literature review to examine value-chain trade in the three regions in comparative perspective, and in particular from the angle of lessons that could be learnt by ASEAN from experience in other regions.

3. Comparative review: ASEAN vs. Latin America and the Caribbean, and Central and Eastern Europe

This section presents the analytical material that is the focus of this chapter. The first subsection presents a selective review of the available literature on global and regional value chains in CEE and LAC, and the second undertakes a quantitative analysis using indicators derived from the OECD–WTO TiVA data set. The presentation here is positive in nature, not normative. Possible policy implications are discussed in more detail in Section 4.

3.1 Literature review

The literature review is divided into two parts. First, the discussion focuses on the development of value chains in CEE and LAC, noting their spread, intensity, and the type of exchanges that typify them. Next, consideration is given to the economic impacts of value chains in the two regions, taking account of the different development paths followed. Given the large amount of literature involved, it is necessary to be selective in terms of presentation. The approach taken is therefore to focus on a number of key contributions that provide important insights, rather than to attempt to cover the field.

3.1.1 Development of value chain activity in CEE and LAC

Previous work on global and regional value chains in CEE has highlighted the important role they have come to play in the period following economic transition, that is, from the 1990s onwards (e.g., Ando and Kimura 2013b). Machinery is a key sector, with value-chain interactions based on trade and investment relations amongst manufacturing firms.

CEE's relationship with other regions is of particular interest. Although interregional linkages, including with Asia, have grown, Western Europe remains the key source of demand for CEE value-chain production (Ando and Kimura 2013b). This is an expected finding, in light of factors such as physical proximity and low trade costs due to both geography and common membership of the EU (Godart and Gorg 2011). Godart and Gorg (2011) identified a range of additional factors that have made CEE countries attractive for Western European firms – especially German ones – looking to internationalize their value chains. Factor costs and productivity are important, but human capital in the form of an educated workforce also plays a significant role.

The rise of value chains in CEE has coincided with restructuring of the Western European manufacturing sector, particularly in Germany, the manufacturing centre of Europe (Timmer et al. 2014). Following patterns established elsewhere, for instance in North America, high value-added activities have tended to stay in Germany. However, other parts of the value chain that are more labour-intensive have been relocated to CEE. The net result for Germany has been a fall in the overall domestic content of manufacturing, as reliance on intermediate goods sourced notably from CEE has increased. This process can potentially result in efficiency and competitiveness gains due to cost savings, as well as specialization according to comparative advantage. However, results are sometimes striking: for example, it has been estimated that only 30 per cent of the value added of a Porsche Cayenne – nominally made in Leipzig – is German. The car is in fact mostly assembled in the Slovak Republic, a fact that gives rise to concerns in some quarters in Germany (Sinn 2006).

Although structural factors are important in the spread of manufacturing value chains from Western Europe, and particularly Germany, to CEE, Godart and Gorg (2011) also highlighted the important role firm-specific considerations play. The decision to internationalize a value chain is a complex one influenced heavily by economic variables, but issues such as management structure and competence, and market pressures in relation to final products, also come into play. The discussion here focuses on structural factors because of the nature of the empirical analysis conducted in the next section, but in practice it is important to keep in mind that many substantive issues affecting the decision whether to internationalize, and which functions to internationalize where, interact in a complex way at the firm level. One implication is that although country- and region-level studies are useful in understanding the broad dynamics in operation, firm-level empirical analysis – both quantitative and qualitative – still has an important role in explaining the particularities of individual value chains.

Compared with CEE, the available literature seems to suggest that global and regional value chains are relatively underdeveloped in LAC (e.g., CIDOB et al. 2014). LAC is a large region, however, and there is considerable diversity within it. In South America, value-chain participation is typically centred on the supply of raw materials, in contrast to the manufacturing core of CEE value chains. In Mexico, however, there is a greater level of manufacturing, for instance, in the well-known *maquiladoras*. As a whole, however, it is fair to characterize value-chain interactions in LAC as less intense than in CEE, and differently focused. In terms of the capture of value added, production is generally centred on low value-added activities, such as the production of primary materials.

Against this background, Giuliani et al. (2005) highlighted the difficulty many LAC countries have had in moving up the value chain to higher value-added activities. The authors highlighted structural issues as an important cause. In particular, they stressed the region's relative abundance of natural resources, and relative scarcity of labour and capital. Of course, labour is available in large quantities, but in relative terms, free markets tend to lead to specialization by many LAC countries in low value-added parts of the value chain. However, sectoral and country specificities can be important, as the example of manufacturing in Mexico shows.

Just as Western Europe, and in particular Germany, played a key role as a source of demand in the development of CEE value chains, the United States has been an important force in the evolution of value chains in North and Central America, particularly Mexico (Kuwayama 2009). As in the case of Germany, high value-added activities tend to stay in the United States, and lower value-added activities move offshore to destinations like Mexico. Assembly is an example of a relatively low value-added task in the manufacturing sector. Kuwayama (2009) argued that, going forward, it will be important for LAC countries to move up manufacturing value chains towards higher value-added activities. This process entails renewed attention to the services sector, including areas such as engineering, research, and design. The author argued that policy interventions may be necessary to realize this goal.

3.1.2 Economic impacts of value chain development

The development paths taken by value chains in CEE and LAC are substantially different, but similar issues arise, in particular concentration on relatively low value-added activities (although the level of value addition appears to be significantly higher in CEE than in LAC). The broad pattern of specialization is the same in both locations, with the advanced 'anchor' market (Germany or the United States) specializing in high value-added activities, and the surrounding countries specializing in activities with lower levels of value addition (e.g. Timmer et al. 2014).

Damijan et al. (2013) highlighted the important role that internationalization of production has played in restructuring the CEE countries' trade relations, both on the export and import sides. In particular, there has been a notable shift towards imports of intermediate inputs and capital goods. Associated foreign direct investment (FDI) flows have been significantly associated with export upgrading in some countries, but not all. Countries that have been more successful in attracting FDI to technology-intensive industries have tended to experience faster productivity growth.

Productivity gains and economic restructuring are important channels by which value chains can influence economic outcomes. There is the potential for economic benefits to flow even though countries specialize, at least initially, in relatively low value added parts of the chain. Moreover, Kuwayama (2009) pointed out that in the case of LAC, where there is considerable surplus labour, low value-added activities can have significant economic and social advantages. The mechanism at play is the creation of employment in relatively low-skill activities. Many developing countries, including some in LAC, have dual labour markets, with significant excess supply of low-skill workers. An expansion of value-chain activity that increases demand for low-skill labour can therefore take up some of the slack in that part of the labour market, which decreases unemployment and potentially, over time, exerts upwards pressure on low-skill wages.

3.2 TiVA indicator analysis

This section provides a comparative analysis of the TiVA indicators discussed in Section 2. The presentation is data driven, focusing on interpretation of the indicators. Policy implications of this set of findings taken together are discussed in Section 4.

The OECD–WTO TiVA data do not cover all countries in the three regions being compared. It is therefore necessary to use those countries included in the data set as proxies. Indicators of regional, as opposed to national, performance are calculated by taking the simple average across all countries for which data are available in the region.

Country coverage in the data set is as follows:

- *ASEAN*: Brunei Darussalam, Cambodia, Indonesia, Malaysia, the Philippines, Singapore, Thailand, and Viet Nam
- *Latin America and the Caribbean*: Argentina, Brazil, Chile, and Mexico
- *Central and Eastern Europe*: Bulgaria, Czech Republic, Estonia, Hungary, Slovak Republic, and Slovenia.

Coverage is reasonably good for ASEAN and CEE, but is quite weak for LAC. Results for that region will therefore need to be interpreted with caution. Nonetheless, much of the region's global value-chain activity is concentrated in Mexico, due to its proximity to the United States and the influence of the North American Free Trade Agreement (NAFTA), so the sample can be regarded as informative for comparative purposes.

3.2.1 VAX ratio

A country's VAX ratio summarizes the proportion of gross exports that is accounted for by domestic value added. Figures 13.1–13.3 present comparative results for the three sectors under consideration – electrical goods, transport equipment, and textiles and clothing. The lowest VAX ratio of the three sectors

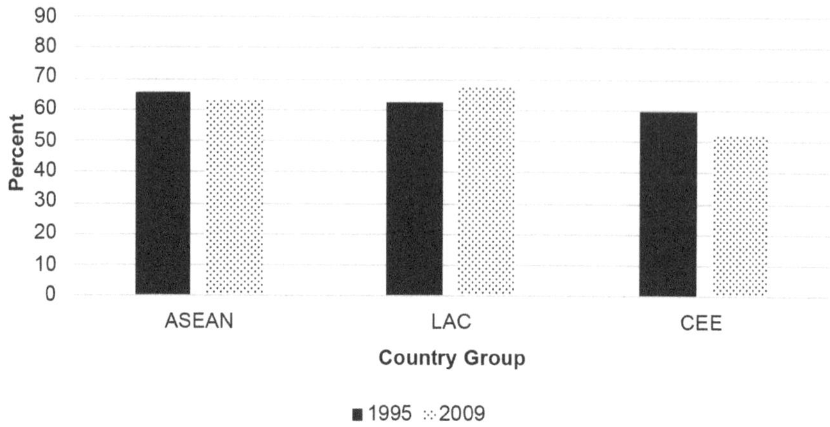

Figure 13.1 VAX Ratio (electrical equipment)

Source: OECD–WTO TiVA Indicators.

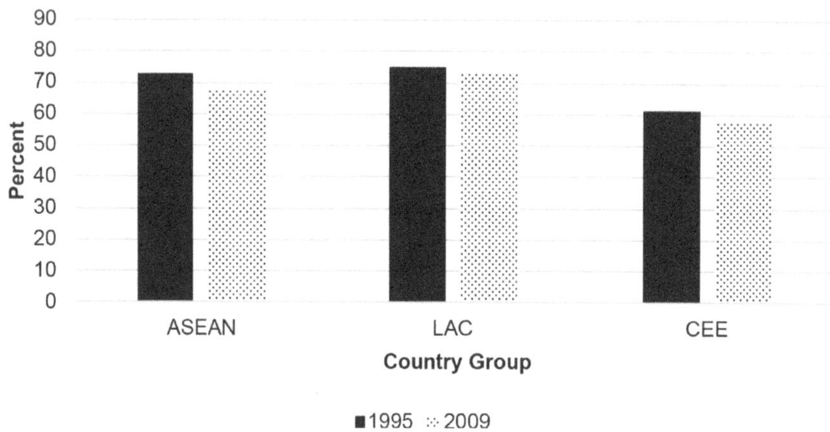

Figure 13.2 VAX Ratio (transport equipment)

Source: OECD–WTO TiVA Indicators.

is in electrical equipment (61.9 per cent for the world average), followed by transport equipment (63.7 per cent), and then textiles and clothing (70.1 per cent). World average VAX ratios fell in all sectors between 1995 and 2009, but the fall was much more pronounced in relative terms in transport equipment (–7.4 per cent) and electrical equipment (–6.3 per cent) than in textiles and clothing (–1.9 per cent). A falling VAX ratio can have a number of economic

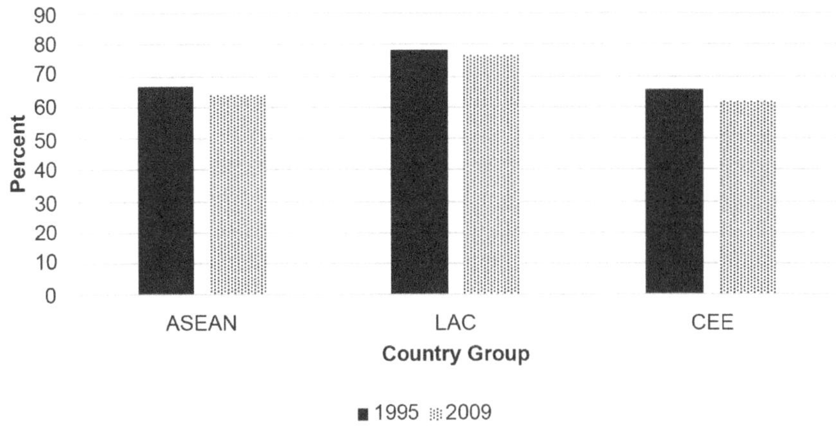

Figure 13.3 VAX Ratio (textiles and clothing)

Source: OECD–WTO TiVA Indicators.

drivers behind it, but a common interpretation is that production chains are continuing to internationalize in all sectors, which means that the ratio of imported intermediates to domestic value added is higher, and thus that the VAX ratio is falling. Under that interpretation, the biggest change in internationalization of production has been seen in transport equipment, followed by electrical equipment, although the latter remains more internationalized (i.e., has a lower VAX ratio) than the former. The level has only fallen slightly in the textiles and clothing sector.

Figures 13.1–13.3 also present comparative regional data for each sector. Those data can be used to compare value chain internationalization, and the dynamics behind it, across regions. For electrical equipment, ASEAN's average VAX ratio is 5 percentage points lower than LAC – indicating greater internationalization – but is 10 percentage points higher than CEE, which indicates the opposite. This is an interesting result, because electronics goods value chains are often associated with Asia, including ASEAN, but this result suggests that the degree of internationalization of production may actually be higher in CEE than in Southeast Asia. Moreover, the relative rate of change is faster in CEE than in ASEAN: the ratio fell by 12.3 per cent between 1995 and 2009 in CEE, but only by 4.5 per cent in ASEAN. It will be necessary to compare these results with more detailed indicators to see if the comparison holds.

For transport equipment, the same pattern of results is evident in levels (Figure 13.2). Indeed, the quantitative magnitude of the differences across regions is nearly identical to the case of electrical equipment. Again, one interpretation is that despite the importance of transport equipment value chains in Asia, including ASEAN, the degree of internationalization is still not as great as in CEE, notwithstanding the lower ratio with respect to LAC. In terms of dynamics, the pattern for electrical goods is reversed: ASEAN is internationalizing at a relatively faster pace than CEE. The former's VAX ratio declined by 7.7 per cent, compared with 5.8 per cent for the latter. The rate of change is much slower in LAC, at just –2.8 per cent.

Results for CEE and ASEAN are closer in the textiles and clothing sector (Figure 13.3), although CEE is still a couple of percentage points lower in its average VAX ratio than ASEAN. LAC has a much higher score than either of the two other regions. In terms of the rate of change, the percentage declines in VAX ratios for ASEAN and CEE are quite similar: 4.1 per cent and 5.9 per cent, respectively. The result for LAC is closer than in the other sectors, at -2.3 per cent.

Interpreting changes in the VAX ratio is not straightforward. At a policy level, there is clearly a mercantilist temptation to become fixated on maintaining a particular proportion of domestic value added in exports. However, there is strong evidence that domestic and foreign intermediates act in part as complements, not just as substitutes: internationalization of the value chain is associated with rapid growth in trade, including domestic value added – which can increase even though its share in gross exports decreases. To make the point, Figure 13.4 presents domestic value added in exports for electrical equipment in ASEAN, where VAX ratios have been falling rapidly. As is clear, domestic value added has in fact increased sharply in all three regions as complementarities between domestic and imported intermediates are exploited.

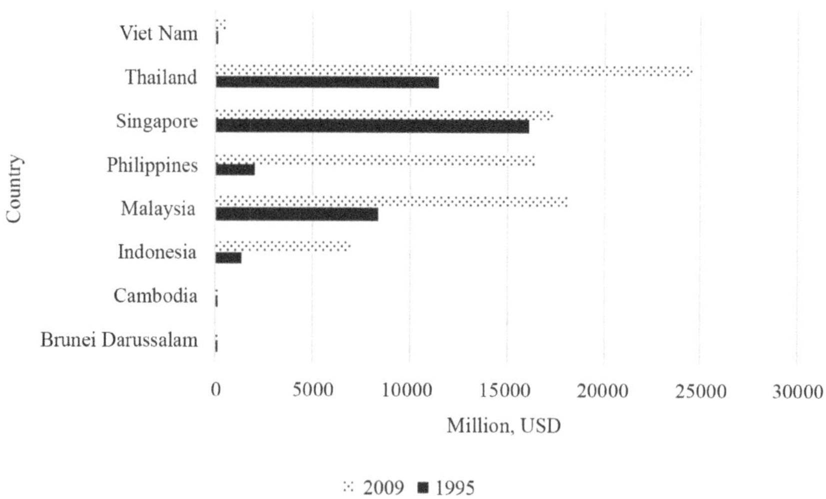

Figure 13.4 Domestic value added embodied in gross exports, ASEAN, (USD million), 1995–2009

Source: OECD–WTO TiVA Indicators.

3.2.2 Number of international production stages

To investigate the hypothesis that internationalization of production is stronger in CEE than in ASEAN, and that it is rising faster, a logical first step is to look at the number of international production stages in each sector and region. Figures 13.5–13.7 present results.

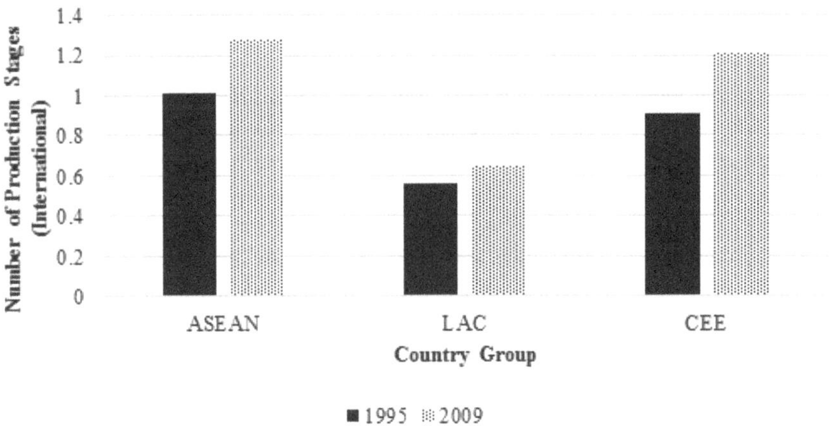

Figure 13.5 Number of international production stages (electrical equipment)

Source: OECD–WTO TiVA Indicators.

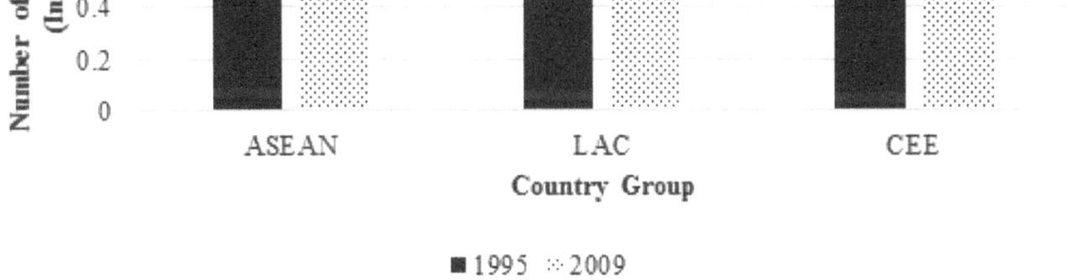

Figure 13.6 Number of international production stages (transport equipment)

Source: OECD–WTO TiVA Indicators.

In terms of the world average, the electrical goods sector has the most international production stages, but results for transport equipment are very close. The textiles and clothing sector has noticeably fewer international production stages. These results in levels are exactly reflected in relative changes over time: the electrical goods and transport equipment sectors have seen the largest increases in the number of international production stages (26.4 per cent and 26.3 per cent, respectively), compared with a much lower rate of change of only 12.9 per cent in textiles and clothing.

Figure 13.5 makes it possible to compare results for electrical goods across the three regions under consideration. Although there are more international production stages in ASEAN than in CEE – a result that seems contrary to the indications of the VAX ratio – the rate of change in CEE is much faster. The number of production stages increased by 33.1 per cent in CEE between 1995 and 2009, compared with 26.6 per cent in ASEAN. On the other hand, ASEAN displays a different pattern from the world average in that the number of international production stages is 24.3 per cent higher than in transport equipment; the two figures are nearly identical for CEE. LAC again seems not to be as dynamic as the other regions, with only around half as many international production stages as ASEAN in 2009, and a much lower rate of change over the sample period (14.1 per cent).

Results for transport equipment are in Figure 13.6. In this case, CEE has considerably more international production stages than ASEAN, the difference being 13.2 per cent. In this case, the evidence from this indicator accords well with preliminary findings from the VAX ratio to the effect that production is more internationalized in this sector in CEE than in ASEAN. It is not just the number of international production stages that is higher in CEE: there is also a faster rate of change, 28.9 per cent between 1995 and 2009, compared with 16.3 per cent

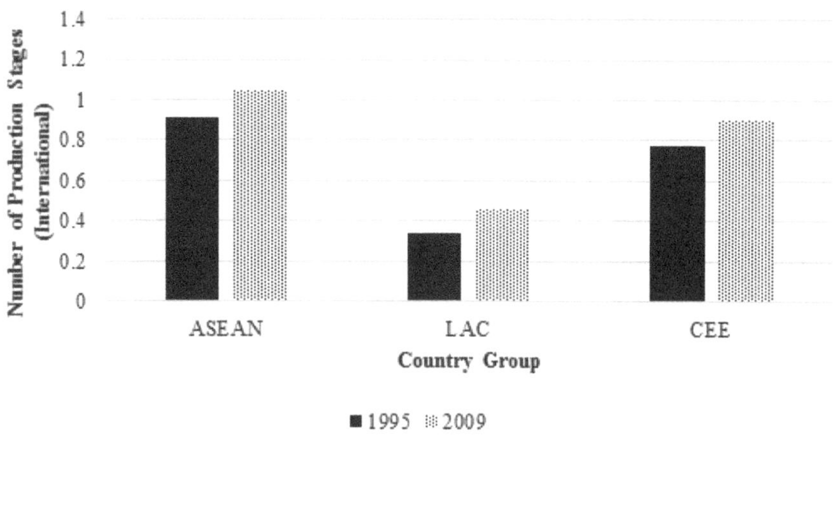

Figure 13.7 Number of international production stages (textiles and clothing)
Source: OECD–WTO TiVA Indicators.

in ASEAN. Although LAC has fewer international production stages than either of the other two regions, its growth rate is actually higher than ASEAN's at 21.9 per cent.

The textiles and clothing sector (Figure 13.7) displays some similarities to the other two sectors, despite its generally lower number of international production stages. ASEAN has the most stages, with a score 16.3 per cent higher than CEE. LAC has far fewer international production stages: its figure is only 43.8 per cent of ASEAN's. However, the picture becomes more interesting when rates of change are considered. On this occasion, LAC has actually seen the fastest growth in the number of international production stages (36.7 per cent), compared with very similar but much lower rates of change in the other two regions (16.5 per cent in CEE and 15.0 per cent in ASEAN). Part of the reason for this result is undoubtedly that LAC started from a much lower baseline in 1995, but the rate of growth is nonetheless impressive.

3.2.3 Backwards and forwards participation

The two participation indices focus on trade in intermediate goods. The backwards index captures the proportion of a country's exports that is accounted for by foreign intermediate inputs. The forwards index captures the proportion of a country's exports that are used as intermediate inputs for other countries' exports.

Figures 13.8–13.10 present results for the backwards participation index. Changes in this indicator, and comparisons across groups, are more striking compared with the indicators previously discussed. In terms of the world average, electrical equipment has by far the highest index: it is twice as high as the index for transport equipment, and more than three and a half times higher than the index for textiles and clothing. In terms of dynamics since 1995, however, it

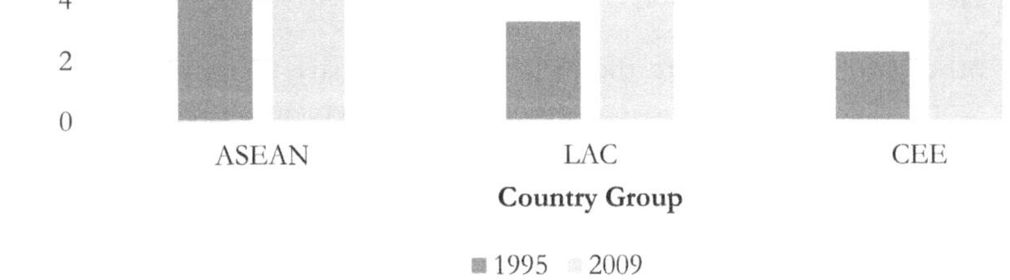

Figure 13.8 Backward participation index (electrical equipment)

Source: OECD–WTO TiVA Indicators.

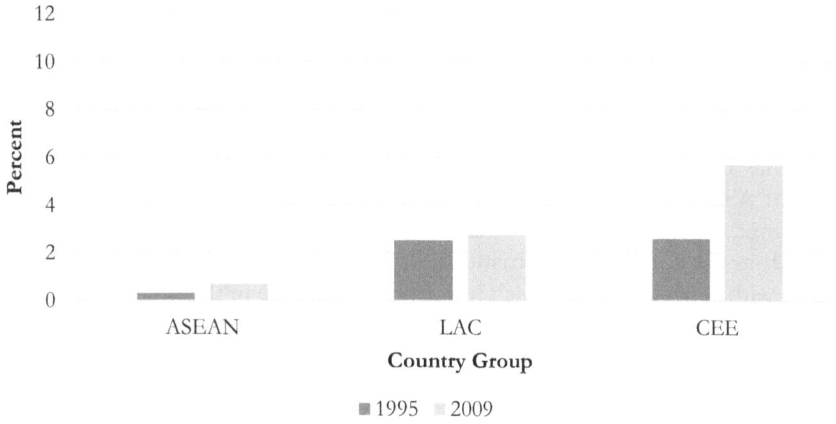

Figure 13.9 Backward participation index (transport equipment)

Source: OECD–WTO TiVA Indicators.

Figure 13.10 Backward participation index (textiles and clothing)

Source: OECD–WTO TiVA Indicators.

is the other two sectors where the biggest changes have taken place in terms of the world average. Transport equipment saw an increase of 39.0 per cent in its score, compared with 13.7 per cent for electrical equipment. Two possible interpretations are open: either value chains have tended to internationalize less over the time period, thereby reducing reliance on imported intermediates, or domestic value added as a proportion of the final goods price has on average increased. Given the extensive evidence already marshalled to show that value chains have generally become more, not less, internationalized over time, the second explanation is the only plausible one: domestic value added has tended to increase in relative terms, particularly in transport equipment, and to a lesser extent in electrical equipment. By contrast, the opposite dynamic is true for textiles and clothing: the world average score fell by 38.3 per cent during the sample period, which tends to suggest that value chains have become more focused on lower value-added activities, such as assembly from component parts.

Just as the world average varies considerably from one sector to another, so too do the results for ASEAN and the two comparator regions. ASEAN's highest score by far is in electrical equipment (10.1); scores in the other two sectors are much lower (4.3 in textiles and clothing, and 0.7 in transport equipment). The pattern of scores across the three sectors for ASEAN is interesting because it does not reflect what has been observed for other indicators, where there has been a reasonably stable ordering of either electrical equipment or transport equipment at one end of the scale, and textiles and clothing at the other. The surprising result for transport equipment is all the more notable because of the contrast with CEE, which has a score 8.2 times higher, despite similar relative growth rates in both cases (131.9 per cent in ASEAN compared with 120.6 per cent in CEE). Although particularly stark in this case, the result for transport equipment suggests that the value chain is not as internationalized in ASEAN as it is in CEE, for example, and that it is less internationalized than the electrical equipment sector. This interpretation is consistent with results for the other indicators, discussed previously.

The comparison between LAC and CEE is interesting for the case of transport equipment. The two regions start from similar baselines in 1995 (2.5 for LAC, and 2.6 for CEE), but subsequent growth rates are very different. CEE has experienced explosive growth of 120.6 per cent in this sector, compared with only 9.6 per cent for LAC. Activity in the transport equipment sector is significant in both regions, although it is more localized in LAC in terms of its involvement in international value chains, being centred on Mexico.

All three regions experienced a fall in their backward participation index over the sample period for textiles and clothing. The fall was most pronounced in CEE (–56.0 per cent), followed by LAC (-28.5 per cent) and then ASEAN (-15.7 per cent). In terms of the pattern of scores, ASEAN has the highest score in both periods, and is separated from the other two regions by a considerable margin. One interpretation of this pattern of results is that there is still considerable assembly activity going on in this sector in ASEAN, although it is declining in relative terms in all regions. It is, however, more persistent in ASEAN than elsewhere, which is reflected in a greater overall importance of imported intermediates in producing exports.

To provide a point of contrast, forward participation indices for the three sectors and regions appear in Figures 13.11–13.13. In terms of the world average, electrical equipment has by far the highest score: it is over four times as high as the average score for transport equipment, and over six times higher than the score for textiles and clothing. Taking results for the two indices together clearly suggests that trade in intermediate inputs is particularly important in this sector,

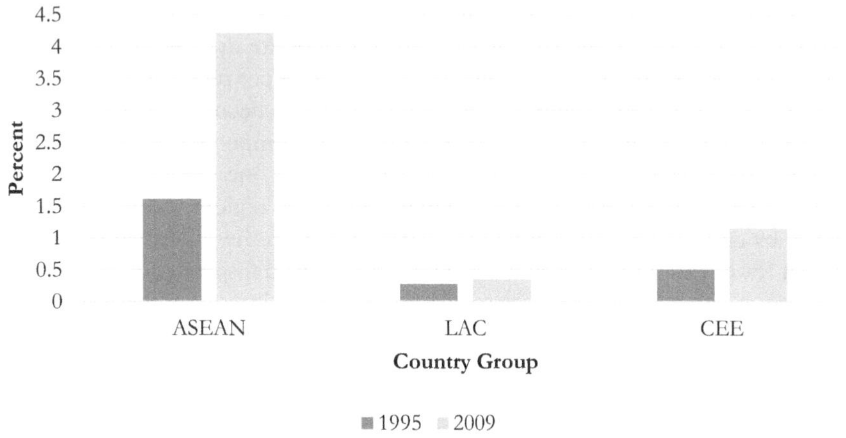

Figure 13.11 Forward participation index (electrical equipment)
Source: OECD–WTO TiVA Indicators.

Figure 13.12 Forward participation index (transport equipment)
Source: OECD–WTO TiVA Indicators.

which is in line with the important role that value chains play in production. This pattern of results is also reflected in an aggregate sense in a dynamic comparison of scores in 1995 and 2009: growth in forward participation has been much faster in electrical equipment than in transport equipment; the world average score has, by contrast, fallen noticeably in textiles and clothing.

ASEAN has the highest forward participation index score of any of the three regions under consideration for electrical equipment. The number has also

Figure 13.13 Forward participation index (textiles and clothing)

Source: OECD–WTO TiVA Indicators.

Figure 13.14 Outward FDI Stocks of China, Japan, and Korea in Southeast Asia, and Germany in CEE, 2011 (USD million)

Note: Southeast Asia is defined as elsewhere in the chapter, i.e. Brunei, Cambodia, Indonesia, Malaysia, the Philippines, Singapore, Thailand, and Viet Nam. Similarly, CEE is defined as Bulgaria, Czech Republic, Estonia, Hungary, Slovak Republic, and Slovenia.

Source: UNCTAD Bilateral FDI Statistics.

undergone rapid growth in recent years (162 per cent between 1995 and 2009). Exports of intermediate inputs for use in other countries' exports is more important in ASEAN than in the other regions, although CEE's level of participation has only grown at a slightly slower rate, albeit from a much lower starting point.

Transport equipment again sees the order reversed: CEE has a considerably higher score than ASEAN, and its growth rate (180.7 per cent) has been much faster than ASEAN's (47.3 per cent). Whereas LAC could be considered relatively dormant in terms of growth of intermediate inputs trade in the electrical equipment sector, this form of trade has actually become less important in transport equipment. There are two possible explanations: either LAC production processes have become less internationalized, or the countries in the sample are exporting more goods for final consumption and fewer intermediate inputs. Given that there is evidence of increasing internationalization in this sector, it is more likely that the second explanation holds.

As was the case for the backwards participation index, the forwards index for textiles and clothing contracted in both ASEAN and CEE over the sample period; in LAC there was slight, almost imperceptible, growth. Interesting, CEE had a higher index score than ASEAN in 1995, but by 2009 had a lower one. The interpretation is that exports of intermediate inputs to be used in other countries' exports have fallen in relative importance in both regions, but the fall has been much sharper in CEE.

4. Policy implications

The TiVA data and the literature review have disclosed substantial differences in the ways in which global and regional value chains have developed and spread according to sector and country. In general terms, ASEAN is a strong performer in comparative perspective, particularly in the electrical equipment sector. Value chains in Southeast Asia have undergone considerable internationalization in recent years, and this is evident in the dynamics of key indicators. The only sector in which continuing internationalization has been rather limited in ASEAN is textiles and clothing. Given that that sector is intensive in low-skill labour compared with the other two sectors under consideration – electrical equipment and transport equipment – this finding arguably suggests movement within ASEAN towards higher value-added activities. More broadly, ASEAN countries have typically been quite successful in joining GVCs in the sectors considered here. The challenge going forward is to move up within those GVCs to higher value-added activities that can have significant economic spillovers, such as research and development, and even to assume a leadership role in terms of strategic decisions at the business level.

In terms of the cross-regional comparison that is the analytical focus of this chapter, one finding stands out in sharp relief: value chain development has been very intense in CEE, but relatively limited in LAC. This result needs to be interpreted with caution, because the TiVA data are much more complete in their coverage of CEE than of LAC. Based on the available data, however, it appears that internationalization of production in LAC remains considerably less than in ASEAN or CEE, and the dynamic of change is much slower.

Another important finding that emerges from the literature review relates to the different natures of value chains in CEE and LAC. In the CEE countries, as in ASEAN, value chain activity is focused on manufactured goods. Although there are also some manufacturing value chains in LAC, particularly in countries such as Mexico, a more typical set-up is that LAC countries supply primary commodities that are transformed elsewhere, with corresponding high levels of domestic value added in primary sectors. The pattern of comparative advantage is therefore perhaps more similar in CEE and ASEAN than it is in LAC and ASEAN.

For these two reasons – greater internationalization and faster change, as well as reliance on manufacturing – it seems clear that the main potential source of policy insights for ASEAN is CEE rather than LAC. A comparison of the two regions, CEE and ASEAN, is instructive for many reasons, but the most obvious one is the different ways in which value chains have internationalized and evolved in the two more sophisticated sectors – electrical equipment and transport equipment. ASEAN value chains are more internationalized than those in CEE in the first case, but the opposite is true in the second case. Composition effects might be at play, but the pattern is nonetheless interesting. Backward and forward linkages are noticeably better developed in electrical equipment in ASEAN than they are in transport equipment, so it is important to examine the potential reasons behind that development compared with CEE. Part of the answer may be found in the special role played by Germany with respect to the CEE countries. Another part of the answer lies in firm strategies in particular sectors: for example, in the automotive sector, Volkswagen favours outsourcing different elements of the production process to different countries, and having a number of assembly locations, whereas Toyota prefers a single location assembly strategy. Advantages and disadvantages of both approaches are possible, but they clearly give rise to differences in the nature and extent of production internationalization in the regions under consideration. Similarly, differences in the way agglomeration economies work in the two sectors contribute to the overall differences seen in the numbers. A final set of operative factors relates to policy barriers. In electrical products, the tariff environment is relatively free thanks to the WTO Information Technology Agreement, and liberalization of air transport markets has facilitated the emergence of value chains by reducing the cost of rapid transport for high-value-to-weight products, such as components in this sector. In the transport equipment sector, by contrast, many countries have used activist policies of one sort or another to try to encourage development of domestic, rather than international supply chains, and the numbers in ASEAN perhaps still reflect in part the lingering effects of policy barriers.

Another striking fact that emerges from a comparison of value chain indicators for CEE and ASEAN is that there is some evidence that internationalization of even the electrical equipment value chain has been more rapid in a dynamic sense in CEE than in ASEAN. For example, the CEE countries increased their backwards participation index by 237.3 per cent between 1995 and 2009, compared with only 18 per cent in ASEAN. The index remains higher in ASEAN than in CEE, but only because of the huge difference in 1995 baselines.

5. Conclusion

This chapter has compared value chain experiences in ASEAN, CEE, and LAC. Three sectors have been considered: electrical equipment, transport equipment, and textiles and clothing. The data reveal different levels of internationalization according to region and sector. In general, the electrical equipment and transport equipment sectors are more internationalized than textiles and clothing. In regional terms, ASEAN and CEE both have much more international and dynamic value chains than LAC, although the data for LAC need to be interpreted cautiously due to a small sample.

From a policy perspective, the comparison between ASEAN and CEE is most instructive. The CEE countries have seen rapid internationalization of their production processes, in line with the economic transition and the path to membership of the EU. They have undertaken deep trade and investment liberalization not only amongst themselves, but also with the large, developed markets of Western Europe. Germany plays a particularly important role as an anchor market for the CEE countries: it is both a vital source of demand for their value-chain production, and also a source of investment through production outsourcing, which can act as a vector for technological change in the host countries. A range of factors lie behind the decision of German manufacturing firms to build productive capacity in CEE, and include the region in international value chains, but amongst the most important are the free movement of goods and capital, geographical proximity that reduces trade costs, differences in factor abundance and therefore relative prices, and levels of human capital. The last factor appears to be particularly important in the ability of some core CEE countries, such as Hungary, to move up manufacturing value chains to higher value-added activities.

The experience of the CEE countries with the EU, and with Germany in particular, has implications for ASEAN's processes of intra- and extra-regional economic integration. The AEC already represents a commitment to the free movement of goods and capital within the region. The CEE experience suggests, however, that the 'ASEAN+' initiatives may be of particular importance in terms of the region's ability to interface with potential anchor economies. Although those agreements have made notable progress in terms of freeing up goods trade, there is still much to be done in terms of promoting and facilitating investment with extra-ASEAN partners. Free movement of goods is important from the perspective of intensifying value-chain trade, but external investment, including technology transfer, is key if ASEAN countries are to move up manufacturing value chains to higher value-added activities.

On a macro-policy level, ASEAN will need to pay renewed attention to the 'ASEAN+' initiatives from this perspective. There is a strong case to be made for prioritizing integration with a regional anchor economy. The regional dimension is important so as to keep trade costs low, which facilitates value chain interactions. Although China is an important source of demand for ASEAN exports – including within value chains – it is still in a process of catch up with respect to the global technology frontier. Japan is much better placed from this perspective, but its economy has been in considerable difficulty for an extended period of time. A potential anchor that offers both technology and dynamism is Korea. It has the advantage of already being a global hub in two industries that are of particular importance for ASEAN value chains, namely, electrical equipment and transport equipment.

Comparing experiences across the three regions considered in this chapter, it is clear that national and regional policies interact to provide a conducive environment to value-chain trade. Getting the regional stance right, particular as regards investment, is important. However, so are national policies in areas such as human capital formation (i.e., education and training). Movement up the value chain requires access to an educated workforce, so investments in this area are a necessary complement to free flows of goods and capital with key partners. The experience of the CEE countries – which started from a relatively strong basis in human capital, and then liberalized movements of goods, services, people, and capital – should be highly instructive for ASEAN going forward, as the emphasis in most countries shifts from joining GVCs to moving up.

Notes

1 The author is grateful to Marwa Abdou for excellent research assistance, and to Hal Hill, Lili Yan Ing, and Deborah Winkler, as well as seminar participants, for helpful comments and suggestions.

2 For the purposes of this chapter, a *value chain* is defined as 'the full range of activities that firms and workers do to bring a product from its conception to its end use and beyond. This includes activities such as design, production, marketing, distribution and support to the final consumer' (https://globalvaluechains.org/concept-tools).

3 Clearly, the 2009 data may be affected by the aftermath of the global financial crisis. However, as of writing, no later data are available. Avoiding the effect of the crisis entirely would require using 2005 data, which are now more than 10 years old. In the interests of keeping the analysis as relevant as possible, this chapter therefore uses the 2009 data, subject to the caveat that they may be influenced to an unusual degree by broader macroeconomic and financial factors.

References

Ando, M and Kimura, F 2013b, 'Production linkage of Asia and Europe via Central and Eastern Europe', *Journal of Economic Integration*, vol. 28, no. 2, pp. 204–240.

Antras, P, Chor, D, Fally, T, and Hillberry, R 2012, 'Measuring the upstreamness of production and trade flows', *American Economy Review*, vol. 102, no. 3, pp. 412–416.

CIDOB, ECSA, and IWE 2014, 'Study on Latin America, the Caribbean, and Central and Eastern Europe: potential for economic exchange', report, EU–LAC Foundation, Hamburg, Germany.

Damijan, J, Kostevc, C, and Rojec, M 2013, 'Global supply chains at work in Central and Eastern European countries: impact of FDI on export restructuring and productivity growth', *Discussion Chapter* No. 332/2013, LICOS, Leuven, Belgium.

De Backer, K and Miroudot, S 2013, 'Mapping Global Value Chains', *Trade Policy Chapter* No. 159, OECD, Paris.

Godart, O and Gorg, H 2011, 'The Role of Global Value Chains for German Manufacturing', in *Trade Policy Research 2011: Global Value Chains*, ed. A Snyder, Department of Foreign Affairs and International Trade, Ottawa, pp. 335–366.

Guiliani, E, Pietrobelli, C and Rabelloti, R 2005, 'Upgrading in global value chains: lessons from Latin American clusters', *World Development*, vol. 33, no. 4, pp. 549–573.

Johnson, J and Noguera, G 2012, 'Accounting for intermediates: production sharing and trade in value added', *Journal of International Economics*, vol. 86, no. 2, pp. 24–236.

Koopman, R, Powers, W, Wang, Z, and Wei, S–J 2010, 'Give credit to where credit is due: tracing value added in global production chains', *Working Paper* No. 16426, NBER, Cambridge, MA.

Koopman, R, Wang, Z, and Wei, S-J 2014, 'Tracing value added and double counting in gross exports', *American Economic Review*, vol. 104, no. 2, pp. 459–494.

Kuwayama, M 2009, 'Quality of Latin American and Caribbean industrialization and integration into the global economy', *paper* No. 92, CEPAL, Santiago.

Saslavsky, D and Shepherd, B 2014, 'Facilitating international production networks: the role of trade logistics', *Journal of International Trade and Economic Development*, vol. 23, no. 7, pp. 979–999.

Shepherd, B and Archanskaia, L 2014, 'Evaluation of value chain connectedness in the APEC Region', *Document* No. 214–SE–01.28, APEC.

Sinn, H-W 2006, 'The pathological export boom and the bazaar effect: how to solve the German puzzle', *World Economy*, vol. 29, no. 9, pp. 1157–1175.

Timmer, M, Erumban, A, Los, B, Stehrer, R, and De Vries, G 2014, 'Slicing up global value chains', *Journal of Economic Perspectives*, vol. 28, no. 2, pp. 99–118.

Index

Italic page references indicate figures and tables.